DISASTER CITIZENSHIP

THE WORKING CLASS
IN AMERICAN HISTORY

Editorial Advisors
James R. Barrett, Julie Greene,
William P. Jones, Alice Kessler-Harris,
and Nelson Lichtenstein

*A list of books in the series appears
at the end of this book.*

DISASTER CITIZENSHIP

SURVIVORS, SOLIDARITY, AND POWER IN THE PROGRESSIVE ERA

JACOB A. C. REMES

UNIVERSITY OF ILLINOIS PRESS
Urbana, Chicago, and Springfield

© 2016 by Jacob A. C. Remes
All rights reserved

1 2 3 4 5 C P 5 4 3 2 1
♾ This book is printed on acid-free paper.

Library of Congress Cataloging-in-Publication Data
Names: Remes, Jacob A. C., 1980–
Title: Disaster citizenship : survivors, solidarity, and
   power in the Progressive Era / Jacob A.C. Remes.
Description: Urbana: University of Illinois Press, 2016.
Series: The working class in American history
Includes bibliographical references and index.
Identifiers: LCCN 2015019369
ISBN 9780252039836 (cloth : acid-free paper)
ISBN 9780252081378 (paper : acid-free paper)
ISBN 9780252097942 (e-book)
Subjects: LCSH: Disaster relief—Social aspects—
   United States—History—20th century. Disaster
   relief—Social aspects—Canada—History—20th
   century. Fires—Massachusetts—Salem—History—
   20th century. Halifax Explosion, Halifax, N.S., 1917.
   Working class—Massachusetts—Salem—History—
   20th century. Working class—Nova Scotia—
   Halifax—History—20th century. Solidarity—Social
   aspects—History—20th century. Power (Social
   sciences)—History—20th century. Salem (Mass.)—
   Social conditions—20th century. Halifax (N.S.)—
   Social conditions—20th century. BISAC: POLITICAL
   SCIENCE / Labor & Industrial Relations. SOCIAL
   SCIENCE / Disasters & Disaster Relief. HISTORY /
   United States / 20th Century.
Classification: LCC HV555.U6 R46 2016
DDC 363.34/8097309041—dc23
LC record available at http://lccn.loc.gov/2015019369

For Mari

CONTENTS

Acknowledgments ix

Introduction 1

1 "Organization without Any Organization":
Order and Disorder in Exploded Halifax 21

2 "A Great Power Had Swept Over It":
Politics and Power after the Salem Fire 54

3 "It Is Easy Enough to Establish Camps":
Geographies of Community and Resistance
in Burned Salem 78

4 "The Relief Would Have Had to Pay Someone":
Halifax Families and the Work of Relief 105

5 "A Desirable Measure of Responsibility":
Halifax's Churches and Unions Respond
to the Progressive State 132

6 "The Sufferings of This Time Are Not Worthy to
Be Compared with the Glory That Is to Come":
Salem Workers Build Power in the Church and Factory 165

Conclusion: Cities of Comrades 189

Notes 201

Bibliography 245

Index 277

ACKNOWLEDGMENTS

The book that follows is about community, social ties, and solidarity in two cities. In the cities that I have called home and where I have worked on this book, I too have relied on bonds of solidarity and mutual aid. Indeed, in the decade I have been working on this book, I have racked up incalculable debts—material, professional, political, and emotional. Without friends, colleagues, and family, this book would not exist or would be much impoverished.

In New Haven, where I was first reared as a historian, I hope that Anders Winroth, Glenda Gilmore, and Jim Scott will be pleased to see their continuing imprint on my thinking and writing. In Durham, where this book was conceived, my first thanks are to Robert Korstad, Sarah Deutsch, John Herd Thompson, and Gunther Peck. Bob's friendly gregariousness and helpful advice are always a welcome pleasure. Sally has the uncanny talent to listen to me ramble and then distill my argument into a few pithy sentences; she has often known better than I what my project is. My writing has been improved immeasurably by John, who will no doubt be utterly horrified at this sentence's wordiness and its use of the passive voice. From the beginning, Gunther has pushed me to ask harder and more specific questions, and has, in the years since graduate school, remained an important mentor, teacher, and friend. I could not have done the research required without funding from Duke University's Center for Canadian Studies, under the leadership of Jane Moss, J. J. Thomas, and John Thompson and the friendly administration of Janice Engelhardt; I was also funded by Duke's graduate school, its Franklin Institute for the Humanities, its University Scholars Program, and the Josephine de Kármán Fellowship Trust. Thanks to Margaret Brill, Robin Ennis, Cynthia Hoglen, Tori Lodewick, Katharine French-Fuller, Alisa Harrison, Paula Hastings, Max Krochmal, Gordon Mantler, Orion Teal,

Anne-Marie Angelo, Kathleen Antonioli, Eric Brandom, Jenny Wood Crowley, Daniel Elam, Kelly Kennington, Pam Lach, Bart Scott, Liz Shesko, Ben Hayden, Sarah Heilbronner, Irene Liu, Abhijit Mehta, and Mitch Fraas. I would have had nowhere to write without the staffs of Alivia's, Copa Vida, Joe Van Gogh, and Beyu coffee shops, and I would have had no books with which to write without the staffs of Lilly and Perkins Libraries. Thanks especially to Fred Smolioky, who was always cheerfully ready for everything.

In Halifax, Salem, and environs, I am grateful to the archivists who built, protected, and maintained the collections that are the lifeblood of historians' work, and who helped me find what in those collections I needed. Thanks especially to Darlene Brine, Philip Hartling, Garry Shutlak, and the rest of the staff at what is now rebranded as the Nova Scotia Archives; to Lynn-Marie Richard and Lynda Silver at the Maritime Museum of the Atlantic; to Barbara Kampas and the rest of the staff at the Peabody Essex Museum's Phillips Library; to Judi Garner at the American Jewish Historical Society New England Archives; and to Susan Edwards at the Salem State University Library's archives and special collections. Nonarchivists Debbie Amaral, the late Richard Adamo, Maurice Umble, Deacon Robert Britton, and the late Bishop Colin Campbell also preserved, found, and gave me access to records. Myron and Robbie Rosenberg, Jon and Becca Bijur, Shayna Strom, Kevin Fogg, Pam and Chad Gaffield, and Nancy Wright hosted me in their various cities. Thanks to Dan Rainham and Tara Wright, and to my archive buddy, Erin Morton.

Okayama, where I wrote much of the first draft of this book, became my Japanese hometown thanks to Teresa and Yasuaki Umeki and especially Kazue and Takaaki Matsuo. Thanks also to the staffs of The Market, Mister Donut No. 0678, the Okayama Prefectural Library, and the Okayama City Library's International Salon.

In Montreal, where this book went to gestate and where I began the hard process of figuring out what it was really about, I was lucky to have an intellectual home at the Centre for Interdisciplinary Studies in Society and Culture at Concordia University. Thanks to Graham Carr, Marci Frank, Kevin Gould, Suzanne Morton, Norma Rantisi, Ted Rutland, Anna Sheftal, and Graeme Williams. Thanks most of all to the American Council of Learned Societies (ACLS), who funded my year in Montreal as an ACLS/Mellon Recent Doctoral Recipients Fellow.

In New York, I am grateful to my colleagues at SUNY Empire State College's Metropolitan Center, most especially Shantih Clemans and associate deans Cathy Leaker and Christopher Whann. My students there have forced me to hone my thinking about disaster studies and welfare history. Dean Cynthia Ward allowed me a year's leave to finish this book.

In Cambridge, where I went for that year, I could not have finished the manuscript without the support of the William Lyon Mackenzie King Research Fellowship and the Canada Program of the Weatherhead Center for International Affairs at Harvard University, overseen by the very able Helen Clayton. Thanks to Alexia Yates, Kirsten Weld, Marie McDonough, Michael Leslie, Alanna Krolikowski, Laura Hartmann, George Elliott Clarke, and Joshua Cherniss. Thanks especially to the Accountability Club, Julia Gaffield and Jillian Powers.

From Urbana, Laurie Matheson at the University of Illinois Press shepherded my manuscript into your hands and coached me through turning it into a book. Anne Rogers was a thoughtful and accommodating copy editor. Thanks to Andy Rothwell for the two maps he created for this book and to Tim Pearson for the Index.

At conferences around the country, my colleagues in the Labor and Working-Class History Association (LAWCHA) have been inspiring models of engaged scholarship and teaching. Thanks especially to Dan Katz, Bethany Moreton, Leisl Orenic, and Heather Thompson, who have become friends and mentors. Thanks to Jim Green for introducing me to LAWCHA and for much else. For reading drafts, chapters, or even the entire manuscript, or for sharing their specific expertise, I am grateful to Julie Byrne, Aviva Chomsky, Nelson Dionne, A. Mitchell Fraas, Donald Friary, Glenda Gilmore, Julie Greene, Linda Gordon, Bartlet Hayden-Heilbronner, Alice Kessler-Harris, Erik Loomis, Nancy MacLean, Suzanne Morton, Deirdre Mulligan, Shaun Nichols, Max Page, Ted Rutland, Daniel Scholzman, James C. Scott, Shelton Stromquist, and the late Richard Winer. Kathleen Antonioli and Margaret Boittin helped me translate from French-language sources, although any mistakes in that regard (or any other, for that matter) are assuredly my own.

In Washington, I thank my family. My sister, Sarah, has been there throughout. My parents, Deborah Carliner and Robert Remes, were and remain my first and most important teachers. I would not be a professor today if it were not for their love and support.

Finally, in Durham, Okayama, New York, Tokyo, and Davidson, North Carolina, where we have been together and apart, there has been Mari Armstrong-Hough. Her political, intellectual, and emotional influence is surely on every page. With thanks and love, this book is for her.

INTRODUCTION

The week before Labor Day 2005, Americans watched their televisions aghast. In a flooded New Orleans, the state and society had, apparently, collapsed like inadequate levees. The first, dominant story we heard in the aftermath of Hurricane Katrina was of a city disintegrating into anarchy, police fleeing their jobs, looters running rampant, and poor people attacking those who tried to help. Hordes of lawless refugees seemed to run through the Superdome and Convention Center, and reporters repeated horrifying accounts of rapes and murders. Other stories told of unknown assailants firing guns at helicopters as their crews attempted to rescue stranded hospital patients.[1] It appeared that without a functioning state, people—yes, even modern Americans—reverted to a state of nature in which brutality reigned, cruelty went unabated, and violence became unchecked. The *Washington Post* banner headline that Friday summed up the lesson by linking two states that were, it implied, intimately related: New Orleans was now "A City of Despair and Lawlessness."[2]

Only a week later, a counternarrative emerged from New Orleans that inverted the roles played by authority figures and those stranded in the city. In a piece first published in the *Socialist Worker* on September 9 that spread quickly around the Internet, two visiting paramedics wrote of spontaneous organizing to find food and a way out of the city. The same weekend, the public-radio program *This American Life* aired an interview with Denise Moore, who sheltered at the Superdome. She denied the stories of murder, rape, and mayhem. "I kept hearing the word animal, and I didn't see animals," she said. Instead, she saw self-organized activities by "gangster guys," who broke into abandoned stores, found fresh clothes for those who needed them, "juice for the babies, water, beer for the older people, food, raincoats so that they could all be seen by each other."

In these stories, the authorities were the brutalizers, the people who in a time of crisis could not be trusted. The paramedics told of police lying to their group and blocking by force their escape from the city. "We were hiding from possible criminal elements, but equally and definitely, we were hiding from the police and sheriffs with their martial law, curfew and shoot-to-kill policies," they wrote. Moore said that authorities' caprice and cruelty fed rumors that the National Guard was planning to massacre those at the Superdome.[3] In this counternarrative, when the apparatus of the state disappeared, people overcame other divisions and organized to provide for their basic needs. Indeed, in this vision, it was the state that attempted to prevent such organization through violence and delegitimation. Its failures notwithstanding, the state was not, as it seemed at first, as fragile as its levees. But neither were the citizens of New Orleans.

This book offers a social history of the tension, so explicit in New Orleans, between the state's often bumbling attempts to help and control, and citizens' work to receive that help and reject control during disasters. The themes of this book—the perceptions of order and disorder in the immediate aftermath of disaster; the organic and mutual self-help that families, neighbors, and friends build in times of crisis; the relationships between working-class organizations and the state; migration and diaspora—resonate in the stories that emerged from New Orleans. How do the state's obligations complement or conflict with expectations for support from individuals or from private institutions? How do citizens' democratic demands intersect with the coercive power of the welfare state? What are we to make of the juxtaposition of informal and mutual solidarity practiced by survivors and the formality, hierarchy, and even violence of government aid?

I ask these contemporary questions through the historical examination of a fire in Salem, Massachusetts, and an explosion in Halifax, Nova Scotia. The Salem fire of 1914 burned a wide swath of the city, destroying the homes and livelihoods of many thousands of people, a plurality of them French Canadian migrants and their descendants. The Halifax explosion of 1917 killed nearly two thousand and maimed, blinded, rendered homeless, and left unemployed tens of thousands more. Three central arguments run through my study of these two disasters. First is that both disasters happened within a transnational region of North America, or to be more specific, in the northeast borderlands of the United States and Canada. This region of shared ideas, migration, fund-raising, and expertise was created by international experts and, more important, by working-class migrants, through their border crossing, their transnational families, and their institutions. The second is to find battles over the control of productive and reproductive labor at the center of contestations over citizenship and gov-

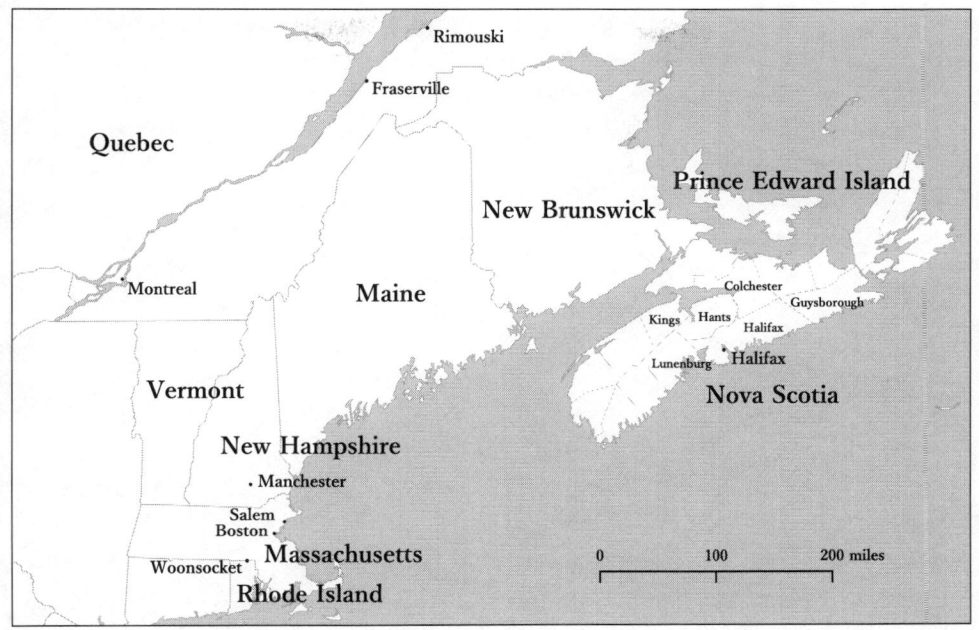

FIGURE 1. The northeast borderlands. Map by Andrew Rothwell.

ernance. Both the fire and explosion were working-class disasters that destroyed workplaces and workers' neighborhoods. Employers and the state sought to trade disaster relief for control over recipients' labor; recipients demanded aid as a right of citizenship but wanted to protect their autonomy and privacy. Third, the basis of the "disaster citizenship" that I describe was a growing sense of solidarity, whether among neighbors who rescued or sheltered each other, workers who organized on more inclusive grounds, Christians who overcame denominational boundaries to rebuild a united church, or residents of Maritimes Canada and New England who cared for each other across the international frontier.

My writing this book was bookended by Katrina, which inspired it, and by Hurricane Sandy, which flooded New York and New Jersey neighborhoods mere miles from my apartment as I was sending my draft manuscript to peer reviewers. Salem and Halifax offer clues about New Orleans, the New York area, and the next North American city to be destroyed. Conversely, what happened in New Orleans and New York informs my understanding of historical disasters, as do other major disasters that have occurred while I have worked on this project, including the Haiti earthquake of 2010 and the Tohoku tsunami of 2011. Historians are properly wary of applying what happens today to previous events, and the institutions and ideologies of disaster relief and reconstruction have changed considerably since the Progressive Era. But as a contemporary American, I necessarily approach my study of the historical experience of disasters in Salem and Halifax with a concern for the future of Sendai, Port-au-Prince, and especially New Orleans and New York.

The chronological context of this study is a historical moment in which individuals, institutions, and the state reimagined, renegotiated, and contested the obligations each had to the others.[4] The Progressive Era saw large-scale cultural and social changes born of ever-increasing industrialization and migration. Families reconfigured members' responsibilities to each other as increased mobility and shifting patterns of wage earning led to greater generational independence.[5] Likewise, workers and their employers renegotiated their relationship and the obligations each owed to the other.[6] Immigrant church members contested the organization of the religious denominations they found when they arrived in North America.[7] They did all this in the context of a greatly expanding role for the state, which acquired rescue as a fundamental purpose—drinkers from the saloons, prostitutes from white slavers, Cubans and Filipinos from their Spanish colonial masters, and child laborers from factories. The state increasingly insinuated itself into relationships among individuals and between individuals and civil institutions, and it took for itself responsibilities that had previously been the bailiwick of civil society.

One purpose of this project is to excavate networks of community obligation and solidarity. A historian has few sources to study the ties that bind neighborhoods together. People rarely record borrowing a cup of sugar from a neighbor or saying hello at the post office. But when those same neighbors rescue each other after a disaster, people take notice and record the event. Disasters thus produce unusual records that document where people turned in times of trouble or crisis. Unions, churches, and mutual-aid societies—never mind more informal connections like families and friendships—were not designed for disaster relief, but how they behaved in disasters shows us something about how they functioned. For this first purpose, then, the disasters function as cameras whose snapshots in time allow us to see social transactions in a particular moment within an era of change.

Understanding how these networks of solidarity and obligation functioned in the Progressive Era is particularly important because they were in flux, challenged both from within and without. From within, new urban, working-class generations experienced the freedom that came with industrial work and adopted new cultural mores. They contested and renegotiated with their elders' previously accepted patterns of obligation. From without, middle-class social reformers worked to devalue the institutions that had resonated in working-class and immigrant populations and to replace them with new structures sanctioned by the progressive state. Exploring responses to the disasters in Salem and Halifax allows us to study how working-class organizations survived and changed—or did not.

In disaster, people not only reenacted everyday patterns of solidarity—by turning to their families, neighbors, and churches for succor—but they also established new expectations. During and after disaster, obligations and relationships between individuals and between them and their institutions were actively renegotiated. The second purpose of this project is to better understand how popular expectations of the state, of civil institutions, and of other individuals evolved partially as a result of disasters. In this mode of inquiry, disasters do more than take a snapshot of a moment; they alter the direction of historical change. Families in destroyed cities had to renegotiate what they owed each other when a brother or a grown child abruptly needed a place to stay for months on end, or when a disabled mother could no longer perform the domestic and reproductive labor upon which her family had depended. Clergy suddenly had to help parishioners find relief even while they were called on to judge their worthiness. Employers and workers contested their obligations to each other when a factory was destroyed or when a massive disaster required rapid reconstruction work. The renegotiations that took place in the aftermath of disasters were built on the structures that had preceded them, but they changed the ways that parents and children, clergy and laity, bosses and workers conceived of each other and their relationships.

The Progressive Era was a period of greatly increasing state intervention, and disasters often catalyzed this expansion. The Galveston Plan of commission municipal government, later widely adopted across the continent, was imposed only after a hurricane destroyed most of the Texas city in 1900. The city's municipal reorganization had been a political project of the city's elite in the 1890s, and the disaster enabled them to pass the plan without a debate.[8] In Japan, the British colonies, and elsewhere around the world in the first quarter of the twentieth century, states discovered that the language and ideology of *emergency* created both need and opportunity: an emergent need for aid and an opportunity for new forms of governance.[9] The Mississippi River flood of 1927 launched a large-scale federal government relief operation; by changing what Americans expected from their government, it set the stage for the New Deal.[10] It is no wonder that New Dealers reached into the history of disaster relief to justify and understand their massive expansion of federal responsibility.[11]

Disasters were moments in which ideas of the state's role in taking care of its citizens changed.[12] As spectacles to both participants and observers, disasters

FIGURE 2. Salemites watch the spectacle of the fire in the Leather District, where it started: men in shirtsleeves, a woman with a parasol, and two girls talking, one using her hand to shield her face from the smoke. Above all these spectators are pillars of smoke. Negative #32828, Miscellaneous collection of circulars, etc., box II, Salem Fire Collection, E S1 F6 1914$_2$, courtesy Phillips Library, Peabody Essex Museum, Salem, Mass.

helped reshape politics. The Salemites in figure 2 who watched the billowing fire were soon joined by out-of-town tourists who poured into the city to gawk at the ruins and by the curious who bought newspapers, postcards, and quickly printed books showing the damage. If the spectacle of baseball helped men comprehend seemingly ruleless industrial capitalism by imposing detailed rules on athletic play, disaster was the opposite: it was a spectacle of rulelessness, unpredictability, and danger. Like other unscripted events, the spectacle of disaster also helped to create a community of watchers and sufferers. The community watching the apparently arbitrary destruction could imagine rebuilding with new rules.[13]

Part of this reimagining was to debate what aid the government owed its citizens when they were rendered destitute by disaster and how to deliver it. Citizens and their representatives had engaged in this debate in newspapers, voting booths, and city council chambers, and through their actions after the Chicago Fire of 1871 and San Francisco Earthquake of 1906, both germinal events in North American disaster history and disaster imagination.[14] After the disasters in Salem and Halifax, these debates played out locally and in their respective national capitals. Both federal states decided that they owed victims more, rather than less. The US Congress had long appropriated relief money for disaster victims, both foreign and domestic.[15] But as we will see in chapter 2, Salem and its allies had to overcome congressional objections that federal aid to unemployed victims of the fire amounted only to support for the merely jobless and was therefore past federal authority. The mass joblessness after the fire also occasioned a test run for what would become the United States Employment Service, which would not officially start until the next year.

Canadians, too, debated a federal contribution to the relief of Halifax explosion victims. "It is pointed out," wrote the *Financial Post*, "that the catastrophe was virtually an incident of the war, and that therefore the whole nation is in a sense responsible for the damage that was done."[16] Linking the explosion to the war added it to the growing list of state interventions that the war had justified or compelled. Between 1914 and 1918, Canadians saw the imposition of federal prohibition, daylight savings, income and corporate taxation, the Anti-Loafing Act (criminalizing men who did not work), and, of course, military conscription. In 1917 alone, the federal government began to intervene in the grain, coal, wool, natural resources, and railway industries.[17] When the Dominion government agreed in March 1918, three months after the explosion, to fund the relief commission with an additional $7 million, it phrased its responsibility as moral, not legal. "No legal liability rests upon the Crown," a Privy Council report insisted. But it admitted that the accident happened because of the war. "These considerations make it incumbent upon the Federal Government to provide reasonable and even generous relief for those who have suffered through the necessities of the war."[18] The necessities of war—and, more to the point, the public unhappiness following the explosion—also required that the Dominion government intrude on what had been a local affair and take over the management of Halifax harbor.[19]

As the list of federal interventions suggests, although Canadian historians rarely refer to this period as the Progressive Era, they see the same trends there as in the United States, albeit often happening a few years later. Canada's mostly Western, agrarian Progressive Party would not arrive until after World War I, for instance. But by then the Manitoba government, dominated by reformers between 1915 and 1920, had already pursued a platform well familiar to American historians: women's suffrage, lawmaking by referendum, prohibition, worker protective legislation, workmen's compensation, a minimum wage, mothers' pensions, and compulsory education in English.[20] And if prairie Manitoba was the center of reform, Nova Scotia, the home of Prime Minister Robert Borden, was no slouch, with reforms including town planning legislation and workmen's compensation.[21]

The Salem and Halifax disasters occurred during a germinal moment in the interlinked histories of the American and Canadian welfare states.[22] Welfare historians have long recognized the Progressive Era as a moment of considerable shift in the mechanism, mode, and ideology of assistance to the poor. Theda Skocpol, for instance, refers to a change in the United States from a paternal to a maternal model of relief, as the primary recipients of benefits went from Civil War veterans to widows, single mothers, and especially their children.[23] In Canada, the maternalist mothers' allowance movement combined with wartime regulation of the labor market to expand the welfare state.[24] Disasters too were part of the process by which the welfare state grew because they created large groups of apparently "deserving" needy who could reasonably make claims on the state and their fellow citizens not on the basis of charity, but on the basis of entitlement.[25] Through their actions, if only sometimes their rhetoric, people who needed government money or services in their ordinary lives or from disaster crafted claims to social citizenship.[26]

As governments at all levels of both federal systems began to take on more responsibility for their citizens' well-being, a key question was how that growing state would interact with organizations that preceded it. Because of the public-private cooperation that characterized welfare systems in the Progressive Era, not all of the state's agents were technically employees of the government. This "franchise state," as Michael Katz calls it, depended on reformers in and out of the state to adopt a common epistemology. Indeed, the professionalization of social workers and the increased role of municipal government in welfare projects muddied the line between private reformers and public governance. The alliance between professional women progressive reformers and their volunteer counterparts was, in Linda Gordon's words, the "defining practice" of the era and explains why "the 'state' often includes more than government."[27]

To expand the welfare system, the state and its agents needed to understand its citizens. Indeed, much of what made actors "state-like" was that they searched and worked for *legibility,* a term used by political scientist James C. Scott to denote the ways that states have "arrange[d] the population in ways that simplified the classic state functions of taxation, conscription, and prevention of rebellion." Scott's key metaphor is the cadastral map, which depicted a geographic region not in all its complexity but rather only for the information that the state found helpful. Moreover, a cadastral map did not just describe; it built a land tenure system by creating and recording legally binding knowledge. As Scott argues, this project of "state simplification" not only facilitated taxation and conscription but also expanded the reach of the state for the aims of public health, political surveillance, caring for the poor, and disaster relief.[28]

This legibility is necessary for modern states to operate and provide services, and the growth of an interventionist, progressive state required an acceleration of the state's project to render its citizens and territory legible. Ironically, the state's dependence on legibility and the erasures the project required left it unable to understand and adapt to the complex and nuanced systems and structures that working-class families and individuals constructed and in which they lived. The illegibility of working-class culture and mutual support hampered state efforts to aid disaster victims, not necessarily because people strove to maintain their illegibility but because the state was unable to read and understand their behaviors and desires.

I seek to do what the state and its agents were unable to do: to describe and understand the alternatives to the progressive state and the formal and informal ways that people rescued each other. The architecture of mutual aid was multifaceted and included families, neighborhoods, friendships, churches, unions, and fraternal societies. I ask how civil society responded to the growth of the progressive state; I offer not only a social history of the state's expansion but of the alternatives people and organizations offered to it as well. *Civil society*, write political theorists Robert Post and Nancy Rosenblum, is a term used so often that "it has acquired a strikingly plastic moral and political valence." I use the term broadly to include formal and informal organizations and institutions in the liminal space between the individual and the state. This includes formal, well-recognized organizations such as churches, unions, and clubs; it also includes neighborhoods, families, and groups of friends. (The Red Cross, a quasi-governmental "instrumentality," straddles the line between state and civil society.) I seek to understand how the various parts of civil society understood their relationships with each other, with their members, and with the state.[29]

Social scientists, long concerned with social capital—"the aggregate of the actual or potential resources which are linked to possession of a durable network of more or less institutionalized relationships of mutual acquaintance or recognition," in the definition of one of its leading scholars, Pierre Bourdieu, or "trust, norms, and networks that can improve the efficiency of society by facilitating coordinated actions," in the terms of another, Robert Putnam—have begun exploring its role in disaster.[30] In a classic work of disaster studies, Kai Erikson showed how the disruption of social networks after a town-destroying flood was a disaster of its own.[31] Through an examination of four twentieth- and twenty-first-century disasters, political scientist Daniel Aldrich has shown that social capital, measured as participation in government, civil society, and the social and cultural life of the community, is a predictor of community resilience.[32] Sociologist Eric Klinenberg explored what he calls "social ecology," emphasizing how built environments of cities and neighborhoods create conditions for social networks and social participation and the protection they provide from the effects of disaster.[33]

Related to these concepts, I discuss what I call *everyday forms of solidarity*. By solidarity, I mean a horizontal, reciprocal care: a care for someone, or a fight for someone, or a connection with someone not out of charity or sympathy but out of identity and empathy.[34] An attention to everyday forms of solidarity helps explain the importance of social capital because it shows how predisaster connections are used in disaster's aftermath. A self-consciously more political term than *social capital*, I use *solidarity* to suggest how these ordinary connections can, with intent and effort, encourage liberatory politics after disaster. It is from an often apolitical predisaster solidarity that disaster citizenship grows.

The Progressive Era was characterized by, in the words of Daniel Rodgers, "efficiency, rationalization, and social engineering," which were based on new forms of knowledge and authority.[35] The rise of the manager, professional, and expert accelerated in all spheres of North American society. Municipal government reform placed experts who were supposedly above politics in charge, replacing or supplementing elected politicians.[36] In factories, the precepts of scientific management emphasized the knowledge, authority, and status of professional managers.[37] Abroad, the experts who directed the construction of the Panama Canal managed material, land, labor, and even government in a complicated dance celebrated as a world-historical feat of engineering.[38] Canada and the United States created technocratic commissions to manage their international disputes far from the corruption of politics.[39] New municipal courts began to move beyond mere arbitration and adopted a sociological jurisprudence that let them intervene in the lives of those who came before them.[40] Doctors, lawyers, and clergymen demanded, built, and maintained professional knowledge, credentials, and power that sought to expel laypeople and their knowledge from newly professionalized terrains.[41] Progressive experts on charity rejected the noblesse oblige of the prior generation, insisting instead on a more scientific approach. "Through constant supervision by visitors of all varieties," writes historian Kathleen McCarthy, "it was hoped that the family could be saved, cured, and improved." Fundamental to this project of supervision was the need to study, know, and understand—in other words, to render legible—the families they sought to rescue.[42]

These projects were not uncontested. Patients, for instance, actively negotiated their authority and knowledge with their doctors.[43] Likewise, working-class individuals and their organizations continued to offer an alternative to the middle-class norms that society increasingly adopted. Sometimes these institutions gave political training to their members.[44] Sometimes they provided space and opportunities to organize explicit resistance. In addition, however, their very existence—their continuation in the face of Progressive-Era changes in culture and governance—itself constituted resistance.[45] In and through these institutions, people made demands on the growing state for aid on their own terms. These demands suggested that potential recipients were attuned to the power relations inherent in progressive rescue. Workers and their families sought to create and preserve spaces and institutions where they maintained their own power and authority in the face of a government, culture, and society that sought to place them under the power of middle-class experts.

The development of the professional expert and the contestation of experts' power happened in multiple countries. They were international, in that they happened in different countries, and they were transnational, in that ideas and people flowed between and among countries and regions. This book is simultaneously transnational and comparative, and these themes and methods run through every chapter. Historian Daniel Rodgers argues that the Progressive Era was a moment of unusual cosmopolitanism in which elites and political thinkers looked abroad for solutions to social problems.[46] Although Rodgers focuses on American links to Europe, the same was true between the United States and Canada.[47] Disaster experts moved across the border and shared their ideas and expertise in both Massachusetts and Nova Scotia.[48]

Ironically, when elites looked abroad for policy ideas, one of the things they shared was immigration restriction. Around the world, the Progressive Era was one of increased concern with and regulation of migration.[49] The border between the United States and Canada gradually thickened, most noticeably in 1906, when US officials started to collect names of Canadians coming south. Even more noticeable were new restrictions for those coming into North America from farther away. Coincidentally but symbolically, in what would become a key moment in the history of Canadian nativism, as Salem burned, Canadian authorities were in the midst of a summerlong standoff with Sikh migrants on board the *Komagata Maru* in Vancouver Harbor. With the help of the nascent Canadian navy, the ship and its passengers were eventually sent back to Calcutta.[50]

Yet no matter the attempts by politicians, bureaucrats, and their soldiers and border inspectors—often egged on, ironically, by trade unions—to solidify boundaries, working-class bodies and ideas continued to cross borders. As immigrants, emigrants, and families of immigrants and emigrants, the victims of the Salem and Halifax disasters experienced them in a transnational context. Only 19 percent of Salem's affected families—those who lost their homes, their breadwinning jobs, or both—were identified by the Red Cross as "American." Irish comprised 20 percent. French Canadians accounted for 43 percent of the victims, more than the "American" and Irish combined.[51] The Salem fire was rooted in a specific, American location, but the population that experienced it was diasporic and transnational. The institutions that supported them and shaped their understanding of the tragedy that had befallen them were self-consciously rooted in Quebec, even while they were deliberately American. Moreover, decisions about relief could not escape the knowledge that most of those who were affected were perceived as foreign. For them, disaster citizenship was also about literal citizenship.[52]

Halifax in 1917 was also a transient city. It was the metropolis of a region undergoing heavy and long-term out-migration to the English-speaking provinces from Ontario westward and especially to the United States. World War I suddenly brought thousands of temporary migrants into the city. Long a strategic port, the British had only withdrawn their garrison in 1905. But the (British) Royal Navy returned, and Halifax was also the home port of Canada's fledgling Royal Canadian Navy. Many of the half-million Canadian soldiers en route to the battlefields embarked there, and its harbor loaded the foodstuffs that sustained those soldiers and the civilian population of Britain. The city thus bustled with stevedores, merchant seamen, soldiers, sailors, and some of their families; wives and children often moved with their husbands and fathers to Halifax for the duration of the war. Many in the city still had family in other parts of Nova Scotia, New Brunswick, Prince Edward Island, or Newfoundland; they also had family who had migrated to central and western Canada or to New England. Haligonian social networks, therefore, reached nearby into the city's hinterland as well as across the continent into its diaspora. In particular, Halifax survivors' ties to Massachusetts, built through generations of migration, shaped the way they imagined their rights to access relief.

As port cities with migrant or transient populations, Salem and Halifax were broadly similar. They were also roughly the same size; Salem had 47,000 inhabitants, and Halifax somewhere around 55,000.[53] Both cities were also firmly embedded in a borderlands between the United States and Canada, a space of migration, cultural exchange, and trade. Part of what made the borderlands was working-class migration; another part was a shared elite political culture of progressive reform. The relief efforts after each disaster were firmly embedded in that progressivism, with people, ideas, and money flowing across the border.

Within a few weeks of the Halifax explosion, the premier of Nova Scotia, George Henry Murray, had written to American officials in California, Ohio, Texas, and Massachusetts asking for information on how they had handled the San Francisco earthquake of 1906, the Dayton River flood of 1913, the Galveston storm of 1900, and the Salem fire of 1914.[54] One of the documents Murray received was the galleys of a new book about disaster relief by a Red Cross executive then being published by the Russell Sage Foundation.[55] Byron Deacon's *Disasters and the American Red Cross in Disaster Relief*, which went on sale the next month, was designed as a manual for charity workers and executives—those who found their cities struck by disaster or those who, as the book urged, wanted to plan ahead.⁵⁶ A compendium of lessons learned from disasters such as the San Francisco earthquake, the sinking of the *Titanic,* and the Dayton River flood, the book suggested principles and procedures for future disasters.⁵⁷ Although Deacon had practical advice for relief workers—for instance, how relief committees should be organized, how national and local organizations should work together, and how to divide stricken cities into districts for canvassing—his major purpose was to describe and prescribe a progressive ideology of disaster relief. Deacon's prescriptions had some weight since by the next year, he was director general of civilian relief for the American Red Cross.⁵⁸ The Red Cross had long made decisions on pragmatic and personal grounds; Deacon's book was part of a shift toward more bureaucratic, professional, and ideological disaster relief.⁵⁹

The first element of Deacon's ideology was the importance of experts and managers. While he noted the "instinctive impulse to help" that arose after disasters, he emphasized that such spontaneous efforts "must be supplanted by reasoned, organized action" by trained experts and led by "prominent people."⁶⁰ The central coordinating committee should represent, he said, "official, business, professional, labor, and philanthropic groups," and should especially include the executives of charitable organizations.⁶¹ Social workers were critical participants; Deacon compared them to physicians and lawyers. Unlike well-meaning volunteers, whom he warned would harm relief efforts, social workers were equipped through their professional training and experience to guide families to recovery.⁶² A failure to centralize control under a powerful and prominent committee or to vest day-to-day authority in trained professionals could lead to failure. Any independent effort would "seriously hamper the execution of more comprehensive relief measures." Worse, disorganized and unprofessional relief could "promote first idleness and then discontent" among the very people the volunteers were trying to aid.⁶³

The second element of Deacon's ideology was that relief should not be intended to make sufferers whole; the job of the relief agency was not "indemnifying the families for the loss of property or of wage-earners, but [rather] equipping them to live healthy, happy, useful, normal lives in spite of their misfortune." "The object of rehabilitation relief," he continued later in the book, "is to assist families to recover from the dislocation induced by disaster and to regain their accustomed social and economic status."⁶⁴ Families—which he argued should be the "unit of relief"—should all be treated separately, with special attention paid to their individual circumstances and needs; while it might appear fair to apply hard-and-fast rules, Deacon argued that the opposite was true, and that it was fairer to examine every case on its own merits.[65] This sort of relief necessarily required the investigation and discretion that Deacon insisted were the sole purview of trained and experienced professionals.

Related was the third, and perhaps most important, aspect of Deacon's relief ideology. For him, disaster relief was simply a continuation of other progressive philanthropy and should thus seek to strengthen families beyond merely restoring them materially. Writing of how relief committees should respond to families in which the husband and father was incapacitated or killed, he emphasized the patriarch's nonfinancial role. "After all, a husband and father is more than an earner of wages," he wrote. He "supports his family in a moral and affectional sense" and "is an exemplar of worthy ambitions." If he could not continue in those roles, the relief organization had to take his place.[66] Deacon's project then, was to use disasters to reshape families into a patriarchal model, or to strengthen that model if it already existed.

"When a disaster relief committee essays to help the victims of calamity, it assumes a responsibility which is not discharged merely by grants of money or supplies," Deacon warned.[67] Although this meant extra work and responsibility, it also led to greater authority and legitimacy. If municipal authorities were unable to protect the health of their citizens, the private relief workers must take over that part of the state. He encouraged relief workers to use disasters to pursue not only municipal reform, but also, as the opportunity arose, to insert themselves into families' legal affairs, their parenting, their education, and their employment. "There will be families to move to cleaner and better houses, housewives to instruct in purchasing and preparing food to better advantage, others to be taught needed lessons in infant hygiene, men and women to arouse from the apathy and despair into which their misfortunes have plunged them and to be heartened to face the future with hope and courage."[68] Deacon urged his fellow progressives to see disasters as opportunities to continue the social reforms they were already working on. Relief work was a time when social workers could establish the "subtle influence of . . . friendly relations" to encourage Americanization.[69]

Deacon's ideology of relief was disseminated not only through his book and its reviews, but also through people.[70] In 1908, the Red Cross had inaugurated a system through which it drafted employees of central, coordinating charitable organizations in major American cities for disaster-relief work. The Red Cross thus became the clearinghouse for information and knowledge about disasters, and it circulated a cadre of professionals who shared a common outlook, training, and ideology.[71] While there was some change in the institutions of disaster relief between 1914 and 1917—notably, the American Red Cross professionalized and masculinized throughout the 1910s, especially after the American entrance to World War I—the key progressive ideology of disaster relief was remarkably constant.[72]

That continuity was partially because the same men who organized Salem's relief helped organize Halifax's. In the explosion's aftermath, a large number of expert men and women converged on Halifax, including three leading American architects of disaster relief: Abraham Captain Ratshesky, a banker from Boston; John Farwell Moors, a Boston broker and social reformer; and Christian Lantz, the general secretary of the Salem YMCA.[73] Ratshesky and Moors had first gotten involved in disaster relief after the San Francisco fire and earthquake of 1906, a key moment for the progressive remaking of disaster relief.[74] They had also both been involved after a conflagration in Chelsea, Massachusetts, in 1908. Moors, a noted Yankee Democrat perhaps most prominently associated with good-government school reform, had been the governor's special representative. Ratshesky, a Jew, was the only non-Yankee on the five-man state commission that replaced elected municipal government.[75] They reprised their roles in Salem in 1914. Governor David Walsh appointed Ratshesky to be one of seven state representatives on the Committee of Fourteen, which ran relief operations, and he sat on the executive subcommittee and on the purchasing committee. Later, he was an adviser to the Salem Rebuilding Commission.[76] Moors, meanwhile, was almost immediately again named the governor's special representative, and he chaired the relief committee.[77] In 1917, Moors and Ratshesky headed a delegation from Boston that arrived in Halifax with the first nonlocal relief after the explosion.[78]

As outsiders and as experienced experts, Ratshesky and Moors had the clout to break through local tradition and personality. They could ignore sentiment and specific, local conditions and impose a standard, centralized relief procedure. A photograph of Ratshesky taken by Richard W. Sears for the Boston newspapers showed him at the bedside of an injured boy named George Arthur (figure 3). Although the caption described him as listening to the boy's "sorrowful story," the banker, still dressed in the fur-trimmed coat his brother sent up to Halifax in a steamer full of supplies, was actually examining a notebook.[79] Sears, accidentally perhaps, captured a telling moment. One of the lessons Ratshesky told newspapermen that he learned from Chelsea and Salem was the importance of local voices. "Salem's problem was a problem for Salem, he said, just as Halifax's problem is one to be handled by its own people," reported the *Boston Globe*'s A. J. Philpott.[80] But the relief ideology developed by the Red Cross, expounded by Byron Deacon, and carried into action by Ratshesky and Moors required that a central and professional logic of "expertise" replace the specifics of local, lay ex-

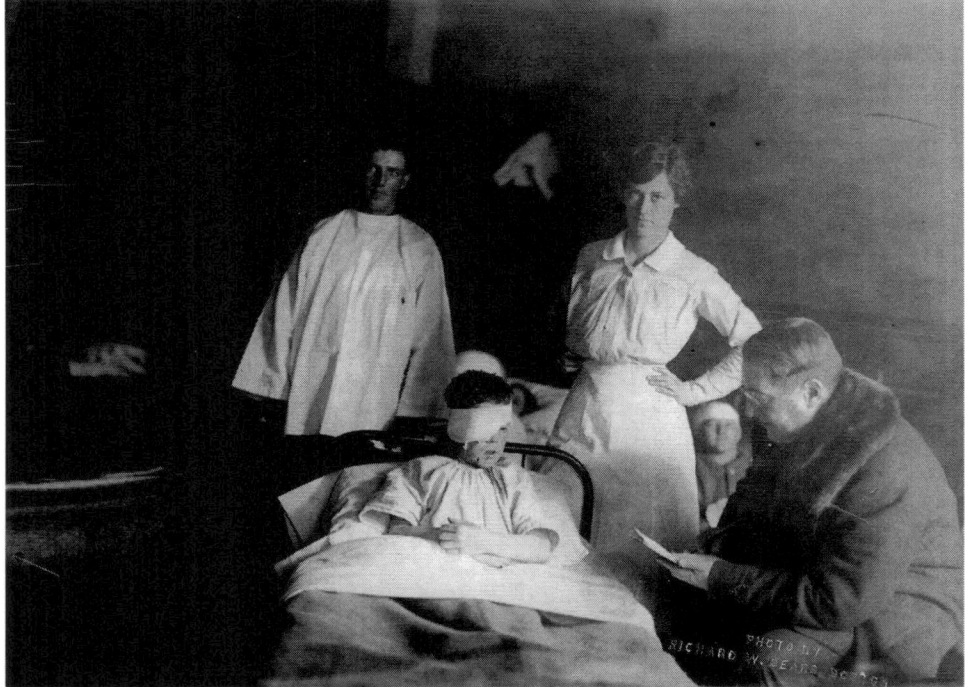

FIGURE 3. "George Arthur, a pupil at the Bloomfield School, who was injured in the Halifax disaster is being nursed back to health by the Massachusetts Relief workers. He is shown in the picture telling his sorrowful story to A. C. Ratshesky. The nurse is Miss Edith Choate of Boston." By newspaper photographer Richard W. Sears. Photograph from RS, from collections of the American Jewish Historical Society, Boston, Mass.

perience. Just as Ratshesky claimed that Halifax's people would handle their own problem, even while "wield[ing] the axe swiftly," so too did he visit the hospital to hear individuals' stories even while paying more attention to his own notes.

Less trumpeted in the newspapers, Moors had perhaps an even greater influence on Halifax relief. When Dougald MacGillivray, a Halifax bank manager and leader at the Board of Trade, found himself not "catching up" to the task of chairing the relief committee, it was Moors who suggested a professional, Christian Lantz, replace him.[81] Unlike Moors and Ratshesky, who both had immense personal fortunes and conducted their philanthropy as hobbies, Lantz was a professional reformer and charity worker, without much personal wealth. Since 1900, he had been the general secretary of the Salem YMCA. It was, as a laudatory *Boston Globe* put it, "as natural in Salem for the people to go to Mr Lantz for a solution of civic difficulties as, so one admirer expressed it, to go to the police when anything is stolen."[82] After the Salem fire, Lantz was appointed by the governor to the reconstruction committee, and from there he was nominated to chair the rehabilitation committee, charged with helping those affected by the fire restart their lives. As the committee's sole executive officer, Lantz made decisions himself about what aid to grant sufferers.[83] "It was an enormous task, for it had to deal with problems of clothing distribution, medical care, furnishing of new homes, getting employment for many who had been deprived of work by the fire, arranging the financial affairs of many families and much other work," recalled a newspaper article a few years later. To keep track, Lantz worked with accountant J. Chester Crandell to develop "a system . . . not unlike a card index system, which simplified to a great extent, the problems of the enormous task."[84] Indeed, when Moors telegraphed Lantz to come up to Halifax, the latter arrived bearing the "paraphernalia" of the system he and Crandell developed.[85] For him, the technology of administrative knowledge and control was inseparable from the ideology it facilitated.

International experts like Lantz, Ratshesky, and Moors show that the construction of a progressive state that depended on their managerial knowledge was a transnational process. So too was the experience of the disasters by those whom they tried to help, through their own migration, the migration of friends or relatives, or the sharing of money. For elites, middle-class experts, and workers, these two disasters existed and were understood in a cultural and political space that crossed the border between the United States and Canada. This did not mean that the specific national contexts were irrelevant; to the contrary, people in Salem and Halifax worked in distinct national (as well as regional) institutional, constitutional, and cultural contexts as well. But a focus on the ways money, people, and ideas crossed the border exposes how ideas about the state and citizens' relationships to it developed.

This book is deliberately and self-consciously transnational, but it is also necessarily comparative. The categories and analysis I develop through the records of one city illuminate the experiences of the other. Yet the differences between Halifax and Salem and their disasters were great. They were, of course, in distinctly separate countries. In June 1914, World War I was approaching on the horizon; in December 1917 it was near its peak. In 1914, the United States was in a depression; in 1917, Canada faced a wartime labor shortage. Salem and Halifax, in their economic, political, and ethnic makeups, were distinct. Most of all, the disasters may seem incommensurate. The Halifax explosion left thousands dead or maimed; only a small handful of Salem's *sinistrés*—a French word for the victims of disaster—faced comparable grief and suffering. "Salem has suffered," wrote John Tivnan, a newspaper editor there who had helped lead relief efforts.

"Halifax is suffering much more."[86] In most of the following chapters, direct comparisons are inappropriate. Instead, the fire and explosion show different ways in which disasters, regardless of cause or size, were key moments in the construction of the progressive state, and how their survivors, in the shadows of such different disasters, worked to build a citizenship that responded to this new form of governance.

• • •

Disaster Citizenship examines the responses of individuals, families, neighbors, and formal organizations to the fire and the explosion, and how those experiences shaped their relationships with the growing state. Each of the six paired chapters covers either Salem or Halifax; each pair grows in time and scope. The first one is about individuals in the first hours and days of each disaster; the second asks how informal communities like families and neighborhoods responded to the disasters and to the state over the span of weeks and months; and the third section looks at formal organizations such as churches and unions, and it extends chronologically for years. Within each section, the order of Salem and Halifax alternate, and the first disaster I examine in depth is Halifax. I begin with the later disaster as a reminder that I do not make an argument here about change in disaster relief between June 1914 and December 1917. My analysis here does not depend on the chronological difference between the Salem fire and the Halifax explosion. Nor are these paired chapters direct comparisons; they ask thematically related but distinct sets of questions.

The first two chapters ask what people did immediately after the explosion, and how individuals' social positions—including class, gender, and geographical location—influenced what they did and how they imagined their cities. They interrogate the categories of order and disorder, showing how each were built into and were dependent on the power structure and society of each city. Chapter 1 asks how people in different parts of Halifax and its society perceived order immediately after the explosion, how they created order for themselves, and how the order they sought differed. Individuals built on their preexisting ties of solidarity, but they also built new bonds and connections. Chapter 2 examines debates in Salem, Boston, and Washington over the state's response to Salem's fire. In both cities, politics were deeply embedded in military and civilian relief work. A theme in both chapters is that when ad hoc and unofficial solidarity was formalized and made hierarchical and official, it became less efficient, useful, accepted, and egalitarian.

The third and fourth chapters start with the greatest point of similarity between the disasters: mass homelessness. They each begin by asking where survivors lived in the weeks and months following each disaster. Chapter 3 describes the conflicts between relief authorities and French Canadians living in a Salem refugee camp. It asks how conflicts about domestic and formal labor played out spatially in fights over the arrangement of the camp and the refugees' presence there. Chapter 4 shows how Haligonians integrated relief aid into their complex family economies. Both chapters explore the balances represented by relief: they ask how the state balanced its need to understand, control, and flatten the complexities of citizens' everyday lives with a fear that it would be saddled with unsustainable material and financial demands. They also explore how survivors struggled to retain their independence and autonomy even while seeking to maximize the money and support they drew from the state. In both cities, questions of productive and reproductive labor lay at the heart of these contestations. So too did mobility, and in chapters 3 and 4 we see how the geographies of the Franco-American and Maritimes diasporas shaped people's access to and practice of aid. This second chapter pair addresses informal parts of civil society: families, neighborhoods, and friendships.

Chapters 5 and 6 move from the informal to the formal; they ask how churches and unions responded to the disasters and to the growth of the state. Chapter 5 explores the role of clergy in Halifax's relief process and asks how churches were rebuilt and reimagined as communities, buildings, and institutions. It compares the relief commission's instrumental use of churches, which emphasized clerical authority, with the ways that lay congregants chose to use churches to come to terms with their grief. It also asks how unions responded to the considerable growth of the technocratic state during the First World War. Chapter 6 combines the stories of the main French Canadian Catholic parish and the textile workers' union at Salem's largest employer, asking how French Canadians crafted an ethnic political culture in church and the workplace. This third section shows how Salemites and Haligonians created formal, explicit political demands and institutions from the informal and implicit politics of relief and aid.

• • •

Writing in Hurricane Katrina's aftermath, Naomi Klein argued that since the 1970s, disasters have been key moments for those who have sought to impose neoliberalism—that is, to commodify all goods and services, radically shrink the social purpose of the state, and simultaneously increase the state's military, coercive, and carceral functions. Disasters are so useful to those who would undermine political and social democracy that if they do not appear naturally, Klein writes, they are worth creating artificially, whether through war or manufactured economic crisis. Central to her book is the metaphor of psychiatric shock therapy, which so loosens and unmoors the patient that it allows a doctor to rebuild a patient's personality from scratch. For Klein, disasters do the same thing: they are destructive, disruptive, and dangerous, and in their aftermath society is so fragile and weakened that neoliberal malefactors can rebuild it in their desired image with minimal resistance.[87]

Klein is right to point to disasters as moments when experiments in new forms of governance can gain increased traction. After the Lisbon earthquake of 1755, the Marquês de Pombal rebuilt using a new type of rational city planning, but he was already committed to the Enlightenment, and the disaster only catalyzed it.[88] Attempts by Southern elites to dismantle or diminish urban democracy, already bloodily on display in the Wilmington coup of 1898, were given legitimacy by the Galveston flood of 1900. And the neoliberals who wanted to privatize public schools and destroy public housing found their opportunity after Katrina. Likewise, as we will see in the chapters that follow, the technocratic project of the Progressive Era—the fetishization of expertise, the privileging of managerial authority, the belief that a disinterested middle class was best able to govern on behalf of "the people"—received a powerful boost from the Salem fire and the Halifax explosion.

More impressive and important, however, is how citizens have challenged and contested these projects. As another public intellectual, Rebecca Solnit, has written, Klein's ideas are "a surprisingly disempowering portrait from the Left."[89] Contrary to Klein's vision of postdisaster societies as pliable and weak, workers and their families in the chapters that follow did not placidly accept the imposition of managerial and technocratic progressive governance. Rather, they worked to shape the new state. Working-class people wanted state support—they wanted to get as much as they could from the preexisting organs of the state and from the new ones that were built in disaster's aftermath. But they also wanted to preserve their own independence and autonomy. This meant subverting the state's demands for power and authority and balancing those demands with their own. In other words, they wanted relief on their own terms: to receive state aid while simultaneously retaining power and dignity.

What they wanted and how they organized themselves were often at odds with the progressive state. Where the state emphasized managerial and technocratic knowledge, working-class disaster survivors shared what they had learned informally. Where the state increasingly thickened borders, survivors built transnational and diasporic politics. Where the state centralized and built hierarchy, survivors preferred solidarity and mutual aid. They were not always, or even often, successful, but working-class survivors subtly created a new form of citizenship for a new era of governance. This *disaster citizenship* is the subject of this book.

1

"ORGANIZATION WITHOUT ANY ORGANIZATION"

ORDER AND DISORDER IN EXPLODED HALIFAX

The morning of Thursday, December 6, 1917, started like any other in Halifax Harbor. It was a clear, sunny day, and the harbor was busy with wartime traffic. The *Imo*, a Norwegian-owned steamer bound for New York to collect supplies for Belgium, had intended to depart the previous evening but had been delayed, so it was rushing out of the harbor. The French-owned *Mont Blanc*, bound for Bordeaux laden with explosives for the front, had arrived late the night before and missed the deadline for entering the harbor, so it too was in a hurry. A bit after nine o'clock in the morning, under circumstances that remain controversial, the two ships collided in the narrows of Halifax Harbor.[1] The munitions on the *Mont Blanc* caught fire and soon exploded in what has been called the largest man-made explosion before the atomic bomb.[2] Two thousand people died.

The Halifax explosion came suddenly, especially in Richmond, a working-class neighborhood in the city's North End, where longshoremen, building tradesmen, railway workers, and their families lived. Some people had been watching the *Mont Blanc* burn, but most Haligonians had not known that there was a ship fire, so they were going about their normal lives: starting school, beginning their workday, cleaning up from breakfast. Suddenly there was a loud noise. For those south of the devastated area, there followed a few moments of confusion, maybe even an hour or so, as people first imagined the damage to be local and relatively minor. Soon, though, they learned just how bad things were, either by traveling north themselves, or by meeting people coming from the North End. For those who started in the North End, it was clear from the beginning that something major had happened. The explosion knocked down houses and sent shards of glass flying like daggers, and as survivors started to escape the wreckage and regain their bearings, they had to contend with a rapidly spreading fire, sparked by the flying munitions and upended coal stoves. In Richmond, on the steep hill overlooking the narrows, what had not been destroyed outright by the shock of the explosion burned down. Even in the South End, a district filled with the gracious mansions of the city's elite and the more modest houses of its middle class, doors came off their hinges, plaster crashed down from walls and ceilings, and windows shattered.

At that moment—whatever moment a person learned that something extraordinary had happened—normality was suspended. Small children mustered uncommon bravery to rescue their parents from burning and collapsed houses. Workingmen abandoned their posts to check on their families. Patients long consigned to the Old Ladies' Home wandered outside for the first time in months or years. Untrained women once squeamish at the sight of blood volunteered for hours of nursing duty at hospitals deluged with the wounded. With the mayor traveling and the city council scattered around town, the city's deputy mayor and province's lieutenant governor essentially ceded political authority to a self-constituted group of local worthies. Nearly a quarter century later, writing the novel that remains the foremost cultural depiction of the explosion, Hugh MacLennan summed it up: "It was all queer; it was a revolution in the nature of things."[3]

The nature of that revolution depended on the perspective of the observer. This chapter examines the perspectives of three different kinds of actors and interrogates their perceptions of order and disorder. First are those I call *relief workers*, mostly middle-class men and women who left their workplaces and homes to help people at the overwhelmed hospitals, rescue the wounded, and clear the dead in the devastated area. Many relief workers extended the roles they played in ordinary times. Volunteer nurses nursed more and in different places. Soldiers worked in the devastated area on their own accord, with little or no oversight. Civilian men and women volunteered their inexpert and unusual labor. What they found, both in the devastated area and in hospitals, surprised some of them. Florence J. Murray, a Dalhousie University medical student, spent the day working at Camp Hill Hospital. She was struck by the "organization without any organization," as scores of people worked alongside each other—trained and untrained, civilian and military—all without direction and on their own authority.[4] This voluntary work was not without hierarchy. Because it was based on people's preexisting social networks or occupational roles, it recapitulated the hierarchy and inequity of ordinary lives. But because their activities threw together strangers, relief workers also negotiated new connections, structures, and hierarchies as they labored.

In contrast stood *relief managers*, rich and middle-class Haligonians whose first instinct was to go to City Hall, where Red Cross leader May Sexton found "total disorganization."[5] For this second group, the city appeared chaotic and dangerous, and they valiantly wielded telephones, typewriters, and pencils in an exhausting battle against disorder and confusion. The chaos they envisioned was a product of their centralized knowledge being insufficient to the task. To these managers, order was by definition created by central committees and the logic of central commands. To some extent, this was a result of a municipal political culture that revolved around the military and its centralized power structure. But it is also the product of any state-based power. The organic, informal logic that volunteers like Murray created was illegible to people like Sexton. It was for relief managers that the events of December 6 were most literally revolutionary, in that they effectively took over government. Yet if the ordinary organs and activities of democratic governance—the city council, the mayor, the upcoming parliamentary election—were absent or suspended, there was little change because by taking over the municipal state, the relief managers also took over the vantage point and organizational logic of City Hall.

Finally, there were the objects of these relief efforts, those who survived the explosion. In addition to the roughly 2,000 people who died that day or in the immediate aftermath, there were 9,000 who were injured and around 25,000 who were made homeless.[6] *Survivors'* ideas and experiences of order and disorder were more complex. They enacted a local, informal order by relying on their everyday networks and practices of solidarity. By going to locations and people that played a central role in community life in ordinary times, they maintained the regular order of their normal lives. In the understanding of civic, political, and military leaders—that is, to relief managers who took control of formal relief efforts—this local order was illegible. To crowd at a doctor's home or to go to the local convent was disorderly and dangerous because it ignored a more regimented, from-the-top relief system. Similarly, helping oneself to medical supplies from a drugstore could be viewed from one perspective as disorderly looting but from another as orderly rescue. Thus North End survivors created an order that masqueraded as disorder.

Survivors were also exposed to the opposite, to disorder masquerading as order. Soldiers canvassed Richmond after the explosion, warning that the fire would soon spread to a munitions depot and forcing survivors to evacuate to open fields. The promised second explosion never came, and the embodiments of the state and its order only spread disorder by delaying rescue efforts and exposing the wounded to prolonged winter weather. One sees in these warnings all the ways in which the state is at best an imperfect relief organization because it is made up of fallible individuals who cannot be omniscient. It is also imperfect because it can only understand things from a centralized vantage point.

Order and disorder are relative concepts, understood only in relation to the subject's position. May Sexton, the relief manager who witnessed "total disorganization" in City Hall, was, from her own perspective, correct. People at City Hall indeed had little idea what was going on in the rest of Halifax. Conditions at headquarters were chaotic—rain and snow came in the windows, city officials wandered around drunk, and key elected leaders remained absent. But what Sexton could not see thanks to her literally central vantage point at City Hall was the informal order created on the ground: Florence Murray's "organization without any organization." This chapter explores and analyzes these differing ideas about and experiences of order and disorder.

• • •

Novelist Hugh MacLennan famously called Halifax a place that "periodically sleeps between great wars."[7] Once a major center of Canadian trade and finance, by the early twentieth century Halifax had lost these functions to Montreal and Toronto. Immigrants to urban Canada preferred the cities in the central and western parts of the country, which became the industrial, financial, and political seats of power.[8] Yet MacLennan's comment was unfair, since it suggested that Halifax's awakening came only at the start of World War I and the influx of new people and money. Rather, Haligonians had long been self-conscious participants in the continental progressive movement.[9]

Two middle-class organizations were the most visible purveyors of progressive reform: the Civic Improvement League and the Halifax Local Council of Women. In 1906 prosperous professionals and businessmen founded the former, initially as a committee of the Board of Trade designed to encourage civic uplift and urban beautification. Frustrated by the failure of the elected city council to resolve what the league perceived as the city's most serious problems, its members shifted their focus to reforming municipal government. They succeeded in creating a hybrid system with both a Board of Control and a city council in 1913. While Halifax was late to reorganize government—by the time the Nova Scotia legislature approved the change, Toronto, Winnipeg, Montreal, and Saint John all had boards of control or commissions—the league encouraged Halifax and Nova Scotia to move more quickly than other Canadian cities and provinces to establish the first city-planning ordinances and laws. The league fell into inactivity in the middle of 1916, and it was formally folded into the Commercial

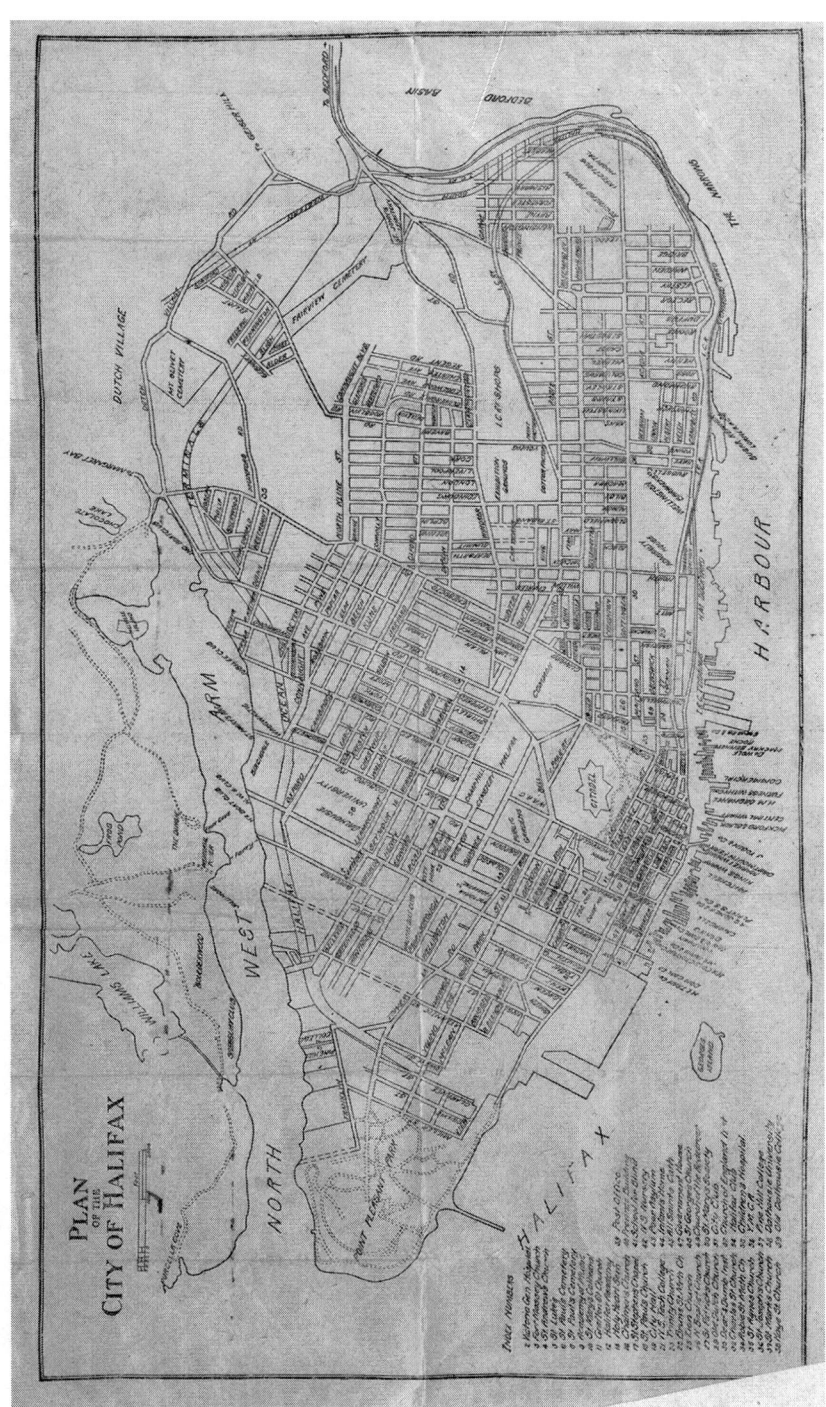

FIGURE 4 Halifax in 1913. "Plan of the City of Halifax, 1913," V6/240—Halifax—1913, Nova Scotia Archives Map Collection.

Club in 1917.[10] While it lasted, the Civic Improvement League fit neatly into a Progressive-Era model: an organization of middle-class and professional men interested in municipal government reform, land use, and city beautification.

The men who ran the Civic Improvement League discouraged women from joining, but their female counterparts had their own organization. The Halifax Local Council of Women organized around questions of health, housing, clean milk, the cost of living, and education. It ran the Women's Department at the provincial exhibition hall, and it helped to found the Children's Hospital, the Anti-Tuberculosis League, and the Halifax Welfare Bureau. It also brought the playground movement to Halifax, creating supervised play spaces in the North End.[11] As leaders in interlocking organizations of women, Halifax's progressive women controlled a large network of middle-class and elite women who knew the city, its people, and its geography. Although their volunteer work meant their knowledge of the city was often detailed and action-oriented, there does not seem to be any instance of their working with poorer women, rather than for them. Their knowledge of Halifax's working-class communities was that of outsiders.

Halifax's progressives, especially women, had particularly close connections to their counterparts in the United States. In some instances these connections were biographical; of the leading progressive women in the city, Eliza Ritchie had studied at Cornell and taught at Wellesley, May Sexton had grown up in Boston and studied at MIT, and Edith Archibald had grown up in New York.[12] But Haligonians continued to build connections even while at home. The Council of Women, for instance, hosted a lecture by Alice Stebbins, the first policewoman in the United States.[13] In 1915, the Civic Improvement League hosted John Sewall, who had led a civic uplift program in Boston, in order for him to spread his ideas to Halifax.[14] Halifax's progressives also existed in a national context, and the Council of Women was the local branch of an organization based in Ontario. Prime Minister Robert Borden, whose government oversaw a number of progressive reforms, represented Halifax in parliament.[15]

Following the progressive impulse to document and find solutions, Halifax's elite hired Archibald MacMechan—an English professor and librarian at Dalhousie, a noted man of letters and Carlyle scholar, and, most important, the neighbor and close friend of the chairman of the rehabilitation subcommittee—to open the Halifax Disaster Record Office. Like Philadelphia's Byron Deacon and Salem's Montayne Perry, MacMechan intended to write a history of the explosion that could serve as both a definitive record and a manual for future disasters. But crippled by depression and faced with waning interest from the relief authorities, MacMechan never finished his manuscript, and it was only published in

1978. Nonetheless, he and his assistant, Dalhousie undergraduate John Hanlon Mitchell, compiled a remarkable archive of letters, reports, and, most important, oral histories. It is this archive that serves as the primary documentary base of this chapter.[16]

After the explosion, middle-class and elite progressives, working on the basis of their predisaster organizations and habits of thought and action, set up a system that valued managerial order: committees and subcommittees, reports and detailed paperwork. In contrast, others, including middle-class relief workers and the survivors themselves, created a spontaneous order based on their pre-existing networks and solidarities. The different ways that Haligonians created and understood order and disorder in their ruined city were often in conflict.

. . .

When the explosion came, most people's first thoughts were of their families. "Workmen employed in the centre of the City whose homes were in the North End hurried from the shops to find out what had happened to their wives and children," MacMechan wrote.[17] Soldiers, sailors, and officers all ran to check on their families, many stopping to ask permission, but not all.[18] Those who were still at home, too, thought first of their families. When M. J. Burris, a Dartmouth physician, first heard the collision, he ran to gather up his wife, daughter, and maid and brought them into the basement, so they were downstairs and away from shattered windows when the explosion came. It is telling that although he told MacMechan about bringing his maid into the cellar, she disappeared from the rest of his story; only his family mattered at the height of the crisis.[19] Parents rushed to find their children. The principal of Bloomfield High School recalled parents running to the school to get their children.[20] Arthur Frye, the foreman at the Nova Print Company, hurried home to his flat on Garrish Street to find his wife "screeching," frantically trying to rescue their baby from the wood and plaster that had fallen onto the bed. Happily, they found the baby under the bed unharmed and their other child blown safely into the backyard.[21]

Not all stories ended as happily as the Fryes' did. Eric Grant was an army lieutenant and the son of the province's lieutenant governor. Home from France on leave that morning, he helped in the devastated area, but there were times he could do nothing. He wrote of meeting a wailing sailor "walking up and down in a dazed and distracted way." "'This is my trouble, Sir,' he said, between great sobs, and taking me around to the other side of the tumbled down house he knelt down beside three prone and lifeless bodies. They were those of his mother, wife and daughter."[22] Throughout the city, men and women faced the same emotional devastation. Worse than their material losses, that was their trouble.

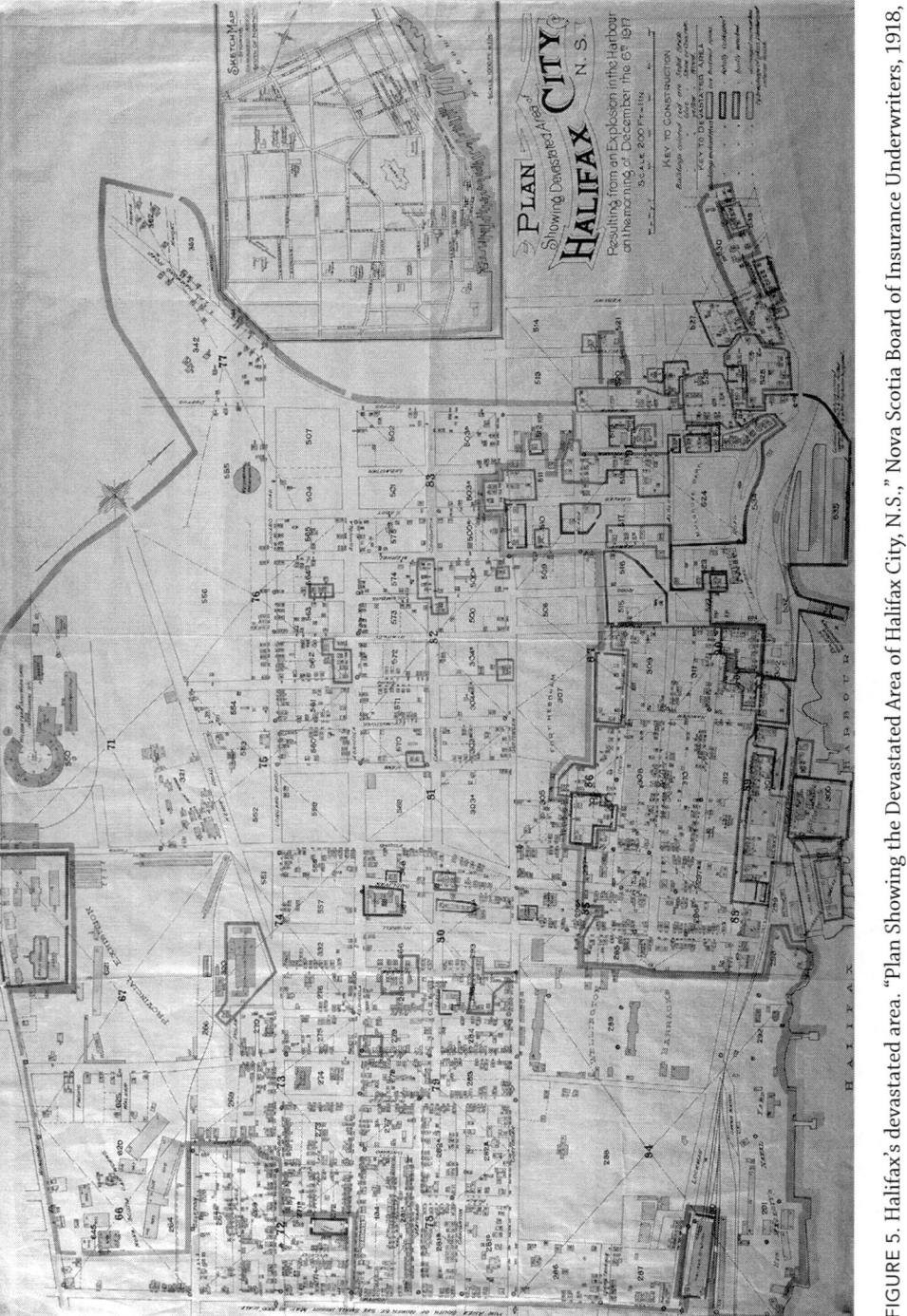

FIGURE 5. Halifax's devastated area. "Plan Showing the Devastated Area of Halifax City, N.S.," Nova Scotia Board of Insurance Underwriters, 1918, V6/240—1917 Halifax, location 4.2.3.2, Nova Scotia Archives Map Collection.

For those who were not injured, who did not have a close family member who needed attention, and who were not distracted by their own overwhelming grief, thoughts could turn elsewhere. *Relief workers* lent their bodies and their labor to help those in need. Often without direction or even suggestion, they went to the devastated area or to hospitals to help the rescue effort. That others who had not been there later derided their work as disorderly and inefficient suggests the way on-the-ground, ad hoc volunteer efforts remained illegible to those in authority. They created order and efficiency without direction. But upon closer examination it becomes clear that even the middle-class volunteers who appear to have had little or no connection with those they were helping were not being randomly altruistic. Rather, they were reenacting their own everyday forms of solidarity in extraordinary times.

When the fire started aboard the *Mont Blanc*, Frank Carew and the seventy-nine longshoremen under his direction were unloading a cargo of flour, oats, baked beans, and "a large quantity of high explosives of a very dangerous quantity" from the SS *Picton*. Though they could not know what was on the burning ship, they did know that any fire was dangerous near the *Picton*'s explosive cargo. "His first thought was for the citizens of Halifax," J. E. Furness, the local director of the shipping firm Furness Withy, wrote about Carew. The crew remained at work and "covered up every conceivable [sic] spot on the 'Picton' so in case of an explosion they would protect as far as they were able, an explosion on the 'Picton.'" Their successful work preventing the *Picton*'s explosion cost the lives of seventy-one men, including Cesare Mariggi, who left a wife and three children. Another six were seriously injured, and four received minor wounds. None of Furness Withey's eighty employees survived unscathed.[23] Although Furness recognized only Carew for his bravery, all of the longshoremen had chosen to sacrifice themselves to prevent an explosion. Carew and his two foremen could not have held the workers there against their will. That not a single one of them chose instead to leave the ship to find safety suggests that their thoughts were not only of the "citizens of Halifax," but also for each other. The solidarity built through long hours of common labor and collective action through their union held strong that day, as each worker stayed with his fellows rather than choose personal safety.[24]

A similar phenomenon was seen among the soldiers and sailors who made up the bulk of the workforce in the devastated area after the explosion. Soldiers rounded up the injured in, to use the phrase of one officer, "wagons, motors, handcarts—anything with wheels" and brought them to hospitals. This could be gruesome work. The officer had experienced the horrors of the front but still found himself overwhelmed by the "awfullness of the situation, burning and fallen houses, the frequency with which one fairly stumbled over dead bodies, etc., etc."[25] All afternoon, soldiers and officers went around in small groups, rescuing the injured and pulling dead bodies out of the wreckage.[26] Others cleared roads so vehicles could get through.[27] In addition to the emotional unpleasantness, this was hard labor. Civilians noted the stamina and the hard work of the military workers. "Them soldiers worked like niggers," exclaimed MacMechan's barber.[28]

Most of the immediate relief work by soldiers was self-directed and voluntary. Mrs. J. G. MacDougall was the wife of a doctor and herself a nurse. She described the work of the soldiers and sailors as "magnificent." They were not working under direction, but on their own initiative, she stressed. They enforced work on each other horizontally, rather than through officers' orders. She saw an American "sailor who crept up on a window ledge and went to sleep," paraphrased MacMechan. He was "dragged out of it by a mess-mate who abused him for disgracing the U.S. Navy."[29] Even officers noted the way their men worked without direction. Colonel Ralph Simmonds was the commanding officer of a regiment made up of local Halifax men that was among the first units on the streets. He praised his men's hard work, noting particularly one private who of his own volition removed his own boots to give them to a barefoot woman, binding his own feet with cloth. (Simmonds was quick to point out that the army replaced the soldier's boots.)[30] Lieutenant O. B. Jones was on a train from Bridgewater that was stopped outside the city. On foot, Jones "proceeded to Richmond and directed soldiers in the work of removing dead bodies," although by his own admission it wasn't clear they needed any direction. "His duty was to see that any soldiers standing about should be given something to do. There were very few standing about. The soldiers worked well."[31] Another officer coming in on the train was Lieutenant Colonel E. C. Phinney. Acting under his own authority, he began to organize soldiers and civilians alike into work parties. "Soldiers came, one at a time," he told MacMechan, and Phinney did not know from where, which means their arrival, like his, was probably not ordered by anyone. They worked so well he was convinced they must have been returned soldiers from the front because they were "very amenable to discipline, more so than those who had not been abroad."[32]

Officers like Phinney may have emphasized the importance of their own leadership, but enlisted men were quite sure that they did the work themselves and on their own volition. One of the very few instances of a soldier telling about his own experiences is the narrative of an anonymous Irish American soldier who had enlisted in November 1917 at a British recruiting station in Philadelphia. He had arrived in Halifax at the beginning of December and was training for the front when the explosion happened. "I wish it to be understood that although we were soldiers, there was no command given to go to the rescue," he wrote. "This was absolutely and entirely outside our line of duty as soldiers. Outside of strictly military operations there is no person or persons having the power to order another person to risk their life. And this was certainly by no means a military operation. It was entirely civilian. Not one of that gallant little band needed a command (even if commands could be given) to go to the rescue of their fellow human beings in that hour of dire disaster, and nobody with a heart could do otherwise!"[33]

But if soldiers and sailors did not work because they were following orders, why did they? Mrs. MacDougall's American sailor dragged back to work by his shipmates offers a clue: like Carew's civilian crew, he was motivated by solidarity with his colleagues. His work in Halifax came from the solidarity and comradeship that his naval training instilled in him. The anonymous American's work was motivated by the same thing, plus affection for the city where he trained. If he already felt solidarity with Haligonians after only a few weeks in town, soldiers who had lived there for longer, who had brought their families there, or who dated local women had even stronger bonds. We can see this in the soldiers' hard work reported by Colonel Simmonds: his men were especially tied to Halifax, and so they were particularly likely to work hard to save their neighbors.[34] Military relief was not without its problems, but when soldiers were working on their own to help save people, they evinced a spontaneous order and efficiency built on preexisting bonds of solidarity with each other and with Halifax civilians.

The doctors of Halifax and its suburb across the harbor, Dartmouth, also worked of their own accord. Like soldiers, whose training and experiences had conditioned them to strong bonds of solidarity, the training and experiences of doctors encouraged them to launch directly into medical service. Dartmouth physician M. S. Dickson was still in bed at nine o'clock and was buried under glass and plaster, but he avoided serious injury. Minutes later his neighbors converged on his house seeking help. Dickson attended to them, aided by his niece Annie Anderson, a Dalhousie medical student who boarded with him. Warned of a second possible explosion, Dickson and other Dartmouth doctors moved outside to the street and continued their work.[35] Perhaps because W. W. Woodbury was a dentist and not a physician, wounded Haligonians did not come to find him. Instead, basing his actions on his preexisting connections, he went to the YMCA, where he sat on the board, and helped with first aid there, mostly binding cuts.[36]

Like their male counterparts, Halifax's nurses and nurse volunteers continued in the work for which they had been trained. Four of Halifax's six visiting nurses in the Victorian Order of Nurses were already in the North End and quickly disregarded their own minor injuries and set about performing first aid.³⁷ "One," MacMechan wrote, "was dragged into a chemist's shop in Spring Garden Road and stayed there all day giving first aid."³⁸ The volunteer members of the Voluntary Aid Detachment of the St. John's Ambulance Brigade (called VADs) had only basic first-aid training, but they all found their way to hospitals or first-aid stations.³⁹

But most women volunteers were neither doing their jobs nor following their training. All around the city, mostly middle-class women, some with basic nursing training, others complete novices, arrived at hospitals ready to work. The Dalhousie student magazine presented an incomplete list of thirty young women "who so quietly went to work, assisting in the dressing of wounds, and ministering to the comfort of patients amid scenes of agony and death to which they were absolutely unaccustomed, and which are known to have shocked the nerves of even those accustomed to surgical work."⁴⁰ As with soldiers, these women were motivated by altruism, but also by peer pressure. By going to the hospitals, the Dalhousie women were enacting their daily patterns of solidarity with each other, in addition to creating new ones with the people they were aiding and those they worked beside.

Jean Ross, a Dalhousie senior who was in the library doing last-minute homework for a philosophy course, was a case in point. At first, she was resistant to helping, she wrote eighteen months later. When "after tea some girl came for volunteers for Camp Hill Hospital," she refused, despite a group of other girls going. "I thought it was ridiculous sentiment and nothing more, and what could girls do there, still more I, who hate all sickness. Besides I was so tired I could hardly stand." The next day, she heard from a friend "who told me of her adventures" at Rockhead Hospital, a severely damaged hospital in the North End. Impressed by her friend's story, Ross decided to go to Camp Hill herself. Still unwilling to go alone, though, she hunted for a friend to go with her. Most of her fellow boarders at Halifax Ladies' College who wanted to help were already there, so Ross trudged through the blizzard to find her friend Abbie Hemphill. The next day, too, Ross went back to Camp Hill Hospital, this time convincing yet another friend to accompany her.⁴¹

The friend who first convinced Jean Ross to volunteer was Margaret Wright, who with Mabel White and sisters Josephine and Helen Crichton, had nearly taken over Rockhead Hospital for two days.⁴² Their stories, too, illustrate the way that everyday solidarities led to the altruism of outsiders helping North End sufferers. The Crichton sisters, who were not Dalhousie students, had originally gone north to check on their relatives. When they reached North Street, they found a slightly injured, eighty-six-year-old aunt, who told them, "Don't stay here, we don't need you, we have our legs and arms, Go North! Go North!" It was around then that the Crichtons met the two Dalhousie students, and the four of them began giving people first aid as best they could with their limited supplies. Around noon, they met a car, Wright told Ross. "The man in it asked what they were doing and they said, 'For god's sake come to Rockhead Hospital!' They consented and he drove them there." Rockhead was a soldiers' convalescent hospital, and when the four women arrived they found it in terrible condition. Ross paraphrased Wright: "The Hospital was entirely cut off from communication, with the city by the burning district the nurses had all been hurt, the medical supplies blown up." Josie Crichton recalled to MacMechan that the convalescent soldiers had all managed to get up to replace the professional nurses who had been injured in the explosion. Besides them, there were only two doctors and some orderlies. "The wounded were simply pouring in and were mostly women and children," Ross wrote. When Wright, White, and the Crichtons arrived, the doctors told them to do whatever they could, so they brought tea to the injured and "looked after them generally," in MacMechan's words. Without any antiseptic or painkillers, there was little else they could do. The floor was wet with water from the broken radiators and blood from the dying patients. "The beds and floor was [sic] crowded," Wright told Ross. "In the course of the night, some beds had dead bodies taken from them two or three times and live patients put in." Eventually, soldiers and others arrived with carts, wagons, and motor vehicles to take the wounded to better-equipped hospitals, but at some point during the night, they grew tired and wanted to stop for the night. Helen Crichton, her sister told MacMechan, kept them on the job "by dint of coaxing and flattering, etc."—another example of informal persuasion, not hierarchical orders, in soldiers' and other relief workers' behavior. When the evacuation was completed on Friday, Wright told Ross that "the last ambulance took three corpses and on the front seat with the driver, Margaret and Mabel. He asked if they minded the load but it seemed a small thing after the night."[43]

When volunteers arrived at the hospitals, be it Rockhead, Camp Hill, or Victoria General, they quickly went to work on a wide variety of tasks. Often they found work themselves, doing whatever seemed appropriate. One woman noted that the patients complained only of being cold, so she went across the street to a home to get hot water bottles to warm them up.[44] A Dalhousie student wrote to a friend of what she jokingly called "my nursing career": "As to what I was doing, it seemed to be a little of everything. I was in tow with a doctor for a day and night, fallowing [sic] his directions which were chiefly to hold the patient while he administered the dressings and other minor duties. The rest of my duties were those of *general waitress*—you know what that would be yourself."[45] D. G. Cock, a Presbyterian missionary on furlough from India, and his wife volunteered at

Camp Hill. They were put to work washing the wounded of the black oily liquid that had fallen like rain after the explosion.[46] Warrena Maddin, another Dalhousie student, walked into Camp Hill "very brazenly." A Red Cross man there asked her if she had any experience or training, and then, MacMechan paraphrased, "kept her busy all afternoon, assisting in dressing wounds, some times would leave her to finish the dressing." She stayed at the hospital past midnight.[47]

The volunteers did this work in unpleasant conditions. The hospitals had sustained significant damage: Cogswell Street Hospital's windows shattered, its doors blew off their hinges, and part of its operating-room ceiling fell down; Rockhead suffered severe structural damage, with its windows gone, pipes broken, and ceilings and even some walls down; and even at Pine Hill, sheltered by Citadel Hill from the main brunt of the explosion, there were broken windows.[48] The four young women who went to Rockhead were so cold they kept their hats and coats on all night, and they had only a single bag of biscuits to tide them over all night.[49] When Marjorie Moir got to Camp Hill, she found "blood everywhere"; the injured who were brought in were "dripping with gore."[50] By the afternoon, Camp Hill was the main destination for injured people. "Patients were brought in in ambulances, carts, waggons, motor cars, or carried in the arms of friends and placed upon the floors in wards, halls, and offices until it was with difficulty that one could pass through the halls," wrote one observer.[51]

All this could look disorganized. After all, untrained workers were showing up, expecting to be put to work in vastly overcrowded hospitals. People arrived with conflicting motives and expectations, and they had to negotiate and renegotiate social expectations and hierarchies while in the midst of difficult, stressful, and unpleasant circumstances. Yet witnesses noted how well people worked, even without experience or direction. Just as the soldiers working to rescue people from the rubble worked on their own volition with little or no direction from their officers, so too did the women who went to help in hospitals. Whether from the bonds of solidarity that relief workers brought with them into the hospitals, or through bonds forged there in stressful times, these workers managed to build well-functioning, if ad hoc, hospital staffs.

Velma Moore, who arrived at Camp Hill after Thursday tea time, was one of those to whom the situation looked chaotic. Like others, she went with a friend, Christine MacKinnon. When they arrived, her "first impulse was to flee," she told MacMechan, because there was "no person to tell you what to do." She had, he wrote, "no experience even of sickness." Yet she stayed and found useful, if unskilled, work: binding wounds, running errands to the dispensary, and finding items that patients wanted.[52] They stayed until two in the morning. MacKinnon, MacMechan wrote, "praised Velma Moore and Gwen Fraser particularly for the way they stuck at their work on Thursday night. They had had no previous experience, had never been up all one night before. It was amateur work, but they used judgement and common-sense, which took the place of training."[53] Mrs. H. Bryant, who—although older than the Dalhousie girls—was also accompanied by a friend as she volunteered, reached Camp Hill around noon "and was struck by the confusion," she told MacMechan. Yet despite confusion, she too found a job washing soot off patients.[54]

After braving the initial confusion, volunteers had to negotiate and define a social hierarchy in the hospitals. Jean Ross, the Dalhousie student who was initially reluctant to volunteer at all, found this difficult. She and her friend were put to work by the matron in Ward L, and they were kept busy, but Ross also noticed "an over abundance of workers." Eighteen months after the explosion, she remained indignant about the hierarchy at Camp Hill, and her place on the bottom of it. "I was pleased when a nurse *asked* me to do anything which they seldom did but I was less pleased to be *ordered* by a V.A.D. when I had a first aid certificate myself and knew it was n.g. [no good] for this purpose. What sent me home however was orders from a woman in an elegant sealskin coat, doling out sweet biscuits and seating herself at each bedside in turn to hear the story. Her car waited at the door. I thought she was just as well able to run after glasses of water as I was. That I was more suitably dressed was a witness to my sense not hers." Worse still was that when Ross and her friend decided to head home, they found only a single car at the door. When the friend asked if they could have a ride home, she "was most indignantly answered that it was Mrs.——'s (our sealskin friend's) car."[55] Yet for every Jean Ross who gave up the social negotiations in frustration, there were more women who stayed despite the indignities.

This appearance of disorder and disorganization—the untrained workers, the volunteers in a huff, the difficulty moving around—masks what was truly happening in the hospitals. Medical student Florence Murray was the daughter of a strict Presbyterian minister in Prince Edward Island. "Inclined to be stout, with very fair, pink and white complexion [and] blue eyes," Murray eschewed cards and dancing and disdained smoking and drinking. Knowing that there would be a demand for her skills, she headed to the civilian Victoria General Hospital, performing first aid on the way. She noticed that people were being brought to the military Camp Hill Hospital, so she stopped there. When she arrived at ten o'clock, an hour after the explosion, she found that the convalescents who were the hospital's normal patients "were working 'like slaves,'" as MacMechan wrote in his notes. But what really struck Murray was what she called the "organization without any organization." Nobody directed the volunteers, but everyone worked intelligently and without argument. The trained, professional staff was so greatly outnumbered that the volunteers had to direct themselves. Further, Ross's story notwithstanding, Murray told MacMechan that volunteers worked wherever they could be of use, whether in the wards or the kitchen, without regard for themselves or their status. Murray especially praised the way volunteers without experience conducted themselves and learned on the job. She knew what that was like: after her day volunteering at Camp Hill, she became the official anesthetist at the YMCA hospital, even though she had but one day of anesthesiology experience.[56]

Related to Murray's "organization without any organization" is the way Christians developed an ad hoc ecumenism even before Protestant ministers from various denominations came together in a formal committee. D. G. Cock, the missionary who washed patients, was unusual. Most clergy in the hospitals performed pastoral duties, and for whomever wanted them. Presbyterian minister Hugh Upham had jumped on a train from Shubenacadie when he heard about the explosion and arrived just past noon. He told MacMechan that patients did not seem to care what denomination he was. "I met a woman in very great pain. Her husband, uninjured, sat at the foot of her bed. She asked me if I was a Roman Catholic priest. I answered 'No, I am a Presbyterian minister.' She said, 'Well, it is all the same now. Would you kindly say a prayer for me?' Her husband was also of the same mind, and there with suffering all around us we invoked the Divine Aid of our common Father in her behalf." Happily for them—and perhaps their eternal souls—the woman recovered. (He also reported a similar interaction in which he was mistaken for an Anglican, a less serious confusion.)[57] In the devastated area, too, Protestants and Catholics worked together. J. P. D. Llwyd was the Anglican dean of Nova Scotia. He went to the North End, where, he wrote, "the scene was horrifying in the extreme." There, "I found a relief party of soldiers taking out the wounded with a young Roman priest helping them. I joined the party and assisted in the taking out of a number of poor crushed and mangled forms, many of whom must have died before they reached the hospital."[58] This cooperation in the first hours after the explosion is particularly notable because only four days later, when Halifax clergymen formally organized to coordinate their duties, only Protestants were represented.[59] When the time came to bury unidentified remains, they were divided between Catholics and Protestants, and each group held distinct funeral ceremonies.[60] In the case of the clergy, "organization without any organization" meant more ecumenism and cooperation than when the system was formalized.

Hospitals and their volunteers succeeded without formal organization. On the first day alone, significantly more than 2,100 patients were seen by emergency hospitals in the city; another 150 were sent by special train to Truro, which had also set up an emergency hospital. This number includes the roughly 1,400 patients treated at Camp Hill, a brand-new hospital that had a capacity of only 250 in normal times. It excludes, however, those handled at Victoria General Hospital, who likely numbered nearly a thousand. Four days after the explosion, concerns that there may have been people who did not make it to the hospitals led to a house-by-house canvass throughout the city looking for untreated injured. The investigators found no such people; everyone had received at least first aid. The only untreated cases they found were eye injuries that would have been hard to diagnose immediately.[61] This is not to say the treatment was perfect. An out-of-town nurse criticized the sloppy job some relief workers did, complaining that they sutured wounds with glass, plaster, dirt, and cinders still inside.[62] Constance Bell, who volunteered at Victoria General, argued although the untrained volunteers had done a good job, it would have been much better to have had more trained people around to begin with.[63] But without any direction or organization, and often without any training, the women of Halifax did pretty well.

Sociologists refer to organic, ad hoc groups like those formed by the Crichton sisters and the two Dalhousie women, or like those who went to Camp Hill, as *emergent organizations*.[64] But these groups did not emerge from nowhere. The frequency with which women volunteered with their friends—Jean Ross and Abbie Hemphill, Margaret Wright and Mabel White, Josie and Helen Crichton, Velma Moore and Christine MacKinnon, Mrs. H. Bryant and Mrs. J. Gillis, Mr. and Mrs. D. G. Cock—suggests the importance of preexisting social networks and bonds in encouraging this volunteer work. Likely, friends provided not only encouragement for the initial decision to volunteer, but also emotional support while there and aid and advice while doing unfamiliar work. This was as true for the soldiers and others who rescued people and uncovered dead bodies as for those who nursed the injured. It was, then, these relationships that structured the outpouring of voluntary aid and gave it order. But these relationships were by their nature private and unobservable by others. The system and order they created thus looked disorderly to people who did not experience it.

. . .

Among those who saw the city as disorderly were people who chose to go to City Hall to help organize rather than go to the hospitals or the devastated area. They had not seen Florence Murray's "organization without any organization," or if they had, they mistook it for chaos. These *relief managers* tended to be managers and organizers in the rest of their lives as well: managers of business and banks, leaders in women's clubs and organizations, people involved in politics

To them, order and organization by definition took the form of committees and subcommittees, each with duly appointed chairs and secretaries and forms to be filled out in triplicate. From their central position—literally in their City Hall headquarters, figuratively at the center of the relief committee—the enemy was a lack of knowledge. Their inability to know, much less understand, what was going on in the city was caused by the physical disruption of telephone lines and the inability of those doing the work to report to City Hall what they were doing. But even had a perfect communication system been in place, those at City Hall could not have known or understood the organic, emergent order in the devastated area and the hospitals because that order was inherently illegible and unknowable to those in a central position.

On the morning of the explosion, Mayor Peter F. Martin was out of town, so authority fell initially to Deputy Mayor Henry S. Colwell. In addition to his elected office, Colwell was a successful merchant, but more important were his memberships in the progressive Anti-Tuberculosis League, the St. George's Society, the Masons, and the City and Waegwoltic Clubs. At age fifty-four, Colwell was a thin man, his face framed by light hair and a square chin, and he sported a wispy mustache and glasses. At nine that morning, he was on his way to his store from his house on South Park Street in the South End. He first returned home briefly to check on his family and, finding them well, only then went to City Hall to direct the city government.[65]

The next thing we know Colwell did was walk up Citadel Hill with the chief of police and the city clerk to meet with Colonel W. E. Thompson, the assistant adjutant general and second in command of the military district that included Halifax. Thompson offered what military men were already doing: rescuing victims, recovering dead bodies, and extinguishing the fires that were still raging in Richmond. He also promised tents, blankets, and mattresses. Colwell also met with W. A. Duff, a railway official, whom he deputized to go out of the city and send telegrams in the mayor's name requesting help from other towns and cities.[66]

At eleven thirty, Colwell convened in the city collector's office a meeting of the citizens who had arrived at City Hall to lend a hand. They were, the minutes reported, "all he could speedily convene"; MacMechan called them "the best brains and strongest energy available in the city." The minutes listed eleven private citizens and seven officials; another four men were probably there unrecorded. All fifteen were rich men from neighborhoods far from the North End. They ranged from James Norwood Duffus, a partner in the English ship line Cunard, to Robert Harris, a judge and perhaps Halifax's most prominent lawyer. Technically, the meeting began as one of the city council, but with only Colwell, two controllers, and three aldermen present—not a quorum—the self-appointed group of citizens quickly took the reins of government. They appointed Nova Scotia's lieutenant governor, MacCallum Grant, himself a prominent Halifax businessman, to chair the meeting, and Colwell served as secretary.[67]

In the forty-five minutes that followed, Colwell reported on his activities of the morning, they briefly discussed "the organization of Committees," and they unanimously agreed to constitute themselves as the Halifax Relief Committee. Having authorized themselves effectively to take over the municipal government, the men at the meeting then appointed each other to committees. They adjourned at twelve fifteen, promising to meet again at three o'clock that afternoon.[68]

Meanwhile, as MacMechan wrote, "to the City Hall, as to the hospitals, flowed a stream of voluntary helpers."[69] So when the second meeting started, there were more people in the room, but they were of the same sort. Joining the civilian worthies that afternoon were General Thomas Benson, the commanding officer of Military District 6, and Admiral Bertram Chambers of the Royal Navy. But by meeting's end the organized relief effort was not much further along. The assembled citizens listened as Colwell read the minutes of the first meeting and then on motion they "duly noted and seconded the action taken" and "ratified and confirmed" the appointment of committeemen. A few subcommittee chairmen made their reports, and at four fifteen they all went back home to the South End.[70]

Why did the "best brains" in Halifax, all of whom had considerable experience with charitable committees, fail to accomplish much in the first days? For one thing, in their central geographic position, people at City Hall could not know what was happening in the North End. The fire chief had been killed that morning.[71] City officials were missing. Of those who were there, one of the controllers was "continuously plastered" and the city clerk was useless; Henry Colwell "tapped his head significantly" when describing the latter to MacMechan. The building was in disarray, with windows broken and plaster down. The citizens' meetings were held in the collector's office because, in Colwell's words, it was "the only room in the building not so badly damaged by the Explosion as to be unfit for the purpose."[72] Mrs. H. Bryant went to City Hall after the explosion to volunteer but found "everything in confusion," so she left. On her way out, she met Dean Llwyd, who suggested that she go to Camp Hill hospital, and though she was "struck by the confusion" there too, at least at the hospital she could be of use.[73] In the hospitals, Mrs. Bryant's initial vision of disorder was deceiving, because volunteers created their own order that did not need to be perceived to outsiders to be useful. At City Hall, organization, control, and direction were the very purpose of activity. Without legibility, there was nothing. The confusion she witnessed had no deeper order because it was all there was.

The second reason the relief managers accomplished so little on the first day was that they were busy taking care of themselves. In the stories they told MacMechan, they frequently spent the morning and afternoon fixing up their houses by boarding up windows and cleaning up fallen glass and plaster. Banker and broker George S. Campbell, for instance, was in his coatroom putting on his boots when the explosion rocked his house. Knowing neither the cause nor the extent of the damage, he went with friends to purchase building supplies. Along the way, he discovered how bad the damage to the city had been, so he went directly to City Hall. Not finding any officials with whom to commiserate, he visited Richmond—not to help, but "to see it for himself." Campbell joined the finance committee. The highlight of his work with the relief committee, he implied to MacMechan, was that the next day he was given a ride home in the blizzard by Prime Minister Borden himself, thus relieving Campbell of the worry of how to get back to the South End in the snow.[74]

And so it went with the others. Dougald MacGillivray, the manager of the Halifax branch of the Bank of Commerce and the chairman of the Board of Trade, was not at either Thursday meeting, busy as he was cleaning up his office. It was only on Friday that he attended a meeting of the finance committee, and on Saturday he attended another "important meeting." He was later "pressed," he said, to take the job heading the rehabilitation committee.[75] Edmund A. Sanders, MacGillivray's secretary treasurer at the Board of Trade, spent Thursday fixing what he called his home's "severe" damage: broken windows and a piece of metal through the door. Experienced in securing supplies for his own house, he eventually went to work on the supply committee.[76] W. A. Major, the Halifax manager of R. G. Dun, had himself driven up to Cogswell Street Hospital to make sure his convalescing son was safe and to see the destruction for himself. Rather than stay and help, however, Major walked back to his South End house. After touring the devastated area (but again apparently not helping the rescue effort), he attended the three o'clock meeting and wound up with the job of trying to round up homeless people and getting them into the ad hoc shelters that were being established.[77]

So it would take until Friday for the "systematic efforts" MacMechan praised to get started in earnest, and the sense of panic, chaos, and confusion at City Hall apparently lasted all weekend, even as relief workers and survivors elsewhere created their spontaneous order. At eleven o'clock Friday morning, in what was apparently the cleaned-up and ready city council chambers, Henry Colwell gaveled to order a joint meeting of the city council and citizens, then quickly turned over the chair to McCallum Grant and took up the secretary's pen.[78] Although Colwell did not record who was there, the meeting was notable not just for having more people—men like MacGillivray had finished repairing the relatively mild damage to their houses and so could lend their attention to City Hall—but for the difference of who was there. Most notably, the meeting on Friday included women. Edith Archibald, Agnes Dennis, Clara MacIntosh, and Jane Wisdom were leading progressives in Halifax, veterans of organizations that had long sought to improve Halifax.

Archibald was a noted suffragist, temperance activist, and women's club organizer; although born in Newfoundland to Nova Scotian parents, she grew up in New York, where her father was the British consul general. Charles, her husband, had owned and managed a mine in Cow Bay, Cape Breton (now Port Morien), but the couple later moved to Halifax, where he was a banker; his business ties to Toronto strengthened her links there, and she often brought ideas from central Canada to Halifax. As a younger woman she had gained fame, or perhaps notoriety, for leading members of the Women's Christian Temperance Union on a raid of three illegal saloons in Cow Bay. By 1917, when she was sixty-three, she had already served as president of the Halifax Victorian Order of Nurses and the Halifax Local Council of Women.[79]

Archibald's distant cousin Agnes Dennis was fifty-eight and had been married for thirty-nine years to William Dennis, who by 1917 was a Conservative senator and the publisher of the *Halifax Herald*, a vociferously prowar, proconscription, and pro-Borden newspaper. She was a leader in her own right, too, working in the same fields as Archibald. After receiving what she later termed her "first training in public work in Temperance Societies," she set about leading nearly every middle-class women's organization in Halifax. She had been the founding president of the Halifax chapter of the Victorian Order of Nurses, had been president of the women's auxiliary of the YMCA since 1910, was the president of the Nova Scotia Red Cross, and was the first executive of the Anti-Tuberculosis League. Her first priority was the Halifax Local Council of Women, which she helped to organize and of which she was president from 1905 to 1920.[80]

Clara MacIntosh had not come to the meetings on Thursday because she was too busy. A little woman, with light hair and blue eyes, she was a nurse and the Lady Divisional superintendent of the St. John's Ambulance Brigade. Her husband was a well-known doctor, but he was not home, so when injured people began to arrive at their house for treatment, Clara put her nursing skills to use. Deluged with the injured—the house was so full she made people wait in the furnace room—she handled both first aid and triage, putting the less injured to work cleaning and, in the case of men, blocking up the broken windows. When soldiers came around ordering people to leave their houses for fear of a second explosion, MacIntosh simply moved the operation outside to the Common. Some patients had to be carried outside, but she laid blankets on the ground and continued her work. She went back inside the house around two o'clock, by which time a VAD from the North End had arrived to help her. They continued giving aid to people until eight in the evening, when they went to Camp Hill to help there. MacIntosh reported to MacMechan that she returned home around midnight, when she grew faint. Nonetheless, because the door was hanging off its hinges, people continued to come in during the night, and she even bandaged a man while she lay in bed.[81]

Jane B. Wisdom was a New York– and Montreal-educated social worker. Born in New Brunswick of Nova Scotian stock, she had come to Halifax nineteen months earlier to direct the fledgling Welfare Bureau, the city's attempt at organized charity. Wisdom was to model professional welfare work, and part of her job was to know the circumstances of Halifax's poor and then judge them with the "objectivity" of an outside expert. Another part was to create more such experts by training a corps of professional social workers. Wisdom stopped by City Hall on Friday morning to see if she could help; finding the meeting already in progress, she went in and participated. But she did so on different terms: as the city's professional social worker, she was the employee—literal or figurative—of the other participants. MacIntosh's brother, for instance, had been instrumental in hiring her. Wisdom was of a generation of professional social workers who sought to differentiate themselves from the volunteer elite women with whom she attended the meeting.[82]

At the meeting on Friday morning, it was the women who led the discussion on how to deal with the suffering masses. The lieutenant governor opened the meeting by reporting on offers of aid from other towns in the province and noted that doctors and nurses had arrived from nearby to help in the hospitals. They talked about committees and appointed more men to them, and there was a report on the financial situation. It was not until Clara MacIntosh spoke up that the meeting's attendees considered how to find out what survivors needed and how to give it to them. MacIntosh proudly announced that all her VADs had been working in the hospitals and offered "to organize voluntary women workers to visit homes from door to door and look after people."[83] Her offer stepped on some toes. Jane Wisdom objected that MacIntosh "took it upon herself without consulting anyone" and had disregarded Archibald, Dennis, and their higher-status organizations. Regardless, MacIntosh's offer was taken up by Archdeacon William Armitage, the pastor at Anglican St. Paul's and a de facto leader of the city's Protestant clergy.[84] MacIntosh turned to Wisdom, who was sitting beside her, and enlisted her help in setting up an office. It was only later in the meeting that Dennis and Archibald offered up the services of their organizations to help MacIntosh's.[85]

The plan was pure City Hall: to send outsiders—middle-class volunteer women—into a neighborhood and have them report back to the center what was happening. It was a difficult task. At noon, MacIntosh and Wisdom took over the mayor's office to set up their operation, and they sent out volunteers armed with typewritten slips of paper telling sufferers where to go for which kind of aid; mostly this meant going to City Hall, where the material relief committee, headed by Wisdom, would see to their needs. If the volunteers found someone who could not come themselves, they were to fill out on a pad of paper what was needed and bring that form back to City Hall. Although Wisdom later described their goal as to "try to fill all urgent needs with a minimum for record," they in fact kept nearly every name and address. MacIntosh reported proudly that the material relief committee had kept a record of all the aid it distributed.[86] They sought to impose legibility on an illegible city.

Dalhousie student John Hanlon Mitchell arrived on Saturday morning to volunteer. Finding MacIntosh and Wisdom's office at "the height of confusion," he "saw that if I wanted a job, I should have to get it myself." A number of women, he wrote, directed by Wisdom, were taking orders for coal, food, oil, and blankets. With a blizzard raging and nary a window in the city intact, blankets went the fastest, and they had to be constantly replenished from the supplies now arriving by train. Many people who came into the office were looking for the registration or transportation departments, which were down the hall. "Most of those who sought food etc were from the less stricken areas. We gave to all without question, though some were patently undeserving. Our only orders were from Controller [John] Murphy to 'Give everything to everybody.'"[87]

Also on Saturday morning, yet another progressive woman arrived. May Sexton, like Dennis a leader in Halifax's Red Cross, was only thirty-seven, young compared to the others. Her husband, Frederick, was the president of the Nova Scotia Technical College; they had met while both were students at MIT, she in chemistry, and married while both were working for General Electric.[88] A suffragist like Archibald and Dennis, Sexton supported the reelection of Robert Borden, who had granted a limited number of women the right to vote. At the time of the explosion, she was in Fredericton, New Brunswick, campaigning for his Union government. After confirming that Frederick was safe, she went about her business, even giving a stump speech. She then left for home, arriving early on Saturday morning thanks to the overnight blizzard. Although she arrived late, she told MacMechan that she had been "prophetic about the conditions at Halifax, i.e., total disorganization."[89]

Like George Campbell, Sexton was concerned with her own convenience, and she was put out that no car picked her up and she had to "skedoodle" into town herself when her train arrived. But she was more concerned about conditions at City Hall. When she got there, she found everyone "dazed" and working in "great confusion." MacIntosh was claiming that her VADs had finished canvassing the devastated area, but no one believed her. Archibald's offer to call out the members of the Imperial Order of the Daughters of the Empire—another women's volunteer group—and the Council of Women had apparently not been taken up. As a result, the women were being "slurred," Sexton thought. Consulting with Archibald, Dennis, and Wisdom, Sexton recalled that the city had already been districted for the election campaign, and she offered to redirect that work toward the registration of disaster sufferers. It was this intervention, Sexton implied to MacMechan, that led to women's inclusion in the relief committee. Until then, in MacMechan's paraphrasing, "there was a distinct feeling not to allow the women to take part in the relief work, or have them on the Committees," perhaps because of disappointment with MacIntosh's perceived slowness. This "distinct feeling" may also have been a continuation of Halifax men's discomfort with civically active women; both the Victorian Order of Nurses and the nearly all-woman Red Cross chapter had been met with resistance from the male establishment. Alternatively, what Sexton perceived as resistance to women may just have been personal hostility stemming from her long history of run-ins with other reformers.[90]

The difficulty in canvassing revealed the inadequacy of progressive women's knowledge of their city. Through political districting and other volunteer work, they knew the city better than their husbands, perhaps, but not well enough. The relief managers of both genders at City Hall rightly saw Halifax as a city in great need: people who required food, blankets, clothes, coal, and building supplies; doctors coming into the city who needed transportation; and supplies shipped in from afar that needed to be organized and distributed. They also saw a major logistical challenge. The city had suddenly become far less legible. The maps and city directories that had been accurate on December 5 now depicted a city that no longer existed. As successful social activists, lawyers, merchants, managers, and politicians—often more than one of these—the men and women who went to City Hall were unaccustomed to the idea that the city could run without them. When they imagined the city around them, they saw chaos, confusion, and disorder. They saw the ad hoc efforts of the relief workers as "higgledy-piggledy," in the words of one of their counterparts in Dartmouth.[91] It was only through their management, they thought, that they could bring order to the city. Their position—both their literal position at City Hall and their social position in the middle class—made it impossible for them to perceive the real order that existed around them.

. . .

When Clara MacIntosh's VADs or May Sexton's Union campaign workers ventured into the North End to find what was needed there, they met people who had spent the previous few days caring for each other. The *survivors*, the group at whom the rescue efforts of both relief workers and relief managers were directed, created their own order. Like relief workers, they built this order on the basis of their preexisting communities, relationships, and networks. But while relief workers used their connections to help others, survivors used them to help each other.

Survivors helped the people they were close to and people they knew. The most basic form of solidarity was with family; survivors often went immediately to their own homes or to the homes of other family members to help them. But people also helped their neighbors and friends—those with whom they had had previous contact and relationships. As with the outside relief workers—soldiers, officers, doctors, nurses—who did under extraordinary circumstances what they were trained for and used to doing in ordinary times, North Enders who were used to helping helped their usual charges. Conversely, people who needed aid knew where to go, so they went to schools and doctor's homes, places where they were used to receiving succor, aid, and support.

The order survivors created and experienced could, from the outside, look disorderly. Digging frantically through a ruined house, trying to rush into a burning building, converging at a doctor's residence, or helping oneself to tools or first-aid supplies could seem disorderly. But in fact it was often outsiders who created disorder. The military devoted substantial numbers of men to policing against looting that never happened and banned civilians from the devastated area.[92] These choices disrupted North Enders' work to rescue each other and distracted from more needed rescue work. Soldiers also feared the fire would spread to the North Ordnance magazine and spark a second explosion. They thus halted rescue work and forced survivors to move south to Citadel Hill to wait outside. Driven by a fear and assumption of disorder and danger, the state and its agents broke up the spontaneous order built by survivors. Order and disorder masqueraded as each other.

Many survivors' first thoughts after the disaster was to check on their families. This was as true for Jack Libby, who ran home and hunted through the burning ruins of his house for his family, as it was for W. A. Major, who left his office at R. G. Dun to check on his wife in the South End.[93] The rush home meant that employers were left in the lurch, which from the perspective of their managers could create disorder. The general manager of the Canadian Government Railways, for instance, complained, "The regular men to man the services were either dead or seeking their dead and temporary arrangements had to be hurriedly made to restore order and carry on the work."[94] From the perspective of a railroad manager, men abandoning their jobs was disorderly; from the perspective of the families they sought out, it was a form of order.

Family members who checked on each other performed an important service. They were often the first people to see their relatives and could judge their physical condition. Helen Cooper was nineteen years old and pregnant, living in the North End while her husband was stationed in Halifax with the army. "Only a kid myself," she joked in 1985, "too young to get married." After the explosion she went to see her husband's father. "That wasn't very far up the street and he was coming down to meet me. And he said to me for god's sake woman, go over and lay down, for your life is a matter of [inaudible]. And he frightened me because I didn't think it was that bad." Her father-in-law had seen what she, in her shock, had not: that a shard of glass was embedded in her neck very close to her jugular. He saw to it that she was loaded in a wagon and taken to the hospital.[95] (In oral histories, stories of glass shards just missing the jugular are sufficiently common they ought not be taken literally. Whatever the specifics of Cooper's injuries, her story implies that in her memory, at least, her father-in-law played a key role in helping her.) Margaret Smith's brother was a seminarian who was in the neighborhood to look at the new organ at St. Joseph's church. When the explosion happened, he went straight to his mother's house at the corner of Kaye and Albert Streets, despite being injured himself. She was stuck under a piano. Because he had checked on his mother, he was able to call out to two passing men, who held up the piano while the son pulled out his mother.[96]

Families' importance continued past the initial rescue; they created the order that structured survivors' lives in crisis. May Gerroir, a fourteen-year-old eighth grader at St. Mary's, went home after the explosion and stayed with her mother, who refused to be evacuated when soldiers warned of a second explosion. This meant that they were both at home when her father, a tugboat captain, and her uncle, who was stationed on the *Niobe*, came home. May, her father, and her uncle went around to the hospitals to find a relative. That evening, the extended family found a place to stay together. "Eventually, of course, my relatives all got together," she remembered sixty-eight years later.[97]

"You went around to all the people that you know that needed help—if you could," one survivor recalled.[98] Although family was the most important community to which people belonged, "all the people that you know" extended to friends and neighbors, too. Sometimes this help was rather mundane: an injured man in the hospital was able to find out his wife's fate by having a hospital volunteer call his mother-in-law's neighbor. The neighbor, he rightly presumed, would know if his wife was okay.[99] One observer described neighbors sharing information about the state of each other's houses and families in the first few hours after the explosion.[100] Other times, this assistance was material. Margaret Nowlan was a schoolgirl just starting the day at St. Mary's School when the explosion happened. Sixty-eight years later, her memory was confused and her story full of contradictions, but she recalled the way friends looked after each other. She recalled waiting at the Citadel for the all clear after the warning about the feared second explosion. "There was another lady up there that knew us and she went home and she got us something to eat and she brought us hot tea. And we ate it up there in the rain, on the Hill." The specifics here—whether she was at the Citadel or, as she had said earlier, another park, or even whether this happened on the immediate day of the explosion—are unimportant. Nearly seven decades later, the memory of neighbors looking out for each other survived when other memories had faded.[101]

Most important, the support lent by friends and neighbors was emotional. At a traumatic moment, they provided camaraderie and a place to sit, often over a meal. Although she lived in the North End, Mrs. Henry Dunstan was sufficiently middle class that she had a maid. "Mrs. D's house was soon filled with wounded people—but there was no place to lay anyone down," MacMechan summarized. Despite the crowds, Dunstan, probably with the help of her maid, fed fourteen people before soldiers evacuated them.[102] The same thing happened in working-class North End homes. Julia DeYoung's father, a munitions factory worker and sometime merchant sailor, had refused to leave the house when sailors and soldiers tried to evacuate him, though the rest of the family went. When they got home from the Citadel, he had tea ready. "We came home—my father had tea made and all the windows boarded up and everything cleaned up. Our neighbors came in and what bread we had and everything—he made everybody something to eat." Meanwhile her mother had gone to Victoria General to help tend to her injured neighbors.[103]

That most of the solidarity shown by sufferers from the explosion was an expression of the daily solidarity that existed in everyday life does not discount the emergent solidarity shown by fellow survivors who had not before known each other. Some of these spontaneous acts of altruism took the form of giving injured passersby blankets or coats. One woman, whose family owned a piano store and who lived outside the devastated area but called her house "shattered," wrote to her cousin about her experiences. A relative of hers had gone out but came running back to the house to warn people about the feared second explosion. "On her way out she gave her coat to a poorly clad woman with two little ones in her arms," the woman wrote.[104]

Similarly, an otherwise unidentified North End woman named Mrs. Moore described several instances of being helped by strangers or near strangers. She had been watching the ship fire with three of her young children while the baby slept in another room. When the blast came, she was knocked unconscious, only to be revived by the small tsunami that washed ashore from the force of the explosion. When she extracted herself from the rubble, she found herself standing in water up to her knees, wearing only her underwear. A man came by, and she pleaded with him, "For God's sake, get my children!" But by then, the house was on fire, and although the man tried, he was unable to save the four children inside. Distraught, she looked fruitlessly for four more children who had been at school. As she wandered, a woman covered her with a blanket, and someone else gave her a sweater coat. She soon found a cluster of other refugees, and they drifted west. As they were walking, a wagon picked them up and took them out to the suburb of Rockingham. The group spent the night in a cottage they found there, and townspeople brought them mattresses, bedclothes, and food. Mrs. Moore had to rely on this sort of emergent solidarity because she had little family left: she lost her husband, five children, two sisters, four brothers, her mother, three sisters-in-law, and twenty-five nephews and nieces, leaving her with only four young children, of whom three were still in the hospital in January, and a married daughter.[105]

Mrs. Moore was apparently too dazed to notice or remember who gave her the blanket and sweater, but it must have been a North End neighbor. If nothing else, they are likely to have seen each other on the street before, gone to the same stores, or had similar, minor interactions. It was perhaps a similar dynamic that motivated a Veith Street shopkeeper. The building ruined, she gave away her entire stock to her neighbors, declaring, "I can go to service again."[106] Her beneficiaries were not likely her friends; rather, they were neighbors and customers, people with whom she had built bonds of solidarity through small interactions on a daily basis. To use urbanist Jane Jacobs's term, she was a "public character," someone who knew many people and who was known by many, but in an impersonal way, and for only a specific purpose.[107] In a time of crisis, however, the role of this public character expanded. Likewise, men and women in caring jobs—notably teachers—expanded their everyday roles. At the local public high school, the principal and teachers evacuated all the children within twenty minutes and administered first aid to the injured ones.[108] At St. Joseph's, a Catholic elementary school, injured teachers mustered their strength to help children escape the wrecked building. "The Sisters, though streaming with blood, cared for the fainting and dying children, securing wraps and coverings for those who could not walk," wrote a proud observer.[109]

Meanwhile, neighbors were arriving at St. Joseph's School and convent seeking shelter and succor. Since it was a preexisting community center and site of refuge, parishioners thought to go to St. Joseph's in a time of crisis. People went to other community gathering places, too. All through the war, the YMCA had been a hangout for soldiers and sailors, and the week before the explosion it reached record attendance. No wonder that people in Halifax knew it as a place to go for help and respite, and that within half an hour of the explosion, "people began to swarm into the rooms, first the hurt and a little later the hungry."[110] Sometimes the gathering points were still more informal. C. Sutherland was a barber on George Street, a fair-skinned and fair-haired man with a big nose and a big forehead but a delicate aspect. He and his colleagues turned their shop into an impromptu dressing station, using the peroxide, cotton wool, and sticking plaster they had on hand.[111] Some people went to drugstores for aid, figuring that they were sites of medicine and that someone could be found there who could help. Medical student Florence Murray had ducked into Buckley's Drug Store to take some supplies when a woman was half-dragged, half-carried in, blood pouring from a cut artery in her face. Murray bandaged her and left her to rest in the store.[112] In all these cases—the school, the Y, the barbershop, and the drugstore—ordinary people went to ordinary sites, places from their everyday lives where they thought they could get help.

Similarly, survivors flocked to the doctors' houses and offices, knowing that they would be able to receive aid there. As we have already seen, many people went to Dr. G. A. MacIntosh's house, where his wife Clara gave them first aid.[113] The same thing happened at the houses of M. S. Dickson and M. J. Burris, two Dartmouth doctors. Dickson worked all morning, assisted by his medical-student niece.[114] Burris, too, had a steady stream of patients through tea time, after which he switched tactics and went around to houses to check on his neighbors.[115] Twenty-two-year-old Catherine Boudreau lived with her family in the North End neighborhood of Merklesfield, on the far side of Fort Needham from Richmond. Her sister was just back from a nursing course in Hartford, Connecticut. "Of course," Boudreau (by now Catherine MacDonald) recalled in 1985, "the local people knew that she was home and she was a nurse so that was their first thought." When Catherine got home from work, "the living room was full of people," and her sister set her to work tearing sheets to make bandages, and then sent her across the street to bandage people who were waiting in a park.[116] C. C. Ligoure, a Trinidadian immigrant and recent medical graduate, was the only doctor in the vicinity of the North End's cotton factory, which burned down. The severely injured came to his office immediately, and he worked all day, ignoring the warning of a second explosion, assisted by his housekeeper and the Pullman porter who boarded with him.[117] Injured people going or being taken to a doctor's office was local order dependent on local knowledge. They had to know where doctors (or in Boudreau's case, nurses) lived. To go there was logical and displayed a sort of immediate order.

The scenes at the Dickson, Burris, Ligoure, and Boudreau houses was repeated elsewhere. Lieutenant Colonel A. W. Duffus described his medical officer, Captain A. McD. Morton, as "immediately besieged at his home."[118] Duffus's choice of words was telling: Captain Morton was "besieged." For people who valued central order or who were dependent on central knowledge, the idea of people flocking to doctors' houses was disorderly. It meant that doctors had to do their work in unusual circumstances, with makeshift tools and bandages. In Dr. Dickson's case, it meant doing his work outside. No records could be kept, no charts consulted, no statistics collected. Worse, it was a spatial transgression: medical work was being done spontaneously, in homes, rather than in hospitals. A medical relief committee report used the same word as Duffus: "Unfortunately for the hospitals many of the local practitioners were besieged in their own offices by crowds of wounded people and as a result for two or three hours the various hospitals were left seriously understaffed."[119] This may have been a serious concern, but it also demonstrates the concern relief managers had for control and legibility.

This was not the only instance of locally orderly behavior appearing as disorderly to those from the outside. All over the North End, survivors "raided" drugstores looking for supplies.[120] They also disobeyed orders, often so that they could continue doing necessary work. Julia DeYoung's father was just one person who refused to leave his house when soldiers ordered him to evacuate.[121] Mrs. Albert Sheppard, who had been a nurse before her marriage, ignored the orders to go south and instead went to the dockyard, explaining, "They need me there, there are no nurses, I'll go."[122]

Most of these disregarded orders were given as soldiers spread the news that there was likely to be a second explosion. This warning and the military orders for citizens to move south and west as a result were the prime example of the way survivors experienced disorder that masqueraded as order. The warning, a YMCA official wrote, "resulted in something approaching a panic that cannot easily be described."¹²³ Nonetheless, some attempted description: "The whole street was alive with people running south," wrote a lieutenant returned from the war. "Mothers badly burnt were hugging babies, old women, who perhaps had not left their beds for years, had in the moment of fright rallied their feeble strength and were bobbling along among the frenzied crowd. Everyone was fleeing clothed as they had been at the time, some had grabbed the thing nearest to them and were carrying articles which were of no value." Later, as he went past the Citadel, he saw thousands of people huddling on it, fearing a second explosion.[124] A schoolteacher wrote to a friend about the "soldiers [who] came galloping down the road ordering people out of the houses immediately as they expected another explosion any minute. . . . For two hours and a half on a frosty winter morning, every living person, even the sick and dying had to stay in the open."[125] The military may have been right to worry about a second explosion—the fire was indeed near the North Ordnance magazine—however, at least one officer "ascertained that there was no danger" but was largely unsuccessful in calming the alarm.[126] Regardless of whether the fear was reasonable, the warning had dire consequences because it delayed rescues, forced injured people to stand outside in the cold, and disrupted the networks and relationships on which survivors relied.

Soldiers and officers imagined themselves to be efficient, orderly, and indispensable. But when twenty men from Wellington Barracks arrived at Bloomfield High School ready to rescue the children, they found that ordinary members of the community—the principal, teachers, and parents—had already done so.[127] Civilian Haligonians, too, viewed men in uniform as purveyors of official, correct information. One returned lieutenant went into the streets to help bandage people but found that "this was difficult work as everyone was so nervous and anyone in uniform was at once surrounded and asked dozens of excited questions at once."[128] Military men knew the power of their uniforms and used it to their advantage. Another returned officer intentionally dressed in what remained of his old uniform in order to assume its borrowed authority.[129] Because officers were fallible but trusted, any decision they made or information they passed on was amplified.

But of course soldiers often did not know any more than civilians. Although many soldiers and sailors in Halifax were locals, many were short-term residents with even less local knowledge than the relief managers. Around midnight the first night, Warrena Maddin was walking home after volunteering at Camp Hill Hospital, when she was stopped by soldiers who offered her a ride. She accepted but discovered they did not know the city, and she had to direct them.[130] One can imagine that their ignorance of Halifax rather hampered their ability to give

Great Halifax Explosion.—Utter Desolation and Devastation so Complete that this Picture might have been taken on the Battlefield of France. —Copyright Underwood & Underwood NY.

FIGURE 6. "Great Halifax Explosion.—Utter Desolation and Devastation so Complete that this Picture might have been taken on the Battlefield of France." The caption—together with the uniformed figures outlined by hand on the far right of the photograph—emphasized how the explosion was understood as part of the war. Postcard published by Novelty Manufacturing and Art Co., Ltd., Montreal. Photograph MP207.1.184/28, negative 20915, Halifax Explosion Photograph Collection, Maritime Museum of the Atlantic, a part of the Nova Scotia Museum.

relief during the day. Alternatively, though Maddin identified them as soldiers, they may also have been American sailors who landed to help patrol the city that night. If that was the case, the story should remind us that outsiders who could not find their way around the city were unlikely to do a very good job policing it.

Soldiers were also fallible because for them, as for civilians, this was a new and unfamiliar situation, and they did not know what to do. Sometimes, as with the warning of a second explosion, they made mistakes. For instance, soldiers did most of the identifying of bodies at an ad hoc morgue in the first week, and the professional who arrived a few days later to take over complained that they had made lots of errors, making his work of identification harder.[131] Soldiers were no better suited for this unfamiliar work than anyone else, but their uniforms vested them with greater authority.

For good or for ill, in a military city during wartime, the explosion was seen as part of the war. "At last the war has come to us, I'm glad," the army's second-in-command admitted three months later thinking at first. "People will know that we're in it."[132] Postcards appeared comparing Halifax's destruction to that of battlefield towns (figure 6). Even battlefield veterans saw a connection. The explosion reminded Lieutenant Rod Macdonald of the war not only because of the physical destruction but because of people's responses as well. Haligonians showed bravery like their sons, husbands, and brothers in the trenches; like those soldiers, they had built spontaneous organization on the basis of preexisting bonds of solidarity. "It was not until some days after the explosion that I was able to view the devastated area as a whole," he wrote. "I climbed Fort Needham and looking from there I thought of Vimy Ridge and the villages to the other side,—Givenchy, Petit Vimy, Vimy, and Avion. It most certainly was as desolate a scene as I have ever seen in France and I am sure that the Nova Scotian lads over there will be proud when they hear with what great fortitude the home folks bore their suffering and bereavement."[133]

2

"A GREAT POWER HAD SWEPT OVER IT"

POLITICS AND POWER
AFTER THE SALEM FIRE

At about one thirty on the afternoon of Thursday, June 25, 1914, a fire broke out at the Korn Leather Company in the Blubber Hollow neighborhood of Salem. No one could say how it started; Charles Lee, the one worker who was burned, had other things on his mind than explaining the fire's origins because he had to jump out of a window to save himself, breaking both his legs. But it was miraculous that a fire had not started sooner; to make patent leather, workers there painted leather with a solution of scrap celluloid film in alcohol and amyl acetate. Then they stained it with lampblack and wood alcohol. All this flammable work was done in a rickety, four-story wooden building with no sprinklers. By two o'clock, the fire had spread to fifteen surrounding buildings, and three hundred workers had already fled their workplaces. Aided by dry conditions and high winds, the flames traveled in a crescent shape across Salem, and by seven o'clock they had crossed into a working-class neighborhood called the Point. "The rush of the flames through the Point district was the wildest of the conflagration," reported the *Salem Evening News*, "the flames leaping from house to house with incredible rapidity." By the time they reached the Naumkeag Steam Cotton Company, the flames were too hot to be stopped by its modern antifire devices. The mill burned, and thousands of jobs went up with it. Overall, the fire destroyed 3,150 houses and 50 factories, leaving 18,380 individuals homeless, jobless, or both.[1]

The Point was a dense neighborhood clustered around the Naumkeag factory, which was sometimes called Pequot Mills for the brand of sheets and pillowcases produced there. About six thousand people lived packed into "loosely constructed

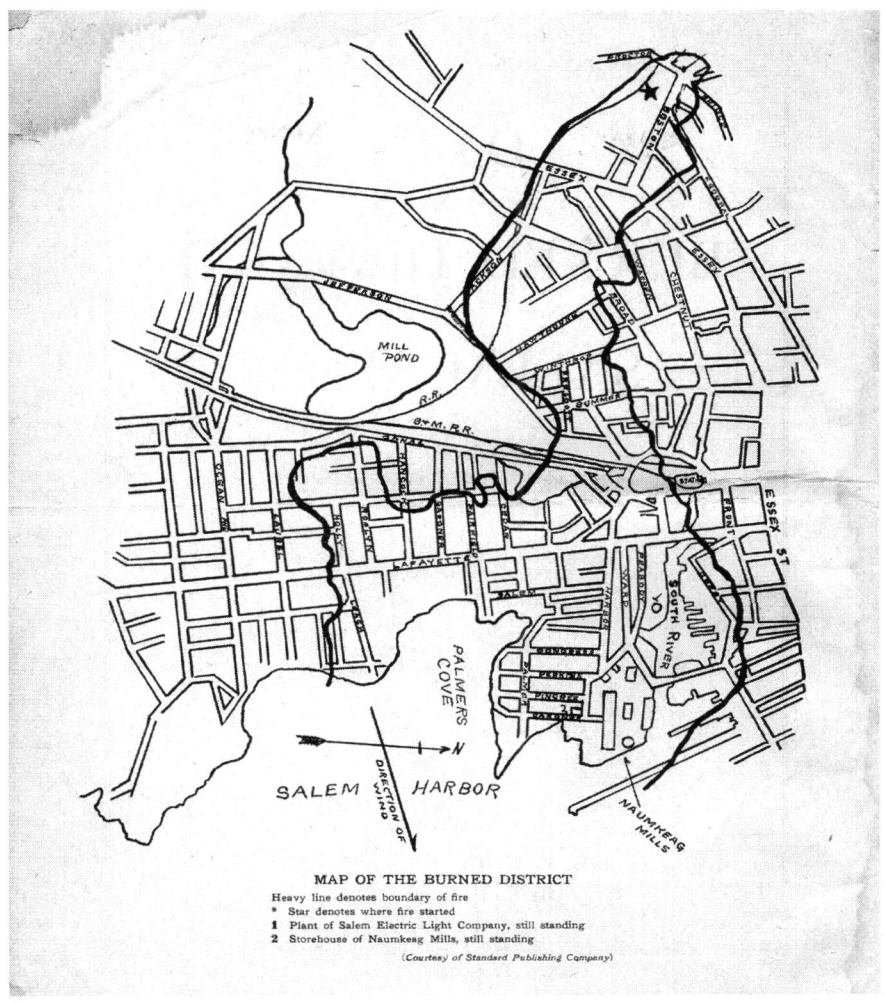

FIGURE 7. Salem's burned district and surrounding areas. Salem State University Archives and Special Collections.

houses, most four and six stories high, all of them wood and closely congested."[2] Three years earlier, two housing reformers had written that the Point was dangerously and unhealthfully overcrowded. "Double four-story wooden tenement houses are common in this district, and the buildings are literally piled one on top of the other," Selksar Gunn and Samuel Schmidt wrote. The tenements were so close together that little ventilation or light reached the bottom floors, and the overcrowding was not only of buildings but within them as well. Only 27 percent of the population in their study slept in rooms by themselves or with one other person; 36 percent slept in rooms with more than three people. All this led, they wrote, to filthy shared toilets, rooms, and yards.³

They were most concerned about fire. There were "practically no building ordinances," they complained, which put the residents there in great danger. Although the law required that buildings more than three stories tall have back stairs, most of these were wooden and thus were themselves flammable. "There is no doubt, that if a fire should ever get a good start in that district during the night, many lives would be lost owing to the lack of proper fire escapes," they warned. "The dangers from conflagration at the Point are very great."⁴

The city Gunn and Schmidt presciently feared would burn was no longer Nathaniel Hawthorne's archetypal Puritan village or a center of the East Indian trade. "Founded in 1629 by sturdy Puritan pioneers and intimately identified with the events which led to the independence and up-building of the American colonies," wrote the national director of the Red Cross, "it is with something of a shock that one finds Salem today the home of a teeming population of immigrants of such recent importation that a large proportion of them is unfamiliar with the English language and without knowledge of, or interest in, the stirring history of Salem herself." Despite its colonial and early republican heritage, Salem was now a firmly industrial city sitting amid the textile and leather cities of North Shore Massachusetts.⁵

Textile work, mostly at Naumkeag, was low paying but year-round; leatherwork and shoe work paid better but only for seven months. Salem residents also commuted to iron foundries in neighboring Danvers and Beverly and to the United Shoe Machinery Company in Beverly. A committee appointed in 1826—the same year Irish immigrants founded the city's first Catholic parish—had warned that building factories would perforce attract foreign workers and create substandard foreign quarters in the city.⁶ Indeed, when Naumkeag was first opened in 1847, its workforce was primarily Irish. As in other parts of New England, French Canadians gradually replaced the Irish as the dominant working-class ethnicity, and it was these French Canadians, plus the ethnic professionals and small businessmen who serviced them, who lived in the Point.⁷ There were also smaller but sizable populations of Poles, Greeks, Italians, and Eastern European Jews; the Greek and especially Polish communities were growing particularly fast in the second decade of the twentieth century.⁸

At the start of the twentieth century, this industrial immigration threatened Salem's political order. In 1901, voters elected Salem's first Irish mayor, John F. Hurley, an immigrant landlord and teetotaling former saloonkeeper. His election changed municipal politics from a sleepy, friendly business to a more rough-and-tumble affair. Between 1901 and 1911, he was elected to four one-year terms, and in 1912 he was reelected for a three-year term, but each election was hard fought and marked by bitter ethnic politics. Yankees accused Hurley of corruption, and the Irish accused his opponents of bigotry. One major bone of contention was patronage, especially in the police department, which under Hurley was increasingly dominated by Irishmen. In early 1914 he had decisively alienated Franco-Americans by firing a police matron and police-car driver. It was Hurley's lax enforcement of Salem's liquor laws that precipitated his final battle. At the end of December 1914, six months after the fire, Hurley was deposed in the first modern recall in New England. He was replaced by prohibitionist Matthias J. O'Keefe, a leather manufacturer backed by the Better Government Association. Despite his name, O'Keefe was native born and probably Protestant.[9] The Francophone *Courrier de Salem* ran an exultant editorial: There would reign "contentment and happiness for citizens with peace, order, and liberty; a magistrate who honors his post by dispensing justice to all, without distinction of language, race, or religion; who imposes respect for the laws for the great as well as the small, rich as well as the poor, friends as well as enemies."[10]

Only 19 percent of the families affected by the fire—those who lost their homes or their breadwinners' jobs—were classified by the Red Cross as American. The rest were foreigners of some sort: 43 percent were French Canadian; 20 percent were Irish; 6 percent were Poles; 5 percent were Italians; 4 percent were "Hebrews"; 2 percent were Greek; and 1 percent were "miscellaneous."[11] Although many of these people may have been legal citizens, having been born in the United States or naturalized, their experiences were marked by immigration. Immigration marked them to others, too. Montayne Perry was a fiction writer, onetime YMCA worker, and past director of Salem's night schools for immigrants. In October and November 1914, she published a serial in the *Salem Evening News* describing the fire and especially the relief efforts, and she republished her reporting in book form the following year.[12] Immigrants' difference was imagined as a fundamental part of the disorder Perry and others observed in the wake of the fire. A National Guard officer wrote to his superior that fall, complaining of the way immigrants had used the toilets provided for them at a refugee camp: "It is needless for me to state that the urinals were used by people of all nationalities a number of whom had no knowledge of the proper use of the modern sanitary toilet."[13] Whether the officer was correct that refugees did not know how to use toilets was less important than his implication that their ignorance was a national characteristic. The question of citizenship was crucial to the city's relief and rebuilding. In the fire's aftermath, politics—in the armory, city hall, statehouse, and US Capitol—would determine order and relief. This chapter traces those politics, civilian and military, at the local, state, and federal levels.

∙ ∙ ∙

The fire had arrived in residential neighborhoods with warning. The flames may have been furious, but the five and a half hours between the start of the fire and its arrival in the Point allowed people to plan and react, and only a small handful of people died. So the tasks for relief workers were quite different in Salem than in Halifax. The emphasis was on saving possessions, not lives. There was not the same need for volunteers to dig the dead and injured out from rubble or to tend to the wounded in hospitals. The closest thing there was in Salem were people who looked after and helped transport families' possessions. Most of those people were those with the ability and inclination to move things—that is, teamsters and others with access to motor vehicles. "Automobiles proved themselves of wonderful value in fighting the fires and in relieving people who suffered from the fire," editorialized the *Salem Evening News*. "Thousands of stories can be told of how the automobile accomplished things that never could have been done by horses and wagons, trolleys, bicycles or any other means of transportation or communication." The fire came at the end of the first period of the automobile's popular growth and just before motor vehicles were to prove their military usefulness in World War I. Car owners were predominantly middle class and suburban or rural.[14]

Among them were Richard E. Eagan, the owner of the Bay State Creamery, a company that owned ten "trucks, ice cream teams, and ice teams and [a] pleasure car." The first squad of militiamen responded to the fire on foot, but when they met Eagan in his car, he offered to carry them to the fire. He spent the afternoon ferrying militiamen around the city and put the rest of his vehicles to work, at one point volunteering his entire fleet.[15] Eagan and his drivers were joined by others who continued in abnormal circumstances what they did in normal times; that is, carrying loads in their trucks. One family rescued its bed frame, sewing machine, and trunk in a coal wagon.[16] Edward Dunbar Johnson, who wrote a report on the fire a year or two later, remembered the help of laundry teams and delivery wagons.[17] The evening newspaper described a nurseryman sending a truck "to the poorer district in the path of the fire, to render aid to those in distress." The crew saved the possessions of forty-six families.[18]

The coal wagon's driver, the nurseryman and his crew, and Eagan and his dairymen did not, on other days, make a habit of carting families' household goods, especially not for free. In that sense, what they did was unusual or extraordinary. Yet in each case they were extending their usual activities in a logical way, offering help within the patterns set by their everyday lives. Moreover, as in Halifax, relief workers created informal, spontaneous organization without any organization. Peter Chase, a militia captain, arrived in Salem on Friday afternoon and tried to take charge of transportation. He found several trucks already on duty, seemingly without direction. "All of these trucks had been provided before I came to Salem and had been at work for one or more days, by whom they were provided I do not know, but I kept them all at work, them and as many more as I was able to get."[19] Nobody appeared to have kept track of who used whose truck, and indeed it seems likely that teamsters and truck owners offered their services and equipment without anybody requisitioning or even asking them. The next day, the newspaper several times spoke of vehicles "pressed into service in the removal of goods."[20] The passive voice suggests that the reporters did not witness a particular person conscripting the trucks or their owners, probably because in fact the assistance was voluntary and not "pressed." As we will see, when the military later attempted to impose external order and create a record of which officers used what truck, the local order that allowed families to save their possessions was lost, and all that remained was confusion.

Even the militia's aid was, as in Halifax, offered unofficially and without direction. "Almost without exception the commissioned officers of the local militia were absent from the city, in attendance at the Service School session in West Newbury," Montayne Perry wrote. "But from every part of the city privates and non-commissioned officers hurried to the armory to report and equip themselves for service."[21] Without their officers, the enlisted men must necessarily have been acting without orders. The postcard in figure 8 shows a family on Roslyn Street being helped out of their three-decker by a militiaman. The fire is only two buildings away, so they have little time left. A woman looks on, her right hand resting on her ample chest as if to calm herself; perhaps she is calling to the man on the second floor, urging him to hurry. The assistance the soldier offered this nameless family was given on the basis of something else besides military discipline. He necessarily worked on his own volition because there were no officers in the city to have given orders.

Although most of the relief work in Salem was directed at saving possessions, there were some injured people. Nineteen people were taken to the hospital with injuries ranging from broken bones to smoke inhalation; they later had to be moved when the flames threatened and then consumed Salem Hospital.[22] Another twenty-one were treated at a field hospital established at the Salem Armory.[23] These forty patients and their relatively minor injuries did not necessitate the large-scale volunteer effort that kept Halifax hospitals running, but they did require people to extend their ordinary roles. The first group of soldiers driven to the fire by Richard Eagan comprised eight members of the hospital corps, plus a sergeant.[24] Perry recognized the self-supervision of these enlisted

FIGURE 8. Postcard: "Militiamen Helping Remove Household Goods, Salem, Mass." Postcard, courtesy of Salem State University Archives and Special Collections.

soldiers, though her analysis was rather different: "The first aid training of the squad demonstrated its value throughout those hours when they worked alone, anxiously hoping that their surgeon, on his way now from the Service School, might find them very soon."[25] In fact, the success of this first-aid station suggested that commanding officers (and their skills) were unneeded.

As in Halifax, civilians with medical training volunteered their help. Perry wrote of a "physician who chanced to come upon" the first-aid station and "promptly offered his services."[26] The *Evening News* described the "valiant service" of a Beverly nurse named Mary Reed, who on her way home for the evening stopped off at Salem Hospital. After the hospital was evacuated, she moved to the armory, where, like her future counterparts in Halifax, she stayed up all night.[27]

In both these instances, the doctor and the nurse were enacting in unusual circumstances what they did in ordinary times. The growing professionalization of doctors and nurses in this period encouraged practitioners to think of being a physician or nurse as an identity, not just a job, and thus something to be enacted whenever the need came.[28] To the extent that the militiamen at the first-aid station followed directions from the civilian doctor, they were voluntarily respecting the authority granted by his training, experience, and professional status, creating spontaneous order based on their positions before the fire.

Even as relief workers proceeded without formal supervision, those who valued central knowledge and organization perceived, in Perry's words, the city as "a scene of seemingly hopeless confusion, with the imperative, insistent need for men of recognized authority to dominate the situation, to assume control, and compass, so far as possible, the protection of life and property."[29] Perry's language is telling: First, what Salem needed was the "protection of life and property," betraying her emphasis on property, rather than community, as paramount. Second, she imagined a particularly male form of power, one whose authority was instantly recognized and that dominated its surroundings. Volunteer women who created less recognizable or dominating informal order were not good enough. Luckily for Perry and those who shared her perspective, both Governor David Walsh and Adjutant General Charles H. Cole—the commander of the state militia—arrived during the evening and lent their sympathy and authority, imbued with precisely the right sort of masculine domination.[30]

On Friday, June 26, as the ruins still smoldered, the relief managers started their work. Early that morning, Cole called a few leading citizens to the armory for a meeting to discuss the civilian relief operation. No minutes remain, but it is likely that John F. Moors, the Boston broker and celebrated disaster-relief expert, attended as the governor's representative. At ten, the city council met and decided to call, in the words of a local newspaper, a "big meeting with all classes of citizens" at noon to consider forming a relief committee.[31]

The use of the word *citizens* is suggestive, given that so many of those directly affected by the fire were aliens. The same article had earlier referred to the "French citizens," so the term does not appear to have been intended to exclude those marked as newcomers. Rather, it suggested a political argument about the nature of relief and rescue. As members of the Salem community, "citizens" in need deserved relief from their community, whether manifested in the state or in an ad hoc committee. Likewise, it was citizens who should make up that committee and exercise control over relief efforts. Referring to *citizens* was an acknowledgment that relief organization was an inherently political task, to be carried out in the context of Salem's preexisting political terrain. How citizens should be heard and which citizens' voices counted were up for debate. Already, some were considering a wholesale takeover of the city government. "It may be necessary to make over its form of government as was done both at Galveston and Chelsea," the *Boston Transcript* wrote.[32] The thought was not unanimous, as the *Salem Evening News* cautioned. "Others said that commission form of government, such as Salem has, originated in a catastrophe such as Salem now labors under, and that as we had just that form of government we had all that was necessary."[33]

Disaster relief is inherently political, in that it is about the distribution of material resources and power within a community. In Salem, these politics played out on a stage constructed by the decade-and-a-half battle between Irish mayor John Hurley and his Yankee opponents. In June 1914, Hurley was halfway through a three-year term, the first since the city charter had been reformed to give the office more power. Friday's noon citizens' meeting, ostensibly to form the relief committee, was a distinctly political affair. At it, a "unanimous and aggressive" audience derailed the agenda in order to make demands about municipal politics. A few days earlier, Thomas Lally, the city's Irish-born, socialist director of public health and a strong advocate for municipal reform, had fired all the members of the city's public-health board. Using their fear of an epidemic as an excuse, the assembled citizens loudly demanded the board's reinstatement. Speakers berated Lally until he finally gave in and promised to reinstate the board. With that important question of local politics solved, the meeting could finally turn to creating a committee to help those who had been burned out of their homes and jobs. Newspaper editor John B. Tivnan declared, in a reporter's paraphrase, that "time for talk was passed and it was time to get to work."[34]

After some discussion, the meeting voted to appoint a Committee of 100, and Mayor Hurley "retired with representatives of various organizations to select" it, Tivnan's *Evening News* reported. The committee, which actually numbered about 110, was remarkably diverse. Judging from names, about 10 percent of its members were Franco-Americans, and seven members were women.[35] Of the total, ninety-one are identifiable on the 1910 or 1920 census, in the city directory, or through descriptions in newspaper articles. The committee was mostly Protestant, including two former mayors and the man Hurley had defeated in the previous election, but it also had a healthy number of Irish Catholics.[36] The chairs of transportation, housing, and employment subcommittees were all Knights of Columbus.[37] Three told census takers that their native tongue was Yiddish.[38] Compared to the people who came to early meetings of the Halifax Relief Committee, if not to the sufferers from the fire, the Salem Committee of 100 was a much broader cross section of the city.

This apparent ethnic and religious diversity should not deceive, however. The committee was overwhelmingly middle-class and elite.[39] Including the mayor and the city directors, there were eleven politicians. The president of the board of trade, who owned a plumbing business, was on the committee.[40] Five members were factory owners or executives; they were almost all placed in charge of a subcommittee.[41] Seven were bankers.[42] There were seven store owners and another five managers, clerks, or salesmen.[43] Fourteen worked in real estate, insurance, or brokerages.[44] Three were lawyers.[45] The biggest profession represented was clergyman, with at least twenty-one, of whom eight were Catholic.[46] The general secretary of the YMCA, the secretary of the local Associated Charities, and the clerk of the city's Overseers of the Poor were all included.[47] Some of Salem's fanciest families—like the Phillips, Batchelder, Gifford, and Rantoul families—had two members apiece.

Moreover, a funny thing happened to the Committee of 100 and its diversity. Montayne Perry explained: "One Hundred was too large a body to do effective work. These men elected twenty of their number, who speedily reduced themselves to seven, to act with seven men sent by the Governor. Thereafter, these fourteen men were the ruling power, a board of directors known as the Committee of Fourteen."[48] In the winnowing process, the original Committee of 100's diversity was discarded. The local portion of the Committee of Fourteen was more Yankee, much more elite, and entirely male, although it maintained a mix of Catholics and Protestants. Edmund Longley was the general auditor and solicitor of New England Telephone and Telegraph and a director of Merchants National Bank.[49] John B. Tivnan, the local committee's chair, was a newspaper editor and had long served on various public boards and commissions; he was Catholic and a local leader of the Knights of Columbus.[50] E. G. Sullivan, elected secretary of both the Committees of 100 and Fourteen, was secretary of the board of trade, and John F. Cabeen was its president.[51] Brahmin John Saltonstall's family had a long history in Salem, but he lived in neighboring Beverly with his cook, butler, coachman, and chambermaid; he was a State Street banker but mostly lived off his family fortune.[52] Eugene F. Fabens was president of the Naumkeag Trust Company.[53] John E. Spencer, a retired colonel, was the owner of a machine factory and the chairman of the food relief committee.[54] Longley and Saltonstall had not even been on the original Committee of 100. The governor's appointees were more diverse, if only by ethnicity and religion. They included a Jew, A. C. Ratshesky; three Irish Catholics, including former Boston mayor John F. Fitzgerald, and Father Michael Scanlon, director of Catholic Charities; and three Yankees, including John Moors and state Red Cross treasurer Gardiner M. Lane. That Scanlon was the official Catholic charitable representative was telling: his organization had been founded to centralize power within the male and Irish archdiocesan hierarchy and to remove authority from nuns and from local, ethnic parishes. His presence thus highlighted the absence of French Canadians.⁵⁵ Also included, as an adviser, was Ernest Bicknell, the national director of the Red Cross.

To the committeemen, the most salient differences were not those of ethnicity, religion, or class, but of location. In public, the local, state, and national Red Cross leaders praised each other and proclaimed that they were working with utmost cooperation.⁵⁶ Privately, though, they fumed, and by mid-July, tensions were so great between the local committee and the Red Cross that they needed a memorandum spelling out that the local relief committee was not under Red Cross authority, that the latter was just there to help, and that while mistakes had been made, they were no worse than in other disasters.⁵⁷ Rumors reached the national press—and were strongly denied—that the relief work was plagued by inefficiencies, graft, and other problems.⁵⁸ The Red Cross complained that the Salem committee wanted too much independence and that it refused to listen to good advice. Mabel Boardman, the chair of the National Red Cross, complained in a confidential letter that "there is a self-sufficiency about this community not altogether unwarranted" that made working with them difficult.⁵⁹ Bicknell and Boardman blamed the local committee for fund-raising failures. "First, they would not appeal outside the state and did not care for a Red Cross appeal," Boardman complained, "then they did appeal outside the State with little success; some of them did not like my effort to stimulate interest by an appeal, so that, of course, we did nothing further in that line and besides it was too late."⁶⁰

Salem relief managers were indeed jealous of their independence. This was most evident when, less than a week after the fire, they began debating the organization and makeup of a Salem Rebuilding Commission to oversee new building codes, changes to the street plan, and permits in the burned district. The question was wrapped up in the same long-term political debates about who should control the city that had nearly derailed the first meeting the day after the fire. The original plan had been to give the commission its own authority to raise money; "in other words," as the *Salem Evening News* reported, "it was proposed to make the mayor and council subordinate to the special outside commission. . . . The whole tendency on the part of some, indicated a lack of confidence in some of the present city council." Regardless of the powers of the committee, the newspaper reported unanimity that the commission should be made up of Salem residents. Proposals for a commission with either a majority or minority of outsiders were seen as an insult, "a confession that Salem was not capable of self government."⁶¹

Such a confession could be complexly dangerous in the context of Salem's ongoing political fights between Yankees and Irish. Starting in the mid-1880s, just when Boston elected its first Democratic, Catholic mayor, the legislature curtailed much of Boston's home rule. By 1910, unelected, state-level commissions controlled Boston's police, debt and taxation, public-school spending, liquor and restaurant regulation, and finance and appointments.[62] In 1908, at the behest of a committee of Yankee finance and manufacturing interests, a Republican governor had appointed an all-Republican Board of Control to take over all of Chelsea's government after a fire there. Of the five members, only three lived in the city.[63] A state-appointed rebuilding or relief committee with its own fiscal power—this time under a governor who was himself a Catholic Democrat—would, it seemed, put Salem in Boston and Chelsea's subordinate position. It could not have escaped notice that the governor's first delegate, John F. Moors, was a leader in Yankee Bostonians' campaign to take over public schools through the good-government Public Schools Association.[64] Thus the question of who would control Salem—Hurley's Irish or his Yankee opponents—would play out unpredictably at the state level, and if Hurley won, the city could lose the independence and self-reliance Boardman had noted were the sources of such pride.

The solution served dual purposes: keeping out outsiders and keeping power away from Mayor Hurley. The final legislation provided for a powerful commission of five Salem residents, to be picked by the governor on the advice of the mayor and city council.[65] Included were two members of the Salem contingent on the Committee of Fourteen, bankers Edmund Longley and Eugene Fabens; a major clothing merchant, landowner, and active member of the Committee of 100, Dan A. Donahue; city solicitor Michael L. Sullivan, who also maintained a large private practice; and Emile Poirier, a Franco-American who was both a medical doctor and, probably more important, a major landlord who had lost many of his properties in the fire.[66] These choices are significant: They marked a considerable shift from those who had been active in the earliest relief organizing while maintaining the dominance of business and financial interests.

An early proposal had called for the committee to include an "expert engineer" and "expert architect" as members. "Some think that these experts ought to be hired outside the commission," the *Evening News* wrote, "otherwise their strong desires to build monuments to themselves, and their idealism, might lead to extravagance beyond what Salem can stand, and that the other three being laymen might not be able to counteract their influence."[67] Eventually, Governor Walsh's final legislation excluded experts as committee members. Instead, the commission hired a paid secretary and held an "informal conference" with architect C. J. Blackhall and A. C. Ratshesky, who had been a leader on Chelsea's board of control after its fire.[68] Blackhall would eventually be hired as a consulting architect, and at times he worked nearly full time planning the rebuilding of Salem.

As Blackhall's role implies, even with laypeople as commissioners, the project was steadfastly technocratic in its outlook and ethos. Inspired generally by new ideas of town planning and specifically by Franklin Wentworth, a former city councilor and the president of the National Fire Protection Association, the commission sought to rebuild a city that would be scientifically sure not to burn again. It heard public debate on new building rules, but it was guided by technical concerns. That more flammable, cheap apartments were outlawed in the rebuilt area shows how technocratic concerns so often mirrored those of landlords.[69]

As politicians in Salem and Boston jockeyed for power, in Washington politicians debated the federal government's proper role in disaster relief. President Wilson quickly proposed a federal appropriation of $200,000, and it sailed through a unanimous Senate.[70] It met more difficulty in the House of Representatives. The national Red Cross, jealous of its role as the nation's official relief agency and still smarting over its disagreements with the local relief committee, lobbied against appropriation. Salem only needed federal money, it and other opponents argued crossly, because it had failed to raise enough voluntary contributions, and because the Massachusetts state government was not doing its share. A federal appropriation would set a dangerous precedent, and every community would now refuse to do its own fund-raising. Worse, fiscal control might be transferred away from the trained and experienced officers of the Red Cross to a "liberal officer" chosen by the local congressman, who might disburse it on political or patronage lines.[71] The request for money first came into trouble in the House Appropriations Committee, which, after considerable lobbying from Massachusetts politicians of both parties, only barely passed it.[72]

In the full House, the concern was whether the victims of the Salem fire qualified for disaster relief at all. Legal historian and sociologist Michele Landis Dauber has described a long history of uncontroversial federal disaster relief. Members of Congress gradually enlarged the definition of *disaster,* and one of the markers of a controversial expansion, Dauber writes, was the appearance of a chart, first used in 1870, demonstrating the precedents for past disaster relief.[73] Indeed, Salem congressman Augustus Peabody Gardner introduced that chart in preparation for the debate.[74] Salem's controversial expansion revolved around unemployment. Michael Phelan, a Democrat from neighboring Lynn, argued that the Salem fire was worse than predecessors because it had burned both homes and factories, thus causing both homelessness and unemployment. But it was this very fact that worried opponents. Potential aid recipients in Salem, opponents claimed, were simply unemployed; that their unemployment stemmed from a fire was immaterial. "It is a new doctrine that the people should be fed by the Federal Treasury when they are out of a job," Mississippi Democrat Thomas Sisson declared. John Fitzgerald, a New York Democrat and the chair of the appropriations committee, feared that the money would be used to keep men on the dole until factories were rebuilt. "It is not a part of the functions of the Federal Government to take care of destitute persons," he intoned. Gardner agreed with Sisson's and Fitzgerald's characterization that relief would help families destitute because they had no jobs but argued that because their unemployment stemmed from the fire, they were worthy recipients.[75]

Because the country was in the midst of a depression, and because radicals around the country had been organizing for relief and free rent for the unemployed, Sisson's and Fitzgerald's arguments may have been particularly important, but they were nevertheless unsuccessful. Phelan made the winning argument: "Where people are in distress and suffering, we ought, without drawing fine distinctions, to give them aid unhesitatingly and ungrudgingly."[76] After only half an hour of debate, the appropriation passed 102 to 64, with 6 abstentions and 201 congressmen absent.[77] But the legislation followed the normal pattern of granting authority over the relief funds to secretary of war Lindley Garrison, who wrote the Red Cross that he "sympathize[d] entirely with [its] views."[78] In his disapproval, he delayed releasing the money. When the money had not been disbursed by mid-August, the chair of the Salem Relief Committee inquired when they could expect it, but he was told, according to a Salem newspaper, that thanks to the "European war situation they could give no time to the matter at present."[79] In November, the War Department claimed to a delegation visiting Washington that Salem no longer needed the money.[80] It was not until December that Herbert M. Lane, a major in the quartermaster corps—the army's logistics managers—arrived to give out the promised funds.[81] This meant that the money would not be controlled by a "liberal officer" enmeshed in politics, as the Red Cross had feared, but by an apolitical expert in logistics and supply.

As Congress decided to expand its traditional ideas of disaster relief to include help for the unemployed, so too did the fledgling US Department of Labor. Terrance V. Powderly, most famous for having been grand master workman of the Knights of Labor, had since 1907 run an office in the Immigration Commission that tried to direct immigrants to parts of the country with higher demand for workers. The project was mostly a failure because immigrants refused its assistance and union leaders tried to sabotage it for fear it would recruit strikebreakers. When the office shifted to the Department of Labor in 1913, the new department worked to remain neutral in labor disputes, even as it faced increased pressure to facilitate the sharing of employment information in the midst of a depression.[82] The Salem fire proved a perfect test run of what the Division of Information could do because it was an event that suddenly threw many men into unemployment but was not a strike.

In early July, Committee of Fourteen member John Saltonstall wrote to labor secretary William Wilson proposing what amounted to a new project. "I trust you may be able to give us some information which would lead to the finding of reemployment for some of these people," he wrote. In response, Wilson or Powderly dashed off telegrams to 95 shoe and boot manufacturers and 218 cotton mills, asking how many displaced Salem workers they could take in; they wrote letters to a total of 1,300 more potential employers from Maine to Maryland. Powderly himself then traveled to Salem, where he worked with the state employment bureau and the relief committee to find work for the newly jobless. As one might expect from Powderly, he tried to work with labor leaders—he solicited the help and advice of the presidents of the Boot and Shoe Workers' Union and the United Textile Workers—but neither man showed any interest in participating, and he made no note of speaking to any local union officials. Through the administration of the local relief secretary and the Massachusetts State Free Employment Bureau and the publicity of the Labor Department, Powderly reported, they found jobs for more than 1,200 skilled workers.[83]

Powderly's work in Salem set the stage for a much larger project to find jobs for unemployed workers and workers for understaffed factories. In January, soon after Herbert Lane arrived in Salem, Powderly's Division of Information was renamed the United States Employment Office; the office expanded especially during the subsequent wartime labor shortage, when it sought to rationalize the nation's manpower.[84] Powderly's and Lane's work were responses to Progressive-Era debates on the role of the state and whether the state, especially the federal government, should aid jobless workers. In both cases, the Salem fire helped find an affirmative answer. They also—like the participation of Ratshesky, Moors, Bicknell, and Blackhall—were part of a Progressive-Era emphasis on the managerial knowledge of experts and professionals.

In a disaster, one reason that expertise was required was that without it, many feared, things would dissolve into chaos. The day after the fire, the *Salem Evening News* described two scenes of perceived disorder. "At the outbreak of the fire," it wrote, "the men and women employes of the Korn factory were thrown into a panic." Later, "the residents of the flame-stricken sections of the city became panic-stricken as the wave of flames pursued their mad rush and gained in size."[85] For some observers, this disorder was gendered. Women, Montayne Perry wrote, were "faint with fear or half-crazed with excitement" and, like children, had to be "cared for and sent to places of safety."[86] If women were infantilized, men were emasculated. Through the voice of a militiaman, Perry told the story of an "old fellow out in his back yard with a short length of garden hose, trying to send its little feeble stream into the face of the fire." Though enfeebled and rendered impotent by his age, the man insisted on "sticking to his post with his little piece of hose." Perry's militiaman character "had to pick him up bodily and carry him out of danger, and him cursing me up and down because I wouldn't let him save his house."[87]

Others likened the fire's victims to animals. George Heustis was the superintendent of the Malden Electric Company, but he happened to be working at the Salem electric plant the night of the fire. When people gathered themselves and their possessions in the plant's walled yard, Heustis, as a male, managerial, middle-class engineer, could see the danger he believed they were blind to. "I will admit I was using some pretty strong language," Heustis wrote in his company's magazine. "It finally became necessary to drive them away, the same as you would drive cattle."[88] Perry used a similar metaphor to describe the way militiamen warned residents to flee the flames. "In the Point district where the conflagration gained its greatest speed, people were literally driven in swarms from their homes, barely in time to escape the flames."[89] Her depiction stands in stark contrast to figure 8, showing a militiaman taking some care to help the residents of a three-decker rescue their possessions.

Observers, and even participants, could signal their difference from the "swarms" by expressing disapproval of what survivors chose to save. Rescuing certain, emotionally significant possessions—or even pausing to save any items at all when the fire was bearing down—appeared frivolous, foolhardy, or disorderly. The importance of what people saved was not necessarily discernible from outside their own families or communities, but their salvage allowed people to maintain some sort of emotional order and control over their lives. The *Salem Evening News* took a tone that was at once maudlin and mocking: "There was a grotesque side, notwithstanding the pitiableness of the general situation, in the sight of frenzied workers lowering to the yard, by a clothesline, some inconsequential article, while friable goods of value were tossed from windows, naturally, to be smashed," it wrote the next day. "One woman might be seen carrying a cheap picture, the glass broken in the crash; another member of the family would follow with an equally unimportant article. Here and there, in the residential sections, curtains in homesteads were drawn. From the front door would emerge occupants carrying sundry cherished heirlooms."[90]

Looking back on the fire fifty-six years later, John Porter Sumner agreed with these critics. "Most of them," he wrote of other refugees, "seemed more anxious to save their belongings than themselves. We had seen them carrying some of their things a short distance, dropping them, and then hurrying back for still more things." Yet when it came to his own family, the order was quite apparent and he was able to understand the salvaged items' emotional, not material, importance. When Sumner, then eight years old, wanted to put down the tea table he was told to carry, his mother explained that it was a family heirloom and must

FIGURE 9. "Salem Fire Ruins, Salem Mass.: Homeless on Common, June 25, 1914." Victims of the fire relax on and near their salvaged possessions. Postcard, Salem State University Archives and Special Collections.

not be abandoned. Sure enough, when his mother died forty-three years later, he inherited the table he had saved, now imbued with even more value thanks to the memory of the fire. Similarly, he and his father went back to the ruins of the house a few days later to find a childhood memento of his father's.[91]

Figure 9, taken that day, shows homeless people camped out on the Salem Common amid what they had managed to salvage from their burning homes. The fire was traumatic, and these people had all lost their homes and most of their possessions. But once safe in the park, they remained with their families, friends, and neighbors, keeping each other company, comforting them, and relying on and strengthening the relationships they had before. The presence of familiar household possessions would have been further comfort.

Survivors relied on these informal bonds of everyday solidarity. The day after the fire, a Mrs. Green of 5 Skerry Street, feared that her neighbor Mrs. Maguire would worry after her, so she placed a notice in the evening newspaper "that she is in Marblehead and is all right."[92] There is no record about what Maguire had been doing, but Green at least expected her to be concerned and looking for her. Perry described friends and neighbors who "paused to call out 'your wife's up in the pastures, Bill, and the kids are safe with her'; [and] . . . women with a half dozen little hands clinging to their skirts [who] bent to lift and comfort the tired, sobbing child of a neighbor."[93] John Sumner likewise remembered his father "helping the neighbors next door, who were attempting to put out a fire on their roof. They were standing on ladders leaning against the house, beating out the flames with brooms. There was very little pressure in the water pipes, and the hoses were almost useless. There was only enough water to fill some pails, and my father got up on the roof of our house to throw water on it."[94] Sumner's recollections were very similar in content to Perry's scene of the old man with his "feeble stream" and "little piece of hose," but the story takes on a very different valence when it describes interneighbor solidarity rather that the laughably futile efforts of an old man.

In Perry's story, recall, the old man represented the disorder of desperate survivors, and order was restored when a soldier "pick[ed] him up bodily and carr[ied] him out of danger."[95] Order was imagined to flow from the military, and it erased the informal order represented by Sumner's father and neighbors. Perry saw the burning city as "a scene of seemingly hopeless confusion, with the imperative, insistent need" for a military commander. "An intelligent commander cannot fail to get the desired results," she wrote. The leader who arrived—unnamed but presumably Adjutant General Charles Cole—was, she said, "of rank so high that his authority was absolute, of personality so calmly commanding that his dictates were unquestioned even in thought, of disposition so kindly and sympathetic that instant confidence was inspired."[96] Hers was a progressive ideology of expertise and leadership, in which order came from a great individual who could rescue citizens and perform wonders of organization and rationality. But the order the military imposed was based on central knowledge and military discipline, and it worked to erase the local knowledge of survivors.

By Perry's account, the militia went through the Point to empty it, "searching every house from garret to cellar, even when assurance was given by the householder that the place had been emptied."[97] The *Evening News* credited "police and citizens" with the work.[98] No matter; the effect was the same. People may have been saved, though given the depictions of the "stifling clouds of smoke, the intense heat, the terrifying rain of embers, the roar of insatiable flame, the crash of dynamited buildings, [and] the incessant, sinister wail of the fire alarm," it is hard to imagine anyone needed outsiders to order them out of their houses.[99] The other effect was to separate families and help destroy the local order maintained by families like the Sumners. The day after the fire, there were reports of lost family members and neighbors, like Olive Fecteau and Mrs. Green of Skerry Street. It took until July 2 for most, but not all, of the missing Franco-American residents of the Point to be accounted for.[100] Perry described the fear of one father and husband, though her language betrays her belief that part of his problem was his foreignness, and perhaps a Catholic insistence on too many children: "'How can I finda my wife?' demanded a swarthy son of Italy with a bright-eyed youngster clinging to each hand. 'The fire it came so quick—she taka da babe an' da two little ones and she run—I snatch a few things and I taka da two big ones and I run ver' fast—but she is out of sight. I search and I search all night—I finda her not.'"[101] In their eagerness to empty the Point—perhaps justified—the outsiders who imposed their military order broke up the local order of families. The *News*'s "Man about Town" described precisely the problem: "Women went into houses to save their belongings and the firemen had to drag them out. Families were separated and mothers and children spent anxious hours hunting for each other."[102]

For the military, keeping order—their order—was a primary concern. "One of the first problems was the mat[t]er of police protection," the newspaper reported the day after the fire, and the task required not only policemen loaned from cities all the way to Boston but also four companies of militiamen.[103] The form of disorder feared most was looting. "Our men were given strict instructions to shoot on sight any looters or burglars," the adjutant general declared. "And the orders carried shooting to kill. We don't propose to stand for any lawlessness in the burned area while we are in control."[104] Yet the fears of looting were obviously overblown. The newspaper, even on the same page as it printed Charles Cole's threat to shoot looters, acknowledged that theft was not a problem: "There did not seem to be any disposition to disturb articles left unguarded and everybody seemed anxious to help others."[105] The perceptive "Man about Town" in the *News* credited the police with keeping "good order," but simultaneously praised the populace for "courage and calmness in the disaster."[106] Moreover, there was nothing left to steal. "If looters were out to loot in the burned district, they would find little to seize upon," the paper acknowledged in an editorial. "The flames swept the field clean. In hundreds of places where happy homes once stood every vestige of contents has disappeared."[107] The militiaman in figure 10 could not have prevented theft, for there was clearly nothing left to steal in the ruins. The soldier's gun and bayonet, like Cole's bluster, were symbolic: they conveyed authority and power, that the militia was now in control.

More than imagined looters, Salem's order was threatened by the thousands of tourists who flocked to the city. These were outsiders, without the networks of solidarity that bound together Salem's victims. "From trains, trolley-cars and automobiles thousands of people poured into the city," Edward Johnson wrote a year or two later.[108] "It was probably the biggest congestion of motor cars seen in New England. They jammed the roadways for miles. There were quite a few collisions and many narrow escapes," the newspaper reported, breathlessly es-

FIGURE 10. A young, slightly injured militiaman stands guard in the ruins. Negative #32584, courtesy Phillips Library, Peabody Essex Museum, Salem, Mass.

timating the number of tourists at a million.[109] They filled the streets, got in the way, gawked at sufferers. Some of them helped, either directly the night of the fire, or in later days by donating to the relief fund.[110] Others were less sensitive, and the mayor had to specifically ban fireworks on the Fourth of July for fear that visitors would upset Salemites' jangled nerves.[111]

The military forces who were assigned to control the actual disorder of the tourists and the imagined disorder of crazed survivors and rampant looters were not value-neutral enforcers of simple order. The 669 men and 96 officers deployed the night of the fire brought with them their own ideas of what order looked like and how to create it.[112] While no record remains of precisely how many of these soldiers were from which units, the first unit to deploy and the last to leave was probably also the best represented: the Second Corps of Cadets.[113] This is significant because the corps was one of two elite, independent units in the Massachusetts National Guard. The officers were local executives, businessmen, and the scions of Salem's old Yankee families, and even the enlisted men were college-educated, middle-class clerks and professionals. Their armory, tucked amid the mansions and museums on Essex Street, had long served as a clubhouse for Salem's elite men, hosting musical reviews and dances. The corps' last call-out had been among the twenty-four-company force that occupied Lawrence for over a month in 1912 trying to suppress the Bread and Roses Strike with tear gas, guns, and bayonets. "The police and militia here," Big Bill Haywood complained at the time, "are the worst I ever encountered." As managers and executives in industries reliant on the same immigrant working class that was striking, it had been a personal assignment for many of the Cadets.[114] In the fire's aftermath, the Cadets were called upon to help the people who, in different circumstances, they could have been attacking. Their use to "preserve order" meant that the order created replicated the social hierarchy that Salem's elite preferred.

The night of the fire, Lieutenant Colonel Charles F. Ropes, the grain dealer who commanded the Cadets, ordered Salem under martial law.[115] In the first few days of the disaster, the militia's primary job was policing, and that meant controlling movement. Nobody was allowed into the burned area without a pass until the fifth day after the fire, and even then passes were required to open safes, remove property, clean up private property, erect buildings, or travel on closed streets or the wrong way down one-way streets.[116] Passes were given out at the armory, an imposing, castle-like building with an armed guard standing out front.[117]

The fire, wrote Franco-American journalist Arthur Beaucage, left Salem "enveloped in a savage, sinister silence," a "vast cemetery" where the surviving chimneys appeared as tombstones.[118] They were the only landmarks. Without familiar buildings, the urban geography was unreadable, even by those who knew it best. "Was this Hancock St. or was it Gardner?" Edward Johnson described Salemites asking each other. "The district known as 'the Point' appeared as if a great power had swept over it leveling it quite smooth and even with the ground." On top of this, the city was now turned into a military camp. "Companies were encamped, in what seemed to the sightseer the amusing shelter of little pup-tents, which sprang up like mushrooms on lawns and open spaces of different sections of the city. . . . It became a pleasant and familiar sight to see groups of soldiers at their camp fires in places where a few days before the imagination could hardly have dreamed of them."[119] In Johnson's view, the soldiers were reassuring, but to one who felt threatened, bossed around, or otherwise discomforted by the militia, the new atmosphere was one more of disruption on top of the dislocation and disorientation.

Even Salem's elite found this military order cumbersome, and the martial rules and regulations disrupted not only working-class informal order but everyday hierarchies on which the middle class and rich relied. More than a year later,

Johnson still complained about the "exasperation in the cases where a sentry would not allow a property owner to go to his own house when apparently there was little danger."[120] Insurance adjusters bristled at the "rigid rules" that made it hard to get passes and the arbitrary power that allowed militiamen to ignore even those passes that headquarters had issued. "One well known business man was placed under arrest for attempting to get at his own safe without having the necessary permit."[121] Montayne Perry recalled "the most dignified citizen of Salem" being stopped by a militiaman, who told him, "You can't get by here, we've seen bluffers like you before!"[122]

Complaints about the militia kept coming. The night of the fire, Salem's director of public works, Patrick Kelley, attempted to create a fire line by dynamiting some buildings. He was stopped, wrote the *Evening News*, "owing to the freshness of a militiaman," who refused to believe that he was a city official and slashed his car tires to prevent him from continuing his mission.[123] The civil authorities asked that soldiers stop digging "garbage holes" in the middle of the streets.[124] A week after the fire, Mayor Hurley and Colonel Frank Graves, then commanding the troops, squabbled in the newspaper over the demolition of walls left freestanding in the ruins.[125] The same day, the commander chastised officers for allowing unsoldierly behavior and comportment. "Enlisted men have been constantly reported appearing in public street with blouses unbuttoned, hands in pockets, and walking in a slouchy and unsoldierly manner," he wrote, threatening to have arrested anyone "appearing on public streets lacking in neatness and soldierly appearance."[126] That same week, rumors of militia misbehavior had grown so bad that Graves was forced to deny them publicly. "There is no foundation for the wide stories that have been in circulation about the misdeeds of some of the militiamen," the newspaper paraphrased, though Graves admitted that five soldiers had been brought before courts-martial for disobedience, drunkenness, or disturbing the peace. In the final case, a drunken militiaman had attempted to visit a woman, but he had accidentally gone into the wrong part of the house and been thrown out. He had not, Graves hotly insisted, assaulted the woman.[127]

In later months, the militia became embroiled in disputes about payment for trucks during the initial relief and rescue work. The night of the fire, trucks had been lent, on the basis of solidarity and sympathy. When the spirit of the initial relief work was replaced by the hierarchical order of the militia, the voluntarism and spontaneous order of the first hours wore off. Originally, "wagons of every description, . . . auto trucks, private automobiles and even baby carriages" were offered freely, without any external organization.[128] Friday afternoon, however, the militia's Captain Peter B. Chase, in civilian life an engineer and physics teacher at Tufts, arrived and imposed military discipline on the operation. Now, rather than trucks being offered voluntarily, they were seen as having been requisitioned, and their owners expected recompense from the state or from the relief committee; the affective relations that structured the original action were replaced with rational, military order. But the records kept by the militia were inadequate, and headquarters had to question individual officers months later to figure out who had used what truck when. It was a recipe for hard feelings.[129] "It seems too bad that after giving the use of all our trucks, ice cream teams, and ice teams and pleasure car, at one time we had as many as ten to help the sufferers, that such little appreciation should be shown," a bitter Richard Eagan wrote to Chase.[130]

These many and varied complaints and disputes suggest an irritation with the soldiers from even the elite and middle class who ran the city and its businesses. If for them the militia was a hassle, a source of "little annoyances that have made us at times a little hot under the collar," in the words of an editorial in the *Saturday Evening Observer*, how much worse must it have been for immigrant workers?[131] For them, asking for a pass to search the ruins of their home for surviving items meant entering a fortified building that had been intentionally designed to intimidate them.[132] The man from whom they had to ask for a pass would likely have spoken English only, and he might have been a boss, or at least a friend of the boss. The difficulties of insurance adjusters in getting passes and then having their passes honored were likely even worse for working-class tenants hoping to sift through the ashes of their homes for stray surviving possessions. Unlike the insurance men, who dressed well and could show their credentials, residents likely had few surviving documents to prove their business in the burned area. A soldier who harassed a pass-bearing insurance adjuster, or who accused fancy and well-known businessmen of being impostors, would be unlikely to let through a working-class neighborhood resident. The National Guard was designed as a tool of class control, and this fact would have been understood as much by the workers as by the bosses.

Martial law was an opportunity to remake Salem's municipal governance. In charge was Colonel Frank A. Graves, the commander of the Eighth Regiment, Salem's less-elite militia unit. In civilian life, he had risen from a mere shoe worker in Marblehead to a salesman for F. M. Page and Company in Lynn; in the militia, he had gone from a private to a colonel.[133] In an angry letter to the state adjutant general, he complained bitterly of the civil government's incompetence. "The civil authorities cannot be depended upon to do the work" of public health, he wrote, complaining that a curtailment of his authority "ha[d] resulted in a neglect and defiance to Sanitary Regulations to such an extent that the city is threatened with an epidemic unless the most stringent measures are taken and enforced by the troops on duty here."[134] As was clear from the first citizens' meeting after the fire, public health was a major flashpoint in Salem's long-running political battles, so Graves's complaints must be read in the context of the fights between the city's Yankee elites—backed, ironically, by Franco-American community leaders—and Hurley and his Irish backers. By asserting that the civil public-health authorities were incompetent or powerless, Graves claimed for himself, on behalf of the Yankee middle class and elite who objected to Hurley on the grounds of "good government," the right to run Salem. As in other cities and at the border, public health was a new knowledge paradigm with its own corps of professionals who policed and protected the body politic. It was thus a particularly powerful discourse for Graves to deploy.[135]

"We are not a military nation and consequently the people do not take kindly to restraint imposed by military rule," Graves told the *Evening News*. These restraints were the cost of military aid, he said. "Many times when complainants have been asked whether they wished the troops withdrawn they have begged that the men be kept here."[136] Indeed, many seemed inclined to make excuses for the militia and its mistakes. The *Saturday Evening Observer*, for instance, defended militiamen as having "tried to do their full duty. When they have erred it has been from an excess of zeal." Ignoring complaints of poor behavior from the soldiers, the paper praised their "deportment" and noted "the marvelous freedom from all rowdyism." More important, it said, the militia had been successful in causing "the absence from Salem of all classes of criminals and despoilers that generally appear in every stricken community."[137] The relief committee praised the militia for its "excellent service."[138] In October, Montayne Perry admitted that the "militiaman is sometimes uncouth in manner, lacking in the externals of courtesy to the public." She provided them an excuse: their training. "The soldier is taught how to approach his superior officer, how to address him and how to answer him, but he has no specific training in relation to the public."[139] This was, of course, an odd excuse, since as members of a volunteer, part-time militia, national guardsmen spent most of their lives dealing with "the public" rather than superior officers.

Salem's elites were willing—eager, even—to forgive the National Guard its mistakes, perhaps because militiamen had imposed the right kind of order, and the inconveniences caused by the militia were better than the disorder they feared without it. Graves was correct to argue that the indignities the militiamen caused were the cost of the central order they imposed. In the immediate aftermath of the fire, Salem's burned district became the latest battlefield in the long-running contest over the city's governance. Salem in 1914 was a city with an insecure elite, besieged politically by the likes of Mayor Hurley and demographically by the growing industrial workforce. The fire created a moment when the middle class and elite could unite to aid their social inferiors, even while commanding power in their streets.

3

"IT IS EASY ENOUGH TO ESTABLISH CAMPS"

GEOGRAPHIES OF COMMUNITY AND RESISTANCE IN BURNED SALEM

When the Salem fire reached the Point at seven o'clock the evening of Thursday, June 25, 1914, it created immediate and pressing problems for the people who had lived in that neighborhood: where to sleep and what to eat. It would soon be dark, and there was suddenly a severe shortage of shelter. Similarly, the fire burned not only whatever food people had in their pantries but also the stoves on which to prepare it and the stores in which they would ordinarily have bought its replacement. These were both immediate concerns—people had to sleep that night and would have to eat the next morning—and lasting ones, since houses and grocers' stores could not be rebuilt instantly.

Some survivors went immediately to neighboring North Shore cities. The newspaper described "immense crowds" on the streetcars in the afternoon and evening fleeing to Danvers, Peabody, Lynn, Beverly, and even Boston.[1] Other burned-out families stayed in Salem. Some secured space in hastily opened shelters at the armory, city hall, the high school, and Christian Lantz's YMCA. Those who couldn't find room inside had to camp. Hundreds stayed on the Common, a downtown park surrounded by mansions, clubhouses, and churches. Others passed a macabre night in the Broad Street cemetery.[2] Thankfully, it was a warm night, but the impromptu camping must still have been uncomfortable.[3] People were surrounded by strangers, sleeping out in the open, and often trying to stretch out on whatever furniture they had managed to salvage. "On the Common and in open spaces in the outskirts of the city thousands of refugees tried to find a little rest that awful night," wrote the *Pilot*, Boston's Catholic newspaper. "Many stretched themselves out on the grass; others used mattresses or rocking chairs which they had managed to save from their homes."[4] In the best of times, open-air camping amid a large, noisy crowd could not have been very restful; coming after the anxiety, excitement, and terror of the fire, it is hard to imagine that many people on the Common or elsewhere got much sleep at all.

The next morning, the militia began to establish two large refugee camps. By Friday evening, soldiers had erected two hundred tents in Bertram Field. Each tent had three cots, and at least originally, one family was assigned to a single tent. The families staying there—mostly Protestants and Irish Catholics—were given their rations in the basement of the next-door high school. Nearby was an encampment of the soldiers who policed the area, served food, and did other tasks that fell to them during the period of martial law. At Forest River Park, overlooking the harbor and next to the devastated Point neighborhood, were most of the burned-out French Canadians. By Friday night, when the governor arrived for an inspection, there were one hundred tents, and the National Guard's Ninth Regiment was busy erecting more.[5] Although military and civil authorities would disagree over who, legally and technically, had control over the camps, in practice, soldiers patrolled and regulated them.

Faced with the sudden loss of much of the city's housing stock, the establishment of these relief camps was sensible, efficient, and altruistic. Only the militia had immediate access to the large numbers of tents, cots, blankets, and the like needed to house those who had been burned out. But like all relationships of rescue and relief, the establishment of relief camps, especially under the control of the military, carried with it a politics and a set of power relationships. Camping outside in the mud and the rain, under the watchful and sometimes abusive control of the soldiers, was no one's first choice. People with the social connections or financial wherewithal to stay inside did so. Soldiers and officers strictly regulated behavior in the camp. The refugees' behavior and its regulation, and indeed their very presence in the camps, became elements of the domination of and resistance to the state, the military, and Salem's elite. Bertram Field and, especially, Forest River Park were, quite literally, contested ground.

The fire created new and unusual forms of state power—martial law to police the city, a state-sanctioned committee to judge the circumstances and worthiness of needy families, and militia-patrolled camps of thousands of working-class individuals—and with them inspired new responses to that power. The choices Salem families made about where to sleep, whom to ask for help, and how to perform their domestic and remunerative labor show how they contested the state's growing power and how they sought to imbed their own concerns into the state's project of rescue and relief. Aggregated, these seemingly personal choices created a disaster citizenship. Salemites and outsiders in dominant positions—the military, employers, those responsible for giving out aid—sought to control the intimate choices of working-class families who had been burned out, including where and how they would live, how their domestic labor should be apportioned, and where and under what conditions they would work. In this context, the choice made by most families to find shelter with friends or family or in hastily obtained apartments reveals more than their preferences about where to obtain support. Living outside the relief camps also freed families from the direct military control to which they would have been subject at Forest River Park or Bertram Field. Conversely, the growing campaign by relief officials to encourage refugees from crowded private lodging into the camps appears not only as a concern for public health, but also as a way of forcing working-class families into a physical location where they could be more easily controlled. Moreover, the use of camp space as a field of contestation ran both ways. Refugees tried to refuse placement in the camps in order to resist allowing the military power over their lives, but at the same time they worked to make the camps their own and treated them like home. Elites wanted refugees concentrated in the camps and waged a concerted campaign to make them move in, but they were terrified of the camps' permanence and so tried to force residents out.

Camps of men receiving or demanding relief carried a particularly fraught meaning in the summer of 1914. In the midst of a depression, unemployed men around the continent organized to demand jobs, relief, and free rent. The agitation had climaxed around the country in March. In Sacramento, 1,500 unemployed men, led by a former general in Coxey's Army, erected a camp, only to be attacked and routed by armed militiamen. In New York, a Wobbly busboy led jobless men in a series of church invasions, demanding food and shelter, until the police arrested him. Meanwhile, in Boston, threats to occupy space were largely symbolic. One hundred fifty unemployed workers demanded that the Chamber of Commerce open its building as a nighttime dormitory, and a few days later eight hundred people marched on the elite Algonquian Club and demanded entrance. Wobblies protested on Boston Common and called on unemployed men "to steal food and whatever else they need to maintain their health and welfare." There is no evidence of such agitation in Salem, and indeed by June this sort of organizing among the unemployed had died down. But the specter of unemployed people occupying space and demanding aid hung over Salem's relief camps.[6]

Three days after the camp at Forest River Park was established, Salem's evening newspaper told its readers that it was nearly a paradise. The refugees, it wrote in its headline, "are snug and comfortable in tents." "Everybody about the place appeared bright and cheerful," it reported, "and there seemed to be a general air of making the best of things."⁷ The next evening, residents of both camps enjoyed band concerts. The Anglophones at Bertram Field were entertained by the United Shoe Machinery Band, and the Eighth Regiment military band played at Forest River. Sensing the continued solemnity of the situation, Ellery C. Quimby, the latter's bandmaster, explained the program to the newspaper: "I did not want to give them a rollicking lot of ragtime selections. So I finally compromised on a program of patriotic airs." The reporter was enthusiastic about Quimby's choices. "That nothing could have been more successful was evidenced by the rousing cheers with which the concert was received," he wrote. Refugees sang along to "La Marseillaise," "The Maple Leaf Forever," and "The Star-Spangled Banner," for which the audience reportedly stood and gave a cheer.⁸

The *Salem Evening News* enthused not only about the entertainment but also the food, which it said was "well cooked, palatable, and wholesome and there is plenty of it"; on July 1, the menu at Forest River included bacon and eggs, bread and butter, and coffee and milk for breakfast; beef stew, tomato soup, boiled potatoes, bread and butter, and coffee for dinner; baked beans, boiled potatoes, bread and butter, prunes, rhubarb sauce, and milk and coffee for supper. Moreover, the newspaper claimed, the militia had specially hired African American civilian cooks to prepare these feasts.⁹ Their supposed presence—not mentioned elsewhere and contradicted by other articles that said refugees had been put to work cooking—was telling. At a moment when black workers, long imagined as pleasingly subservient, were displaced by a new ideology of service that emphasized skills only whites were imagined to posses, their continued employment at some hotels and on Pullman cars was deliberately tinged with nostalgia for earlier standards of luxury. By emphasizing or inventing African American cooks, the *Evening News* suggested that the refugee camps were like luxurious Pullman cars, fancy hotels, or even Old South plantations.¹⁰ A few months later, Montayne Perry had not yet emerged from the reverie inspired by military bands, hired cooks, and military tents. "It is probable that many of these tent dwellers had quarters far more sanitary and comfortable than their burned homes had furnished, and food of better quality and preparation than they had known before," she wrote.¹¹

But bandmaster Quimby's selection of "The Maple Leaf Forever"—a song long rejected by French Canadians for its veneration of the British conquest—hinted at trouble in paradise. In fact, the refugee camps were not pleasant places. Though the weather before the fire had been hot and dry, afterward it turned cold and stormy. By Monday, June 29, the camps were so muddy that Forest River Park had to be oiled to prevent mosquitoes from breeding, and all week newspaper readers needed daily reassurance that refugees were comfortable despite the weather.¹² The rain not only made everything damp and muddy; it also heightened the indignities of life in the camps. On Monday, the line to get the evening meal was long—it "started before 6:00 and was long then; it was still long at 8:00," the paper reported—and refugees had to wait for food in the rain, unsheltered. They could not return to their tents to try to stay dry or they would lose their place in line. A sympathetic reporter saw "one woman, hatless and coatless, put her hand over her head to protect it from the rain. In her other hand she held the spoon, tin dipper and tin plate, the utensils with which everybody at the park has been supplied."¹³

The woman's militia-issued kit was a tangible representation of the humiliation of receiving charity in public. "The bread line is a study," the *Evening News* observed. "In it one sees people who still retain all the vestiges of former prosperity but who are now stripped of everything they so lately cherished."¹⁴ Reduced to poverty and dependence, the camps' residents also had to contend with the surveillance of suspicious policemen and curious visitors. The police department deployed officers to "watc[h] the lines of people who wait for supplies to see that no fakirs or repeaters get to work at the expense of real sufferers."¹⁵ If the police treated every refugee in line as a potential fraud, the relief committee's health subcommittee suspected them all as potential carriers of disease. William McDermott, a doctor on the committee, urged that those in line for food and clothing be "casually inspected from time to time, for contagious diseases."¹⁶

Perhaps worse than this official surveillance was the presence of tourists in Salem. Making preparations for what all expected, correctly, to be a massive influx of sightseers over the Fourth of July weekend, the Committee of Fourteen heard complaints that "curious hordes [were] desecrat[ing] the privacy of the present homes of the campers." The committee agreed with a visiting Boston doctor that it was "an outrage and it should be forbidden," but the governor "believe[d] in a limited right to see the camp" by the public who funded its operation.¹⁷ The governor's opinion was representative of the broader idea that those who provided relief—in this case, state taxpayers—deserved some knowledge about and control over the way their money was spent. By extension, this meant that they deserved power over the refugees and their personal decisions.

This power was most clearly personified by militiamen—the same young, poorly trained, and sloppy soldiers that rich Salemites decried as capricious and rude. They held new authority over the intimate and quotidian details of refugees' lives. Many of the demands soldiers made of their "inmates"—as they were occasionally called—revolved around health and sanitation.¹⁸ One colonel bragged to the newspaper of the military order he kept in camp. "Wherever paper or other rubbish is seen about the refugee camp," the paper paraphrased, "the militiaman patrolling that street calls attention to it and some youngster is made to pick it up and put it in waste barrels provided for the purpose."[19] Woe to the adult who refused a soldier's order about garbage; when one man refused to empty a trash can, he was arrested and denied food until he complied.[20] An Eighth Regiment lieutenant saw Peter Levesque "thro[w] paper and other refuse promiscuously about the camping grounds" and arrested him for violating health regulations. By the next day in court, however, the charges had been changed to intoxication. The militia retained power even in the courtroom; Levesque's prosecuting attorney was the regiment's quartermaster, a Boston lawyer ordinarily in private practice. Though drunkenness charges ordinarily brought fines of ten or fifteen dollars, Levesque was sentenced to a month in prison.[21]

This obsession with cleanliness and sanitation seemed, to contemporary observers, inextricable from the camp's military control. Military logic and training necessarily sacrificed autonomy for order, diversity for sameness, solidarity for hierarchy, local knowledge for central. "Throughout the camps military discipline prevailed," Montayne Perry wrote in the autumn immediately following the fire. "Absolute cleanliness was a rigidly enforced rule, and the most sanitary conditions were manifested."[22] Another observer made a nearly identical point about a year later: "Military regulations enforced order and cleanliness, in the care of tents, in the persons and habits of the refugees and in the preparation and the serving of the food." He credited the "vigilance of the military authorities" and the collaboration of the priests from the Franco-American St. Joseph's parish for Forest River's "excellent general moral condition."[23]

Even when the camps were technically civilian operations, they remained under military control and discipline. The militia erected the camps with militia-owned tents, blankets, and the like.[24] National Guard officers at least twice served as prosecutors in cases related to behavior in the camps.[25] When the militia officially departed—the Second Corps of Cadets turned over Forest River Park's camp to civilian authorities on July 7, twelve days after the fire—they left relief in the hands of two officers, Colonel John Spencer and Colonel Charles Cutting.[26] Moreover, the two colonels, who were always referred to with their military titles, were assisted by remaining officers and soldiers. For instance, the National Guard detailed to Spencer, a factory owner who took over feeding the refugees, a captain, a lieutenant, two sergeants, and a private, all handpicked by Spencer.[27] The military aspect of the relief camps—even when formally under civilian control—was particularly meaningful in a city where the National Guard was so thoroughly associated with the economic elite.

In addition to the difficulties and indignities of life in a militarized refugee camp, residents of Bertram Field and especially Forest River had to contend with

FIGURE 11. Domestic life and labor in Forest River Park camp. The photographer, M. E. Robb, must have staged the photograph, of which this is a detail. Note how everyone is posed looking at the camera except for the woman at the front, who is performing what was perhaps the iconic act for refugees: carefully putting trash in the barrel. Note also how, in order to air out the tents, families had to give up the last vestige of their privacy by raising the flaps and exposing their possessions. Negative #4606DetailB, courtesy Phillips Library, Peabody Essex Museum, Salem, Mass.

a relief ideology conditioned by the Salem elite's desire to control their labor. Contemporary writers about the fire emphasized that its defining characteristic was that both homes and factories had been destroyed.[28] Although in public this phenomenon was described in terms of the multiple afflictions of the worker who lost both his job and his house, it was also a deep concern to employers. Salem factory owners were dependent on the labor provided by the very people whom the fire had burned out. Suddenly unemployed, they might leave town in search of work, leaving the city with a labor shortage when the factories finally rebuilt. Perhaps worse, the workers might grow used to life on the dole and refuse to return to their hard work once their labor was needed. The refugee camps simultaneously represented an opportunity to keep workers in the city and under the control of their employers and a dangerous summer camp in which workers could get accustomed to leisure. If employers' domination was about controlling workers' location and maintaining discipline, workers' resistance was conversely located around maintaining autonomy over their personal geography and refusing that discipline.

Salem employers' fears of being abandoned by their workers were most pronounced at a contentious meeting of the Committee of Fourteen the morning of Friday, July 3. Transportation chair Dan Donahue, a businessman, developer, and store owner, reported that his subcommittee had paid for 150 people to leave the city, and he asked the committee for a policy on, in his words, "letting people out of Salem." For the previous week, he said in the newspaper's paraphrase, "if he found an able bodied mechanic he refused him transportation on the ground that this man would be needed here later on when rebuilding starts." This policy was endorsed by John Cabeen, the president of the board of trade and a master plumber, who complained that thirty-six families had already moved to New Bedford because their breadwinners had found jobs there. James J. Phelan, one of the governor's appointees, urged that "efforts should be made to hold the inhabitants here," the latter said, "except in cases of dire necessity." Two others of the governor's appointees, A. C. Ratshesky and former Boston mayor John F. Fitzgerald, objected to this proprietary attitude and thought people should be free to go wherever they could take care of themselves. After much argument, Ratshesky and Fitzgerald won. But the local men had demonstrated their fears of being abandoned by their workforce.[29]

Employers feared the loss of their newly transient labor forces after other disasters, too. In a massive Mississippi River flood in 1927, for example, white landowners in Mississippi forced their black workers into armed camps to prevent their departure.[30] For employers of French Canadians, too, the fear may have been particularly pronounced. Moving to a different town was a characteristic way for French Canadians to withhold labor in times of labor strife.[31] Labor transience was thus marked as implicitly political—as would have been employer attempts to "hold the inhabitants here."

The officers who ran the refugee camps—themselves Salem employers or their allies—made sure that the men held there would not fall into dependence or laziness. On the first Monday after the fire, the labor committee, chaired by shoe manufacturer William F. Cass, was appropriated $5,000 of the relief funds with which to employ two hundred or more men to clean up the ruins beside regular city workers. Another forty were put to work as cooks and laborers at Forest River, with a similar number at Bertram Field.[32] Three days later only 125 men were employed in clearing the burned district, and the Red Cross's Ernest Bicknell predicted that the number would not exceed 175.[33] Men not lucky enough to get a city job were chastised and threatened if they did not look for employment or if they did not accept whatever job they were offered. "Unless you find work you must report each day at our office between 9 o'clock and 11 o'clock in the morning," the employment bureau instructed. "Failure to report drops your name from our list," as did failing to go to work when called.[34]

The afternoon before Phelan and Donahue proposed not "letting people out of Salem," a Captain Blanchard, a Second Corps of Cadets officer at Forest River, took his commanding officer, Lieutenant Colonel Charles F. Ropes, on an inspection tour of the camp, accompanied by a *News* reporter.[35] What Ropes and the reporter witnessed highlights how both domination and resistance centered on Blanchard's attempts to discipline work and workers. "The trip disclosed a large number of men entirely content to go on as they are," the *News* man wrote. "Capt. Blanchard explained that he had other plans for them." As he saw it, his main job was to disrupt and prevent the indolence and dependence that he considered the primary dangers of the camp. "Unless we start these men, this camp will become permanent," Blanchard told his guests. He complained disgustedly that a few days earlier, he had needed carpenters and "called some of the refugees. They wanted $4 a day apiece," the union rate for an eight-hour day.[36] Blanchard may have seen the disaster as an opportunity to break the power of the carpenters' union, just as his Second Corps of Cadets colleagues had tried to break the union during their deployment in the Bread and Roses Strike. His adversaries demonstrated the strength of their union culture by rebuffing his demands. But the story may not have been specifically about unions. To Blanchard, hiring camp residents as laborers was a way of disciplining them and forcing them to sell their labor; his opponents thwarted his attempt at conscription by insisting on a wage he thought too high.

As the tour went on, Ropes and the reporter witnessed firsthand a similar interaction. The group came across a painter named Ovid Pelletier, and Blanchard demanded to know why he was not working.[37] Any man still in camp in the afternoon, Blanchard implied, was a lazy shirker, his work ethic—if he ever had one—sapped by the public charity he was receiving. Pelletier and Blanchard went back and forth, Pelletier offering an excuse why he was not working, Blanchard countering with another reason why he should be. After Pelletier said that he could not find a job, the two men argued about whether a particular firm—one that Blanchard claimed was hiring—was an open or closed shop. (Pelletier did not belong to the union.) To end the argument, Pelletier said that he was not working that week anyway because of an injury. Blanchard closed the encounter by demanding a doctor's note "or you get out and get a job tomorrow." As with the men who demanded four dollars a day for carpentry, Pelletier and Blanchard were in a battle of wills and power, the latter seeking to control the former and the former refusing the latter's authority.

Scenes like that between Blanchard and Pelletier played out several times during the afternoon inspection. Blanchard accosted another man, who responded that there was no work to be had because of the rain. Even the reporter, otherwise sympathetic to Blanchard, had to grant the man that "much of the outside work actually was stopped yesterday by the rain." However, even though Blanchard lost the battles with Pelletier and the second man, he was able to regroup by exercising his coercive power over women "inmates." A woman came to the camp seeking to hire three women to help her move and clean a new home. The women in the camp all withheld their labor until Blanchard "ordered three husky young women out of the mess line for her." It was only by denying food that Blanchard could coerce labor out of camp residents. The other technique he tried was to threaten to eject them from the camp altogether. When the inspection party came across an Italian in bed at three o'clock in the afternoon, Blanchard ordered him to get a job tomorrow or leave camp, just has he had threatened Pelletier.

Battles over formal, paid labor bled into fights about informal, domestic labor. The fear that Salem's working-class men were becoming lazy in the camps seemed to find confirmation in the way families divided their work. Montayne Perry sympathized with women who, she said, had an easier time in the camps. "It does seem good to eat a few meals that I didn't cook myself," a mother of eight sons told her. But men with a similar attitude she denounced as "the shiftless, the improvident, the unfit."[38] To Blanchard, that women did any visible domestic labor at all was evidence of their husbands' laziness. "I found the women paddling all the way across the open field at all hours of the night to get the milk prepared by the nurse for their babies," he complained. That women did this work did not appear to him to be an acknowledgment of women's independence or the proper division of household labor. Blanchard insisted that women ought not be up and about at night; that was the proper province of men. "I made a rule that if a woman came over alone, a sanitary guard would be sent back with her and the husband would have to get up." But the refugee families resisted Blanchard's intrusion in their arrangements. "I had to rescind the rule, for the husbands wouldn't get up. Our night feeding fell off to almost nothing. They would have let the babies starve rather than get up."[39] Resistance to the militia's intrusion into domestic arrangements was not exclusively male. When hospital nurses came to Forest River to give children baths, some families were happy to participate, at least on their own schedule. The nurses gave fifty baths in the morning, but "then there came a lull, as mothers failed to come forward with their offspring," the newspaper reported. Though interest picked up again later in the afternoon, in the meantime, "the nurses went from tent to tent, gathering up such children as they could find, and took them away to the big hospital tent."[40] When authorities tried to intrude on the family, mothers and fathers simply refused to participate. When the authorities used force—as when at Bertram Field they took a boy with tonsillitis to the contagious hospital against the wishes of his mother—parents used other tools to protest. The boy's mother could not stop his being taken away, but she symbolically protested by leaving the camp. "She got past the guards at the gate and went out into the darkness where for a time she was lost," the paper reported.[41]

As Blanchard's threats to eject residents from the camp and the sick boy's mother's escape indicate, presence in the camps became a tactic in the fight between burned-out workers and their employers and allies. Families with the choice preferred to stay with friends and family, away from the watchful eye of men like Blanchard. Relief authorities, though, wanted people in the camps so that they would be easier to control. At the same time, they acted contradictorily and disrupted life there to encourage people to leave.

・・・

The night of the fire, waves of people left Salem. "Many were fortunate enough to find homes with friends in unburned sections or in adjacent cities," Perry wrote. "Doors were thrown open everywhere with friendly welcome to the sufferers."[42] Acknowledging the dispersal of burned-out Salemites, the relief committee established depots in Beverly, Danvers, Ipswich, Marblehead, Lynn, and Peabody.[43] The Cercle Lacordaire, a French Canadian temperance group, immediately suspended its Salem meetings, noting that its members were in "Beverly, Peabody, Lawrence, Lynn, Danvers, Lowell, etc." and could attend meetings there.[44] Eight-year-old John Sumner's family went to Peabody to stay with the boy's paternal grandparents. "That night I slept on my grandfather's horsehair sofa in the parlor. I missed my own bed and our own home," Sumner recalled fifty-six years later.[45] Thirty-six-year-old Selma Florence Bartol and her husband welcomed her displaced parents, even though she was expecting to give birth any day. When she died soon after giving birth, the newspaper blamed "the excitement of the fire and the misfortune of her parents" and added her to the short list of fire casualties.[46]

If people had no nearby family, they could stay with friends. Six days after the fire, the *Evening News* printed on its front page a column listing Salemites who had gone to Danvers "to stay with friends."[47] Less than a week after the fire, Ward 1 residents, the newspaper reported, were already complaining that so many of the burned out had come to stay with friends that the neighborhood was overcrowded and at risk for disease.[48] The *News* praised a working-class French Canadian family in Danvers named Soucy for housing thirty-five refugees. "Mrs. Soucy," the newspaper said, "was completely worn out by her incessant hours of labor in caring for the unfortunate refuge[e]s, but she only said that she wished she could have one more."[49] Since there was no mention of a system by which strangers could find each other, we can infer that all of the Soucys' guests were friends or acquaintances. Staying with friends instead of family could cause confusion, though, and families feared their relatives lost when in fact they were staying with friends. Sarah E. Abbott, reported to be among the fire's few deaths, was eventually found alive, staying with a friend instead of family.[50]

Even more confusion arose when people found shelter with strangers or in public buildings. The day after the fire, the *Evening News* ran a list of notices in which refugees informed families and friends of their present locations. A typical entry announced, "Mrs. Labelle's children are safe at the home of Miss Page, 2 Loring avenue." Similarly, "Mr. and Mrs. Theriault want[ed] their children, Joseph, William, Louis and Henry to know that they are at Salvation Army hall, Beverly."[51] No announcements were needed when families were together; that arrangement was the default. A particularly telling story was that of Theodore Boisvelt, his wife, and their five children. His family missing, a worried Mr. Boisvelt went to the newspaper the day after the fire to ask it to run a notice inquiring about their whereabouts.[52] The next day, it emerged that Mrs. Boisvelt and the kids had been "picked up" in Danvers and taken home by Mrs. M. E. Willis. To reunite the family, Mr. Willis brought them to Lowell, where they stayed with Mr. Boisvelt's father.[53] For the Boisvelts, being together, with family, in a different city, was preferable to being with strangers nearby to Salem. Furthermore, although Mrs. Boisvelt may have been grateful to the Willises, and we ought not discount the latters' generosity, their rescue caused a good deal of upset for the husband, who did not know to where his family had disappeared.

Even when people moved into new apartments, they often did so with friends because unoccupied apartments in the area quickly ran out. Six days after the fire, the chair of the housing committee, noting that tenements were in great demand, requested that all landlords tell him of empty apartments in their buildings; the next day he announced that all spaces had been already filled,

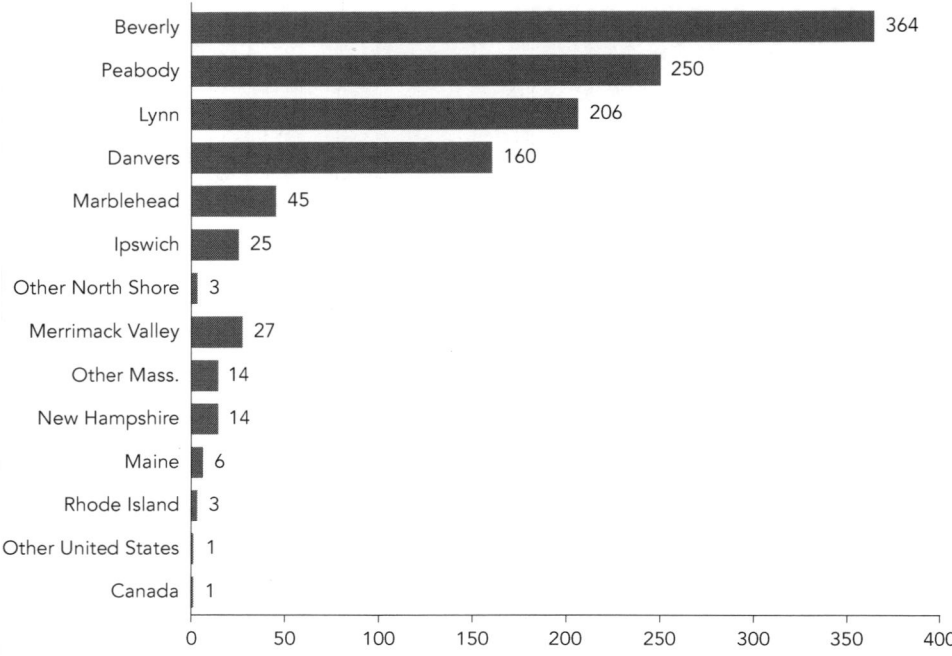

FIGURE 12. Number of families rehabilitated in outlying areas. Other North Shore destinations include Gloucester, Manchester, and Saugus. Merrimack Valley includes Lawrence, Lowell, and Methuen. Derived from undated list, third unlabeled folder, Miscellaneous collection of circulars, etc., 1914, box I, Salem Fire Collection, E S1 F6 1914_2, PEM. Totals here incorporate the handwritten corrections to the typed list.

and that he would start a waiting list for homes in Salem.[54] It is no wonder that twenty-five people crammed into a house on Lawrence Street and that seventy Franco-American families had to fit seven or eight families to a house at Bessom Beach in Marblehead.[55]

Figure 12 is drawn from an undated list showing the locations of families "rehabilitated" by the Salem Relief Committee in places other than Salem.[56] It is incomplete: it does not, for instance, include the Lapointe family, who returned to St. Arsène, in Témiscouata County, Quebec.[57] Nor does it count the couple who, after almost a month at Forest River, went to stay with their uncle in Greenville, New Hampshire.[58] It also does not include the families of at least three workers recruited by M. H. Dumas to work in a factory in Cohoes, New York; those whom another labor agent recruited to Amoskeag Mill in Manchester, New Hampshire; or the thirty-seven sent by the labor committee to farm in Topsfield, Massachusetts.[59] It does, however, provide some indication of the relative popularities of different locations to which refugees moved. The vast majority stayed nearby, in the North Shore cities of Beverly, Danvers, Ipswich, Lynn, Marblehead, and Peabody. Fewer, like the Boisvelts, went to the Merrimack Valley cities of Lowell and Lawrence, both industrial centers of the French Canadian diaspora. Smaller numbers still traveled to other cities in Massachusetts and other New England states. Notably, few families traveled to major Franco-American centers. Of Fall River, Massachusetts; Manchester, New Hampshire; Holyoke, Massachusetts; Lewiston, Maine; Woonsocket, Rhode Island; and Worcester, Massachusetts, the list shows only Manchester welcoming more than one family.[60]

Although the relief committee's list is incomplete, it corresponds to what was reported by Arthur Beaucage, the editor of the *Courrier de Salem*. On July 24, he estimated that 150 French Canadian families had moved to Lynn, and 90 families each to Danvers (comprising 700 people) and Beverly.[61] A week later he greatly upped his estimate of the number of families in Beverly to 400, and on August 7 he said 214 families were in Lynn.[62] His numbers were not necessarily right, especially the second Beverly estimate, which is likely to have been inflated. But they demonstrated that, in accord with the numbers from the relief committee, the bulk of Salemites went to nearby North Shore cities. Beaucage's newspaper occasionally mentioned other towns and cities in Massachusetts, elsewhere in New England, and Quebec, but only as destinations for individual families, never as places large numbers of families went. Most families preferred to stay in the area rather than to venture far either within the French Canadian diaspora or back to Quebec.

Despite being migrants, Salem's French Canadians had built communities, networks, and institutions that kept them in Salem and its environs. Rumor had it, wrote the Labor Department's Terence Powderly, that "French-Canadian business men [were] urging their countrymen not to leave Salem even temporarily for fear that they may not return."[63] Whether Powderly was correct—and, given his longtime hostility to Canadian migration, there may be reason to doubt—the large Franco-American centers in New England do not appear to have exerted much pull on Salem's refugees.[64] Indeed, the philanthropic and informational ties within the *Franco-Américanie* appear to have been quite weak. The Woonsocket-based Union St.-Jean-Baptiste d'Amérique (USJBA) raised only a bit more than $3,300 to distribute to its 254 *sinistré* members, an average of a mere $11.12 for each family. By August 1915, it had not even managed to give out the entire sum collected. And a month after that, the president of the organization could not even remember when the fire—a profound event to his fourth-largest council—had occurred, referring to it as "a year or two ago."[65] USJBA's main competitor in fraternal life insurance, the Association Canado-Américaine (ACA), joined with

Manchester's *L'Avenir National* for its own fund-raising. By the time the newspaper stopped actively advertising for donations at the end of July, it had received only seventy donations for a total of $190.80, of which fifty dollars came from the ACA itself.[66] Meanwhile, French readers far from Salem would have had a hard time learning the details of the fire and the affected community. Although French-language newspapers elsewhere in New England covered the fire, they rarely did so as a communal story. The initial coverage in *L'Avenir National*, for instance, did not mention French Canadians on the front page. When the article finally got to them on page five, it was only in the context of listing which buildings burned down; that is, it covered the burned church building, not the parishioners.[67] Coverage was even slimmer in Quebec newspapers. Montreal's *La Presse* printed news of the fire for three days, then fell silent.[68] *Le Progrès du Golfe*, the newspaper in the largest city in the region from which most Salem Canadians had come, never reported on the fire.[69] There does not appear to have been a fund-raising effort anywhere in Quebec.

Many families who did leave wanted to return home to Salem, even when that home was adopted. One man mentioned in the *Courrier* originally went to Lewiston, Maine, but a few weeks later moved back to Salem, where he lived at Forest River.[70] Elie and Adele Gagnon had moved to Salem from St. Arsène, Quebec, in 1889 as young parents of a single son. By the time of the fire, they had lived almost half their lives in Salem, where they had raised a total of thirteen children, and where Elie worked on the Boston and Maine Railroad. According to family lore, during the fire Adele had sprayed holy water around their apartment in hopes of protecting it, until Elie insisted, "It is time to leave." Homeless, they made their way to Maine, where Adele had relatives. Adele wanted to stay there with her relatives—again suggesting the importance of personal relationships—but Elie insisted on returning to Salem.[71] A pregnant Alphonsine Banville went with her husband and children to stay with her sister in Nashua, New Hampshire. For them, Salem's draw was strong enough to return even though she had lived in Nashua before and her husband was born in Sanford, Maine.[72] Refugees stayed members of Salem organizations to maintain their local connections. When the USJBA lodge elected its six officers for 1915 in December, four of them were living in nearby cities. Two years later, all of the lodge's officers gave their addresses in Salem, including two of the same men.[73]

Nonetheless, some Salem families did leave the city for good, if only to neighboring cities; the Census Bureau estimated Salem's population decreased by nearly 4,500 people between 1914 and 1920.[74] Franco-Americans who moved to Danvers and Beverly were sufficiently stable that during the summer they began agitating for Franco-American national parishes in those two cities.[75] In

September, the *Courrier de Salem*, a vocal agitator for the proposed parishes, noted sadly that three hundred Franco-American schoolchildren in Danvers had started public school—where the author fretted that they would lose their language and religion—because there was no French parish school.[76] Similarly, a good number of members of Salem's USJBA lodge transferred to other lodges, and still more fell off the organization's rolls entirely, suggesting that they left Salem permanently.[77] Of course, those who moved to Beverly or Danvers had stayed in the Salem area and so had not attenuated their local relationships. The French Canadian diaspora, although transnational by definition, was experienced as deeply local.

• • •

Before they resettled permanently, *sinistrés* bounced around geographically, looking for comfortable and safe spaces to live. For the first week after the fire, the population of Forest River Park's camp kept growing, as refugees returned to Salem.[78] It was these people—homeless, unsettled, often uncomfortably overcrowded, and geographically up for grabs—who were at the center of the dispute over where refugees should live. In the first week after the fire, there started a concerted effort to encourage or coerce refugees into the camps.

Five days after the fire, on Tuesday, June 30, the newspaper reported that "isolated" refugees at "more or less distant points from the center of the city," either in outlying camps or making homes for themselves in tenements or friends' houses, were complaining that they could not get food and bedding from the relief committee. The relief committee responded that they held "no disposition . . . to hold back necessities from any of the people," but that if people wanted this immediate aid, they would have to come "to the two concentration camps" at Bertram Field and Forest River Park. In other words, aid was contingent on families being willing to accede to the military authority and strictures of camp.[79]

The next evening, refugees began to receive another message of why they should go to the "concentration camps" when the newspaper ran two stories of overcrowding—the twenty-five people on Lawrence Street and the concerns from Ward 1 residents. The latter article raised the specter of a public-health crisis, suggesting that allowing refugees to stay in overcrowded tenements could lead to an epidemic, not just for those in the building but for the whole city.[80] The cure for this unhealthy and dangerous overcrowding could be found in the next evening's paper: camp life. "In spite of the rain, the refugees in the camps are comfortable," the paper promised on its front page. "They are far better off than herded into buildings, with less liability of sickness or an epidemic."[81] Inside, the paper was even stronger. People's health would be "improved rather than weakened by camp life. Many of the campers are people who have not had a vacation from hard work in years and while their present life is anything but a vacation it in time becomes regarded as at least a relaxation from the grind of years. In many cases more good than evil will result from camp life." Despite rumors to the contrary, the newspaper urged, and despite the rain, it was better to be in tents than in overcrowded tenements.[82]

The publicity campaign continued the next day. The *News*'s "Man about Town," an anonymous columnist on the editorial page, reported that because of the rainy weather, some had suggested that refugees be moved indoors to the high school and normal school. "No worse move could be made," he confidently responded. "Tents, properly pitched and cared for, are as dry as the proverbial bone and perfectly comfortable. I know from years of experience, not only with the militia on all the recent maneuvers, some of which were campaigns in the mud and rain, but from spending my summers for many seasons under canvas." Those unlucky enough not to have ever been camping, the militia would help teach how to stay dry in tents. No doubt remembering his own happy camping expeditions in the woods or the countryside, he insisted that keeping refugees out in "God's pure air" would keep them healthy. "Don't worry about the refugees. They will be all right," he promised.[83] In a news article, the chair of the health subcommittee of the relief committee, Dr. Walter Phippen, appeared to agree. Though he preferred to summer at his cottage on Cape Cod rather than "under the canvas" with the "Man about Town," he urged that the city "close the shelters at the Salvation Army and Father Matthew halls" and send their residents to the camps.[84]

If indeed this is what Phippen wanted, it was a change from what he and his committee had urged only a few days earlier. Three days after the fire—before the *News* began its drive to get refugees to move to Bertram Field and Forest River Park—his committee recommended that "all camps be depleted as fast as possible," for precisely the same reasons the *News* advocated the opposite. "The committee believes that it is better to have refugees slightly crowded in houses, rather than in camps, because of the difficulty of maintaining satisfactory sanitation."[85] The next day, the committee reaffirmed its position: "It was reported that there are scattered groups of refugees in houses, and it seems best to the committee that these persons should remain in houses rather than go to the different camps, but there should be some medical inspection of these groups."[86] We cannot know whether Phippen changed his mind later in the week, perhaps because he perceived an improvement in camp sanitation; whether the newspaper misquoted him; or whether he continued to believe that sufferers would be healthier inside but wanted them in the camps for some other reason.

Whether for reasons of public health, efficiency, or power, Salem's elites wanted refugees to live in the camps where it was easier to control their labor and family life and subject them to military discipline. Refugees, in contrast, apparently preferred to stay with friends and families, even if it meant crowding dozens into a house. Although the elites had the power of the state on their side, in the forms of the militia and the relief committee, refugees were able to use the state's concern for public health and its desire to concentrate sufferers at camps to their own advantage.

We can see this process in the example of Salem's Eastern European Jews, who comprised 4 percent of the fire's victims, or 139 families.[87] (A separate count found between six hundred and seven hundred homeless individuals.[88]) After the fire, between fifty and eighty individuals found shelter on the second and third floors of Rome's Furniture Store on Lafayette Street. The conditions were terrible: the building was old and ramshackle, with steep, narrow staircases and bad plumbing. The water gave out entirely over the weekend. Despite this, the building was dry, it allowed the community to stay together, and it provided a place for kosher cooking. Food was particularly important, and by one estimate, in addition to the people who slept in the building, another twenty or thirty ate there.[89] On Sunday, June 28, the chairman of the general relief committee asked the medical committee to inspect the premises.[90] There is no record of what the inspector found, but there was no attempt to make the residents move. It is likely, though, that the residents asked for help in making their lodgings more comfortable, and especially a plumber to fix the water and the broken toilets, because when none had arrived two days later, one of the residents came to the armory demanding one. Rather than send a plumber, the head of military operations in Salem, Colonel Frank Graves of the Eighth Regiment, made a personal inspection. What he found outraged him: "The building is full of flies and dirt and all its rooms are filthy and nauseating in the extreme. Its toilets are insufficient and filthy, and dirt of all kinds prevails to such an extent that the place is reeking with foul odors."[91] Graves ordered the building closed and its residents transferred to Bertram Field.[92] In exchange, they demanded a separate section of the camp, their own cook, and their own cooking equipment.[93] The community was apparently sufficiently satisfied with this arrangement that in addition to the people forced from Rome's Furniture Store, perhaps forty more came to the camp from Beverly.[94]

We see in this story the way that both the relief authorities and refugees used location to negotiate and contest authority and conditions. Though Graves may have been exaggerating the bad conditions in the building, it is clear that the people living there were dissatisfied too, or they would not have sent the emissary off to demand a plumber. Yet they preferred to live there than Bertram Field, since they could stay together and eat kosher food. When the authorities let them have a separate section of the refugee camp and allowed them to maintain their dietary rules, they agreed to recognize military authority and move to the camp. In order to make Bertram Field acceptable to the Jewish community—that is, to get them to move from the space over Rome's Furniture Store to the outdoor space of the camp—the militia had to accept a change in the camp's geography.

The Franco-Americans at Forest River fought similar battles over space within their relief camp. In both cases, residents worked to make the camp understandable to themselves, and the militia worked to impose legibility from above. The camps, especially Forest River, were crowded and busy. "One can easily lose a comrade there and not find him for an hour, so extensive is the camp and so thick the crowds," a reporter wrote.[95] The density—the thickness—that the reporter noted was not just of people, though it was certainly that: in a space of roughly twenty-nine acres, about 300 families, comprising around 1,700 people, lived in 274 tents.[96] It was also a density of information, of relationships, and of knowledge. The militia, unfamiliar with the customs, the language, and the personalities of the residents, worked to make them legible by shaping the space of the camps; the residents worked to preserve their own local knowledge by resisting the militia's shaping.

The tents in which refugees were housed were erected in strict, straight order. Figure 13 shows the tents, lined up in rows, all indistinguishable. Arranged in seventeen streets—all nameless, so far as the official record was concerned—it must have been difficult to tell one tent from another, never mind to explain to a friend how to find one's temporary home. Finding friends or former neighbors was made still more difficult by the refusal of the Post Office Department to disclose people's new addresses. Because people filed change-of-address notifications with the post office, it was the best source for knowing where burned-out families were now living. Yet postal privacy rules forbade revealing addresses or similar information for purposes besides mail delivery.[97] Salem's postmaster received special permission from Washington to share the information with the relief committee—effectively an extension of the state—but not with anybody else. That meant that an individual who wanted to find a friend had to go to the information bureau in the Now and Then Hall instead of the post office that had been established at Forest River Park's camp.[98] The substitution of regularized, central knowledge for the complexity of local knowledge made the camp easier to govern, but harder to live in. Neat, straight rows of tents were good for patrolling, but bad for finding friends and neighbors, and regulations intended to protect postal customers' privacy made it harder for them to reconstruct their communities.

Identical, neatly ordered tents were an inadvertent erasure of local knowledge, artifacts of the militia's desire for order and regularity. Similarly, the Post

FIGURE 13. Rows of identical tents at Forest River Park. Photograph by Leland Tilford. Postcard, Salem State University Archives and Special Collections.

Office Department's reticence was the unintentional side effect of a long-term commitment to confidentiality.[99] In contrast, some erasures of local knowledge were intentionally designed to destabilize residents' lives. Captain Blanchard of the Second Corps of Cadets warned, "It is easy enough to establish camps; it is more difficult to discontinue them," and he and fellow officers worked to make it easier.[100] One camp commander frequently shifted the rows of tents, making families remove their belongings while the tents were moved and then reestablish their homes elsewhere. "No use of letting some folks get to feeling too settled," he told Perry. "Some of them would be willing to stay here till the judgment day if we made them too comfortable."[101] Similarly, Blanchard objected to refugees building furniture, cooking shacks, or ice boxes because they signaled permanency. Ovid Pelletier, the man Blanchard sparred with over the availability of work, initially attracted the captain's attention because he had built a wooden bedstead, table, and benches.[102]

Blanchard and his fellow officers worked to dominate the camp's space. Arbitrarily forcing families to move their belongings from one part of the camp to another served as a reminder of who was in charge. So too did harassing children to clean up rubbish. Yet residents fought back, claiming the camp and its space as their own. Pelletier's furniture was crafted from what Blanchard termed "camp lumber," an appropriation akin to poaching; not only did Pelletier find the materials he needed to improve his quality of life, he also enacted resistance to the military by taking its property. By personalizing his tent, he rejected officers' attempts to dominate his family's intimate space. Other residents, too, refused to bend to military order and maintained their tents as they had previously lived. The *News* reporter who toured Forest River with Blanchard noted that some tents were "arranged in orderly fashion; more, in the helter-skelter heap which is so familiar in the tenements."[103] The reporter meant this comment as a complaint or an insult, but it demonstrates the way camp residents worked to re-create a sense of neighborhood normalcy after their lives were upended by the fire.

Camp residents also responded to military order by finding ways to create and re-create their own local knowledge. To fight the sameness of their tents, and to make navigation easier, families pinned pieces of paper with their names to the outside of their tents, creating what the newspaper termed "a rude substitute for a doorplate."[104] The *Courrier* advocated that refugees rename their camp Binetteville to honor the assistant priest who, in the absence of the traveling pastor, was the community's spiritual leader.[105] It is unclear if anyone took the newspaper up on the idea—the nickname did not appear again in its pages or ever on the pages of the English-language *Evening News*, which preferred the formal "Forest River Camp" or, once, the folksier "New Frenchtown"[106]—but the suggestion indicates Franco-Americans' desire to claim the space as theirs. So too did residents' description of the camp as home. When Joseph Bérubé and his wife visited his uncle Bernard Thibeault, a former Salemite then living in Greenville, New Hampshire, the *Courrier* described them as "of Forest River Park," as if that was their permanent address.[107] In mid-July it noted that Forest River residents were entertaining visitors in their tents, treating them like the tenements that had burned down.[108]

Sinistrés' work to claim camp space as their own was simultaneously practical and ideological. Pelletier's homemade furniture materially improved his family's life and also challenged the militia's authority. When, one Sunday in the camp, Arthur Tremblay set up a stall to sell beer to Forest River refugees and their visitors, he and his customers were mostly concerned with making money and quenching their thirst, respectively. But they also battled to set communal standards in their space. This implicit battle, too, had material consequences: The police quickly arrested Tremblay for violating Salem's strict—if unevenly enforced—liquor laws. After he was pilloried in the press, Judge George B. Sears dismissed his case for lack of evidence.[109]

Another way in which residents and the police battled was over relief fraud. After the first day or two, when the relief committee prided itself on keeping a minimum of "red tape" and being willing to help whoever asked, authorities became stricter and demanded more proof of a refugee's status and need. Five days after the fire, the newspaper reported the committee's complaint that refugees were sending their friends to register them for aid. "The best way to accomplish results is for the people affected to go themselves" so that relief workers could inspect their documents and help them if they showed "the proper credentials."[110] Meanwhile, the committee promised to crack down on those requesting aid but who were undeserving. "In other words," the newspaper wrote under the headline "To Eliminate Fake Sufferers," "it is the aim of the committee to do more for the afflicted and to cut off the lists all who were not burned out—in short to get ahead of the fakirs."[111]

So began a concerted effort of arrests, publicity, and threats. That day, a burned-out man named Price D'Entremont somehow obtained two relief rations, instead of the one to which he was entitled. He tried to sell the second, asking fifty cents for it—a quarter of what it was worth—but he picked his customer poorly. The would-be purchaser turned out to be a police inspector, who promptly arrested him. He was tried in police court the very same day and was jailed for two months.[112] The police announced a plan of arresting "fakirs" for obtaining goods under false pretenses, a subset of larceny, and publicizing these arrests. "When it becomes known, [that fraudulent applicants would be arrested and jailed, it] will undoubtedly deter many from taking advantage of the relief committee," the newspaper confidently predicted.[113]

When Carlo Caielli, an Italian from Beverly who fraudulently applied for and received a suit of clothes from the relief committee, was arrested, the authorities vowed, "He will be made an example of in the district court." The unmarried sixty-four-year-old told police that he was unemployed and so could not afford the clothes he needed, but this explanation did not sway the court the next day. Caielli was sent to jail, though the newspaper omitted his precise sentence. At the same sitting of the court, Judge Sears sent Felix Richards, a married, forty-five-year-old Salemite, to jail for four months for stealing sixty-nine dollars' worth of clothing. But he also dismissed charges against Harry Pinchuk, a Peabody man infamous for being shot in a fight a few months earlier. Police accused him of fraudulently claiming a Salem address, moving into Bertram Field, and obtaining a pair of overalls from the relief committee, but Sears found there was insufficient evidence to warrant conviction.[114]

Stories like these, like all criminality, might be read as resistance on the part of the fraudsters against the state. The fraud perpetrated by Caielli, for instance, might be the hidden transcript of a poor, unemployed worker, scamming the elite out of whatever he could get—in this case, a new suit, an old coat and vest, a new shirt, a suit of underclothes, three pairs of socks, a pair of shoes, and a felt hat.[115] That the story ends in Caielli's arrest and imprisonment indicates the danger people faced when their hidden transcripts of resistance were exposed to the powerful. Read more expansively, however, the stories of relief fraud and their prosecution demonstrate the broader resistance among the camps' residents as they hid "fakirs" in their midst.

Faced with the difficulty of discerning worthy relief applicants from the fakes, authorities relied on the local knowledge held by police inspectors from the affected communities, French Canadian Frank Pelletier and Pole John Bozek, to watch the lines of relief applicants, exposing, running off, and sometimes arresting those they deemed ineligible. If those policemen knew who was a fraud and who was not, so too must have at least some of the other refugees, who would have been just as well versed in the neighborhood gossip, scandal, and innuendo as the policemen. One of the men Bozek was said to have caught was "known personally to the inspector to be worth $40,000 to $50,000," and another man was said to have $25,000.[116] Knowledge of that type of wealth would have been common knowledge in the Polish community. Authorities recognized this local knowledge when they asked Father Donat Binette to tell his parishioners to turn in the fakes. On the Saturday morning after the fire, Binette warned his parishioners at the camp that "people from Lawrence were here getting supplies by claiming that they were fire refugees, [and he] asked the French people to be on their guard against these vandals."[117] Binette's flock must have disregarded his request. Montayne Perry claimed that fall that a man from Lowell had stayed at Forest River all summer apparently unmolested by his camp neighbors.[118] The newspaper credited Pelletier and Bozek with running off hundreds of impostors, including the rich men.[119] Even if we allow for some exaggeration, it is clear that some number of refugees looked the other way when they saw people they knew were ineligible for assistance receiving it. In so doing, they were casting their lot with the members of their community and against the authorities. Relief authorities claimed that the zealous prosecution of relief fraud was to the benefit of legitimate sufferers. The sufferers rejected this zero-sum logic and created a moral economy that valued solidarity over the marginal potential increase in aid they would receive if they turned in the "fakirs."

This moral economy appears to have recognized the legitimacy of applying for and accepting whatever aid was offered, regardless of need. The camp residents who refused to turn in the strangers in their midst and the people they knew were receiving too much aid demonstrated and created this legitimacy. "What's the use of working?" a refugee asked Montayne Perry. "Aint there millions of dollars pouring into this city from all over the state, for us? I'll stay right here and get my share of it, you bet!"[120] If Perry—an outsider—heard comments like this, we may assume that they were spoken openly among the refugees themselves. In another section of her book, she told the story of a woman who went to relief headquarters explicitly to see if there was any more aid she could get, regardless of her need. "I just wanted to make sure I wasn't missing anything," she told a social worker. In Perry's telling, this attitude arose from "a part of human nature . . . to get one's share of what is being given," and she likened "the impulse which leads a broker to strive to corner the wheat market" to that which "lead[s] the poor but wily carpenter to try for two sets of tools, when they are being given away, that he may have one to sell to his less clever neighbor."[121] But what Perry saw as a competitive impulse was actually a cooperative one. If people requested aid to match what their neighbor or relative got, it suggests they got encouragement, or at least information, from the friend whom they were trying to match. *Sinistrés* shared information with each other and hid information from the authorities so that each could receive the maximum. In building and defending this moral economy, camp residents were not only demonstrating their solidarity with their neighbors and compatriots; they also resisted those who controlled the relief effort. This does not imply that no *sintistrés* were strictly self-serving, but rather that collectively refugees crafted an implicit agreement about the appropriate extent of such behavior.

Even Pelletier and Bozek must have sided with their compatriots to some extent. The *Evening News* may have claimed they ran off hundreds of fraudsters, but they do not seem to have arrested anybody, and the fakers' names were kept out of the newspaper. Only one of the people charged with relief fraud was Franco-American—and he had the misfortune to sell his misallocated rations to an undercover, Anglo cop. The Italians who were overrepresented in the arrests—in addition to Caielli, there was an unnamed woman shopkeeper from Beverly whom police charged with "t[aking] away all she could carry from the relief station" and two children named Sofusca who were let go with a warning after returning twenty-five dollars' worth of supplies to which their family was not entitled—were less protected by a broader community.[122] Like other ethnic groups, Italians had officers in the camp who, after the departure of the National Guard, policed their own community.[123] But that community was numerically smaller than either French Canadians or Poles, and it was less well organized institutionally. Salem supported three newspapers in English, and one each in French, Greek, and Polish; when the health committee wanted to advertise its regulations in the ethnic press, though, it could find no Italian medium.[124] French Canadians had two parishes of their own in Salem, and the Poles had one, and as we saw in chapter 2, Canadian and Polish clergy were on the relief committee. In contrast, Italians had to make do with a small mission, indicating the relative poverty of Italians' social networks and institutions.[125] Their overrepresentation among those arrested for relief fraud reflected that they had a smaller and weaker ethnic community to protect them.

The thinness of Italian formal institutions probably mirrored something similar for Italians' informal community. Only 146 Italian families received any aid at all from the relief committee, which means that in the camp Italians had fewer friends, compatriots, and allies.[126] This left them less protected by the networks of solidarity and resistance that sheltered French Canadians. Forest River's Captain Blanchard bragged that relief authorities had "shipped" five families back to Italy, by way of New York City, at the expense of the Italian consul. "We have only two or three Italian families left and one of them has got to be broke up," he said. The father of the family had been in Buffalo for months and the mother did not know how to reach him. She was too busy taking care of the children to do any other work, and so Blanchard and his colleagues were trying to put the children in an orphanage to free their mother's time for paid labor.[127] With a stronger local community, these Italian families might have avoided expulsion or forced disintegration. The mother with her husband in Buffalo might have been able to share child care with another family, or fellow Italians might have sheltered the family through subtle means of resistance. Without the richness or density of the French Canadian and Polish communities, Italians had less ability to resist the authority of military and relief authorities. Alternatively, the relative thinness of Italian social networks may have made these families less interested in staying in Salem. They may have been perfectly happy, grateful even, for a free ride back to Italy, especially because as a population Italians were considerably more likely to remigrate to their home country.[128]

Whether to Italy or Ipswich, refugees in the camps did indeed leave; Captain Blanchard's fear that camps would be impossible to dismantle was not well founded. The population of the camps fluctuated in the first week after the fire, as refugees and authorities contested their relative power. Soon, though, camp residents started to leave in earnest, and by a week and a half after the fire, Forest River's population had fallen to about one thousand[129] (see figure 14). After that, though, the decline in population stalled. People wanted to leave, but they were stymied because there was nowhere for them to go; the fire had created a housing crisis. The number at Forest River, reported an English-language weekly, "is steadily dwindling and as fast as a man finds employment he is supplied with the necessary articles to enable him to start housekeeping." But the newspaper predicted that the rate of families moving out would slow. "The remainder of the camp is likely to remain some time, however, as there is not one unoccupied tenement in the city so far as can be ascertained. Many of the families are moving to nearby cities and to towns, where they intend to remain only until the Salem mills are again in operation."[130] Moving to another city, even a neighboring one,

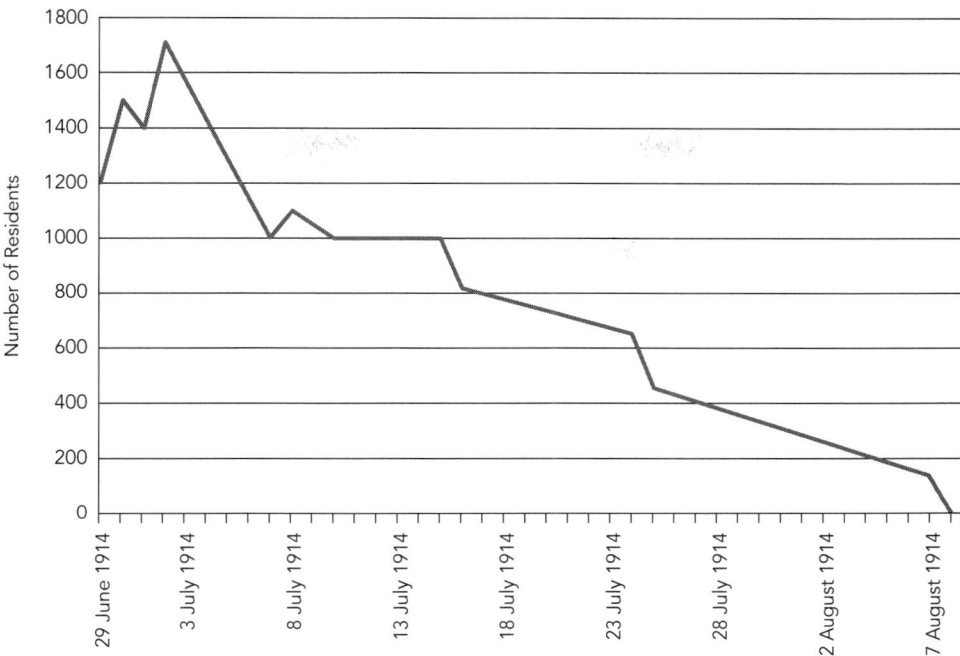

FIGURE 14. Estimated population of Forest River Park camp. Derived from SEN: 29 June 1914, 5; 30 June 1914, 10; 1 July 1914, 10; 8 July 1914, 2; 9 July 1914, 10; 10 July 1914, 5; 11 July 1914, 6; 13 July 1914, 12; 14 July 1914, 10; 15 July 1914, 4; from CdS: 9 July 1914, 6; 24 July 1914, 8; 7 August 1914, 8; from SEO 25 July 1914, 1; Edward Dunbar Johnson, "The Salem Fire," after June 1915, E S1 F6 1914₁₁, PEM; Report of T. V. Powderly to [William Wilson], 28 July 1914, reprinted in United States Department of Labor, *Reports of the Department of Labor*, 1914 (Washington, D.C.: GPO, 1915), 295–97.

was not so easy, however. The same week, the *Courrier* complained that some Lynn landlords were refusing to rent apartments to families with children.[131]

It was around this period that people increasingly began to move farther away. Just as the initial contest over the camps centered on labor, so did families' decision to leave them. When the federal labor secretary had written to hundreds of potential employers looking for jobs for Salem's textile workers and leatherworkers, the *Courrier* warned "that not all of the responses are optimistic."[132] Given the worrisome cast of that news, it must have been a relief when, the next week, labor agents arrived from factories in Manchester, New Hampshire, and Cohoes, New York. Both companies already employed numbers of French Canadians.[133] Two weeks later, the *Courrier* noted three families who moved to Cohoes.[134] Indeed, T. V. Powderly reported that more than 1,200 men had found work through his nascent labor bureau between July 14 and July

28.[135] Although some migrations appeared voluntary, others were less so. Some preferred unemployment and relief to a forced migration to a city where they were strangers. For those people, coercion returned. "They came and yanked down our tent and shoved us into an automobile truck and told us our future home was in East Saugus!" one upset man told Perry. "What do they think I want to live in East Saugus for?" Another family was grateful to have a job and a month's rent paid in Lawrence, but even they appear not have been given any choice in the matter.[136]

And so the relief camps ended as they began, in a struggle over labor discipline, families' and individuals' autonomy, and a dispute over the importance of community. To the man who complained of being forcibly moved to East Saugus, the community he and his family had built in Salem, their freedom to choose for themselves where to live, and—if one is to believe a cynical Perry—their ability to collect more relief were worth more than the promise of employment and permanent housing in an unfamiliar city. Just as some burned-out Salemites had preferred to stay in the discomfort, filth, and overcrowding of their temporary shelters rather than subject themselves to the arbitrary power of the militia in the refugee camps, so too did some want to stay in those camps when the labor committee dictated otherwise.

Neither the relief authorities nor the residents of Salem's refugee camps were unified groups. They acted on competing motivations, and so their actions were sometimes contradictory. Authorities wanted the refugees in camps, where, concentrated, they would be easier to deal with. But they also feared the creation of a relief camp in which families would receive long-term aid without working. Refugees were similarly inconsistent. Although they appear to have preferred to live outside of camps, they also stayed there and avoided being moved out. Both groups of actors had conflicting motivations, and they had to balance them as they made decisions. At the root of their decisions and contestations, however, were questions of power, authority, and control. Refugees fought to retain power over their own lives, whether it was the power to decide who in the family would be in charge of night feeding, what jobs they would accept and for what wages, or where they would live. Relief authorities, particularly the militia, saw their roles as including the right to control the intimate decisions of their "inmates." In Halifax a few years later, relief authorities and explosion survivors engaged in a similar set of negotiations and fights over authority, relief, and power. They exposed the tensions inherent in progressive ideology: Reformers wanted to rescue and relieve people, but they sought to make the cost of that help the loss of autonomy and control. Aid recipients, in turn, worked to maximize what they got but minimize the power they ceded over their lives.

4

"THE RELIEF WOULD HAVE HAD TO PAY SOMEONE"

HALIFAX FAMILIES AND THE WORK OF RELIEF

Even as the fires raged on December 6, 1917, many Haligonians were seized by a sudden generosity. "Citizens came in large numbers 'flocking' to give us places to put people," reported Frank Gillis, an alderman from Ward 2 in the South and West Ends and chair of the relief committee's transportation subcommittee. "One said 'We'll take two families or two women or husband and wife.'" Among those flocking citizens were Lieutenant Rod Macdonald, recently returned from the front, and his parents, who tried to volunteer their house for up to twenty-five refugees. Meanwhile, as in Salem, the army sprang into action to provide rudimentary housing for those left homeless. "Relief bands of military were covering the Common and the slopes of Camp Hill with a mushroom like growth of bell tents, which spring unto being with the passing minutes as if some magical force was behind them," wrote a breathless *Chronicle*. And around six o'clock that evening, the owners of private spaces opened them as public shelters. Homeless survivors were welcomed at "St. Paul's Hall, the Academy of Music, Strand Theatre, Columbus Club, in fact in any place where some warmth and food could be given them," Gillis recalled.[1]

Yet the putative recipients of this generosity appeared uninterested in it. Lieutenant Macdonald was disappointed that "a little lad and one family were the only occupants" of his guest rooms. Without actual sufferers, the Macdonalds instead welcomed twenty-five of the nurses then flooding into the city.[2] Nor did anyone spend the night in the army's tents. "Unfortunately although this cot hospital was complete and very comfortable no patients could be induced to enter it," reported medical relief chief Frank McKelvey Bell. "The patients were stunned and refused to be transferred into the tents at night." Despite his implication that the homeless were unreasonable to refuse the heated tents, he also allowed that the "frightful blizzard," which deposited three feet of snow around the tents by the next morning, made sleeping in them undesirable. Bell's boss, General Thomas Benson, in his own report to Ottawa, blamed "the fact that people with homes shared what they had with the destitute, and Institutions, Halls and Theatres were thrown open."[3] But authorities had to cajole survivors to stay at those halls, institutions, and theaters, too.

It wasn't for lack of need. Eight-year-old Gertrude Hook lived in the North End of Dartmouth, but her father, an auctioneer, worked in Halifax next door to the Old St. Paul's Church Hall. The proximity in everyday life may explain why, after he returned to Dartmouth to gather up his family, they all crossed the harbor again to return to the church hall. "We were way up in the top floor," Gertrude recalled sixty-eight years later. "My father was covering windows with blankets and we were—many children—many people were there—that they gathered up." After a night in the parish hall, they went to friends, where they remained until Gertrude's mother's brother and a half-brother of her father arrived from Ontario and took them there. After that, they moved to Detroit for a year.[4] As the Hooks' friends were rescuing them from the public shelter, Gillis's committeemen were "herding to-gether" survivors, trying to convince people to accept shelter. In the backyard of a house on Kane Street, one relief worker found a sailor and his wife who had somehow fitted up an old henhouse into a shack where they had spent the night. This couple had themselves found eleven people—including a mother and four-day-old baby and two or three small children—and invited them into the shed. Several of these people, including the sailor's wife, were wounded. Elsewhere he found an army sergeant and four women in a cellar, all gathered around an oil stove trying to keep warm. Surrounded by death, these two groups offered not only warmth but also literal conviviality.[5]

To relief workers and managers, the aid they offered seemed obvious: houses were destroyed and uninhabitable, and the army, people, and institutions of Halifax stood ready to help. So they were confused and disappointed that so few people availed themselves of their generosity. The people they tried to help often preferred to stay in their ruined houses, in the overcrowded homes of their friends and relatives, or even in hastily jerry-rigged shacks. The Hook family experience is suggestive. They initially went to an official shelter—indeed, Gertrude's father helped to set it up—but, as soon as they could, they escaped the crowds of people "that they gathered up" and went to a friend's house. That the objects of charity preferred other aid and support does not negate the original altruism. It does suggest, however, an uncomfortable aspect of charity: that even when offered in a spirit of genuine care, nonmutual charity carries with it a heavy burden of hierarchy that often makes it less desirable than mutual aid offered in a spirit of solidarity.

What the transportation committeemen tried to provide in those first few nights after the explosion was the first of many offers of charity—that is, aid given hierarchically by people and institutions that have more money, prestige, status, and power to people with less of these things—increasingly from the state or state-like institutions. This chapter is about how the recipients of that aid responded. The Hooks illustrate many of this chapter's themes: survivors working together, like Mr. Hook and his comrades creating a shelter, to support each other; the family preferring to stay in the private home of friends or relatives; and a reliance on the Nova Scotia diaspora. Haligonians, like Salemites, engaged in delicate, subtle, and often tacit negotiations, seeking to maximize the material aid they claimed from the state while minimizing the autonomy and privacy the state took from them in return. To do this, they carefully inserted this new form of labor—that of applying for, managing, and retaining state benefits—into their preexisting family economies.[6] They also mobilized their relatives in the Nova Scotia diaspora, building a new political power that came from donations from abroad. In the course of understanding these negotiations, we can see how Haligonians managed the broader family economies to which relief was added.

Progressive relief represented a trade for the state as well. In order to adjudicate applicants' claims, the state and its agents labored to make their family economies legible. The job of relief authorities was, in the words of Byron Deacon's manual, to "assist families to recover from the dislocation induced by disaster and to regain their accustomed social and economic status."[7] In order to determine each family's "accustomed social and economic status," trained workers had to investigate each claim carefully, rendering the complex, informal, and illegible family economy into simple, formal, and legible decisions about money, housing, and material goods. A key way the relief authorities rendered family economies legible was by trying to monetize all contributions. But as Christian Lantz, the Salem rehabilitation expert who came to set up the Halifax system, warned, "Rehabilitation does not mean that losses incurred would be made good from the Relief Fund. There is no prospect that the Relief Fund will be sufficient to cover more than a very small percentage of the loss."[8] Had all elements of the informal economy truly been rendered legible and accounted for monetarily, the state would have gone bankrupt. Just as applicants had to balance maximizing the aid they received with their desire to preserve their autonomy and privacy, so too did the state attempt to maximize the legibility of its citizens while still seeking to keep some labor unrecognized.

This chapter relies heavily on a simple random sample of 739 case files of the Halifax Relief Commission.[9] Every family that received aid from "the Relief," as they often called it, was assigned a file, now called "pension files" because their longest lasting use was in administering pensions, into recent decades, to those widowed or injured by the explosion. Although occasionally pension files preserve the voice of an aid recipient, for the most part the documents were written by social workers. Nonetheless, because they contain the case notes of individual families, they are the closest we can come to understanding those families' experiences and choices after the explosion. These 739 families are a cross section of the affected communities. They ranged from destitute to middle class, although because of the geography of the explosion, they were generally working-class families.

Pension files were begun by the Halifax and Dartmouth Relief Committees, which were the independent, volunteer organizations created by relief managers in the days immediately following the explosion. The local committees were superseded in January by a three-man Halifax Relief Commission, created by the Dominion government. The commissioners were chair T. Sherman Rogers, a prominent Halifax lawyer; William B. Wallace, a Halifax County judge who had served on the progressive juvenile court since 1911; and Frederick L. Fowke, a merchant from Oshawa, Ontario. Federal and provincial legislation granted the commission wide powers for rebuilding the devastated area, including apparently absolute control of the funds appropriated by the government and given by most private donors.[10] It operated courts in which property claims were adjudicated; it expropriated land in the North End and redeveloped it; and it decided who got what aid. The latter decisions were made primarily by the staff of the Rehabilitation Department, which was helmed by George B. Cutten. Cutten, trained at Yale University as a psychologist, was then president of Acadia University in Wolfville, Nova Scotia; later, as president of Colgate University, he was a noted eugenicist.[11] But although Cutten was in charge, most day-to-day decisions were made by a staff of women social workers.

It is these social workers whose voices we hear in the pension files, and it is these social workers whom the aid recipients must have seen as the very embodiment of the state. Often, though not always, they were outsiders, experts from Boston, Montreal, New York, Toronto, and Winnipeg.[12] They came to help and also to judge. They carried with them a progressive ideology of state assistance, which often conflicted with the expectations and desires of those they claimed to help. Partially because the investigators who made these files were making judgments about recipients' worthiness and what aid they deserved, the records are not entirely reliable. They contain social workers' unknowable mistakes of knowledge and judgment. Potential recipients were unlikely to reveal sources of income that were illegal or that might have cast aspersions on their respectability. Thus, except for occasional innuendo, criminal activity such as prostitution, smuggling, and theft were absent from the Halifax depicted in these files, although they could not have been absent in a port city overrun with soldiers. Likewise, the emotional toll of the disaster's death, destruction, and dislocation only leaks in around the edges.

Haligonians objected to the relief apparatus, especially in late December and early January, as bureaucratic, slow, and ungenerous. It needed to know and understand the city before it could provide the promised assistance. But the process that social workers and their managers used to understand the city—to make it legible—obscured as much as it revealed. "The fact is that the admirable and infallible managers who are in charge of this relief business are so enamored of their routine and ritual, their paraphernalia of office equipment, their card catalogues, their indexes and all their multifarious apparatus, that they do not seem able to visualize the realities of the situation or get down to close quarters with the actual facts," wrote Ben Russell, a judge, in a letter to the *Herald*. "A little common sense and real human sympathy would be worth vastly more just now than such an overdose of 'BUSINESS EFFICIENCY' and social service pedantry as we have been having for the past five or six weeks."[13]

To help resolve such complaints, the American Red Cross dispatched J. Howard Toynbee Falk, a social reformer based in Winnipeg, to reorganize the rehabilitation office. After meeting with the *Herald*'s publisher to quiet public criticism, Falk promised changes: that the 1,500 families who had registered for relief but had not yet been investigated would only be visited if they asked; that all relief activities would be coordinated by a single office; that a "labor man" would be invited onto the rehabilitation committee; and that an executive committee would meet daily to resolve problems. Notably, the latter two reforms never took place, so the rehabilitation department never gained a working-class voice. The first reform solved the problem of too many unwanted inspections from intrusive visitors, but it did nothing to make working-class families' economies more comprehensible to the authorities. Moreover, despite the fact that much of the rehabilitation staff's job was to understand families' complex domestic arrangements, Falk was insistent that his successor be a man.[14] This may itself have been a sop to critics, since objections to prying social workers often emphasized their gender.[15] To Falk, however, region seemed more salient than gender or even class. An immigrant scion of a family of noted British reformers—his maternal uncle was settlement-house pioneer Arnold Toynbee—he shared the common conception of the Maritimes as clannishly unknowable to outsiders. Thus he designed a system that emphasized professional, technocratic expertise even while searching for local knowledge. The middle-class friendly visitors who made most of the home visits were outsiders to North End, working-class communities, but Falk saw these laywomen (and some men) as experts because they were Maritimers. Trained social workers could then make their objective, technocratic decisions on the basis of fact.[16]

• • •

Haligonians had long built complex and multifaceted family economies. John Oxner earned $12 to $18 a week as a shoemaker, to which his family added the $8 per week a fifteen-year-old son, Louis, made as an apprentice baker under his foreman uncle and the $3.50 a week Louis's brother Fred, a year younger, earned as an errand boy. Outside of the formal labor market, John's wife Margaret took in plain sewing, which contributed a bit more money.[17] The older Gibson family similarly cobbled together survival from a variety of sources. Husband John, age seventy, had spent twenty years working as a blacksmith's assistant on the railroad and then a decade as a teamster. During this time, he and his wife, Ellen, were able to purchase or build three houses, two of which they rented out. The income they took in as landlords supplemented the $12 per week that John had been earning for the past two years as a blacksmith at the dock yard. They also made money from the grocery store Ellen ran from their home. Although we do not know how much money she made from the store each week, when it was destroyed, it contained unsold merchandise worth $400 and carried $100 insurance.[18]

The Oxners and Gibsons represent broader patterns. Working-class Halifax families relied on family members of more than one generation to earn wages in the formal economy. They engaged in a variety of informal and creative activities, like keeping boarders to earn extra cash, keeping chickens or cows to decrease their expenditures, or keeping extra houses to let out.[19] Extended families cooperated, often renting houses or apartments to their relatives. In my sample of 739 families, 94 households included multiple members who earned formal wages; 56 kept boarders; and 30 participated in obvious informal economic activity. Twenty women and six men worked, like Ellen Gibson, as shopkeepers. These numbers are bare minimums—only instances in which the plural survival techniques were recorded by social workers and friendly visitors. Others likely relied on unrecorded wages of multiple family members, and still more relied less visibly on the informal and domestic labor of mothers, daughters, and wives, be it through scrimping, kitchen gardening, caring, or maintaining relationships with neighbors.[20] Most working-class Haligonians survived before the explosion only with the support and labor of multiple family members.

These patterns shifted, stretched, and shuddered in the disaster, but they remained the basis of families' survival strategies after the explosion. People's continued reliance on their families, especially their extended families, is most clearly seen in the decisions they made about where to stay after the explosion destroyed their homes. More than any other institution, survivors relied on family members for emotional and financial support and for access to shelter and food. In my sample of families, out of 194 families that experienced homelessness (about 26 percent of the total), twenty-nine (15 percent of the total homeless) went to the wife's parents; forty (21 percent) went to nonparent relatives of the wife; nineteen (10 percent) went to the husband's parents; thirty-four (18 percent) went to other family members of the husband; and thirty-seven (19 percent) went to the houses of grown children. A further twenty-eight relied on people described as friends—though in fact these "friends" were often distant relatives—and twenty found shelter with their neighbors or rebuilt with them.

When neighbors, friends, and relatives helped each other, the aid they offered was more than just merely material. Frank Brinton remembered that he and his family stayed the first night with their neighbor before leaving the city for several months: "She had a room and she wanted us to come in there."[21] Brinton's recollection that their neighbor "wanted us to come" is telling because it suggests that the neighbor got something—company, emotional support, perhaps their assistance closing up windows and cleaning up the house—from the Brintons. Sarah Ellen Powell took into her mostly undamaged home three of her married sisters and their families, a total of seventeen people. In ordinary times, an unrelated family boarded with her, so she was used to sharing her home with others, though perhaps not quite so many others. But after the explosion, extra families in the house carried a different meaning and served a different purpose. The explosion had killed Powell's husband, a stove fitter, and we can imagine that the busyness and bustle of her sisters and their families may have distracted her and their presence provided her emotional support. This support may have been particularly important to her, both emotionally and practically, because her six-month-old daughter, Hilda, never fully recovered from a cut on her face that she received in the explosion. In April, the baby died. "Mrs. Powell's husband was killed on the day of the explosion which makes it very sad," the friendly visitor, A. P. Stairs, wrote. A month later, a visitor described how "Mrs. Powell [was] very sad over death of baby; says time hangs heavily on her hands at all times." In the face of this understandable grief and depression, the presence of Powell's family became all the more important.[22] The categories of benefactor and recipient were not cleanly distinct.

It was not only families who helped and supported each other in reciprocal ways. Jim Martin's habitual drinking led to difficulty with his family—his pregnant wife sometimes had to call the police to extract money from him—but he apparently had good friends. He and seven other men built a shack on Bilby Street to live in after their houses were destroyed.[23] Gordon Mitchell, a twenty-two-year-old driver, may have been among those men. His house destroyed and his possessions smashed, he built a shack on Bilby Street with "several men of the neighborhood" while his wife and baby daughter stayed in "more comfortable quarters." The shack must have been uncomfortable indeed, since in February the men were still asking for mattresses and beds.[24]

These shacks were physical manifestations of neighborhoods. Men chose to stay in cold, uncomfortable shacks rather than go with their families to warmer, if more crowded and alien, quarters. Staying in the neighborhood, with other "men of the neighborhood," must have been very important to them. We know that male companionship was important to Jim Martin, who later that spring went on a fishing trip with his buddies—which apparently devolved into an alcoholic bender—even though it meant delaying his wife's doctor's appointment.[25] Perhaps this style of homosociality was similarly important to all the men in the Bilby Street shack, and we can imagine that drink would have kept them warm through the winter. Or perhaps focusing on the male neighbors is misleading: Although Gordon Mitchell's wife eventually found indoor quarters, the family had at first built their own shack, which they shared with his parents and a married sister. That the wife stayed in the shack at least through the end of January suggests that there was more going on than masculine carousing.[26] The community networks that the Mitchells and the Martins had built in ordinary times before the explosion took on new and added importance in the crisis after the explosion. Just as we saw in chapter 1, where in the explosion's immediate aftermath people reenacted and reinscribed their daily patterns of solidarity, so too did they here.

A family named the Purcells—Margaret, Norman, and their nine children, ages two and a half to twenty—demonstrates several common things families did after the explosion: splitting up, relying on more or less distant family members for support, spending time in rural areas outside of Halifax, staying in formal shelters with strangers, and working to rebuild temporary shelter near their former houses. Norman was a hostler for the railroad, and the two eldest sons were a fireman and an automobile mechanic. Margaret kept house and looked after the children. In the explosion, Margaret broke her arm, and their seven-room house in Richmond was destroyed. Norman stayed home from work afterward to build a shack in what had been their backyard, using material provided by the Relief. They sent the two youngest daughters to relatives in the country, and Norman and the eldest son stayed in the railroad's roundhouse, which had been turned into a temporary shelter for employees. The auto-mechanic son, meanwhile, stayed home from work to keep house because Margaret's injury prevented her from doing so. The family, used to relying on a variety of members in ordinary times, reshuffled their labor in a time of crisis but maintained the principle of multigenerational family support.[27]

As the building material Norman Purcell used suggests, just as the explosion forced families to rejigger their labor and their locations, its aftermath added a new form of work: applying for, receiving, and maintaining benefits from the Relief. The labor required to obtain the material support offered by the state could be big or small. The Mountain family, for instance, put their thirteen-year-old daughter to work as a messenger to communicate with the relief committee because her father was at work and her mother needed to stay home to mind the other children.[28] Likewise, the Boutilier family relied on the wife's brother to write an angry letter to their social worker demanding more money.[29] Florence Simpson, the widow of a stevedore who died in the explosion, sent for her two grown daughters from a first marriage, Louisa and Mildred, to come from Toronto. Louisa was, wrote a worker at the relief committee, clearly unhappy about being there. Pretty and well-dressed, the daughters resented any implication that they were responsible for their mother's support. Rather, they saw their jobs as advocates. "From their attitude [it was apparent that] they are not intending to work here, but to wait until mother was (as Louise put it) well taken care of, and supported by the Committee."[30]

Few things exemplify better the way relief money was integrated into families' economies than the controversies that arose when, in June 1918, the relief commission put a notice in the newspapers asking that those who had sheltered refugees put in claims so that they could be reimbursed for their expenses.[31] Family members who had happily provided shelter as a matter of familial obligation—indeed who had refused payment when it came from a relative—now discovered a way to increase the money they received. Pearl Morgan and her parents-in-law, with whom she lived normally, went to live with her aunt and uncle, the Hilcheys, after her husband died in the explosion. The elder Morgans and the Hilcheys were barely related, so when the former left for the country after about three weeks, they offered the latter money. The Hilcheys refused any payment. There was probably a similar discussion when Pearl Morgan left a few weeks later to stay with her parents-in-law in Bridgewater. Yet in June, the Hilcheys filed to be reimbursed. Pearl was upset: "When asked if she thought bill was just, and if she would have been willing to pay it, woman said she would not. She knew the Hilcheys would not ask or expect her to pay." Her parents-in-law were less polite. When they spoke to Dorothy Judah, the district supervisor and a social worker who had come in from Montreal, they "appear[ed] mortified and highly indignant at Hilchey claim." Here we can see the two families negotiating what the obligations of extended relations were to each other. The senior Morgans were unsure what the Hilcheys owed them and offered to pay. The Hilcheys at first signaled that the Morgans—the parents-in-law of their niece—were sufficiently family that whatever aid the Hilcheys gave was as friends and relatives, not strangers. Yet when someone else offered money, they were all too happy to accept it.[32] Applying for, accepting, and maximizing payments and in-kind aid became a new and important economic strategy for families like the Hilcheys.

As the dispute between the Hilcheys and Morgans suggests, the presence of money for labor that was ordinarily done informally or for free could change family dynamics, and family members did not always agree about the financial value of certain labor. Jacob and Charlotte Bardsley and their family had a complex system relying on multiple earners in different generations, a plot of land in the countryside, and livestock in the country and city. Jacob worked at the cotton factory—which burned down from the explosion—for just under twelve dollars per month. Son Donald, eighteen, earned considerably more than that at the grain elevator. Daughter Hattie, seventeen, also earned more than her father as a pieceworker at the Moirs chocolate factory; she contributed four dollars each week for her board. Henry, fifteen, worked in construction for two dollars a day. In addition to this formal labor, the family kept two cows and several chickens, which—like the several young children—were Charlotte's domain. Just before the explosion, they had slaughtered one of the cows, but the market-bound meat was destroyed. While in a shelter in December, they managed to find several of the chickens to eat them, but the others died in the cold. They did not "live off" their livestock, Charlotte told an investigator the next summer, "but woman says family find it very hard to manage without them now." Charlotte clearly felt their loss more keenly than Jacob, and she complained quite bitterly about the small amount the Relief gave her for their loss. Jacob, less connected to the livestock and more mindful of the formal wages, undercut his wife and accepted the payment she would not, agreeing with the investigator that she was "very unreasonable."[33]

The differing responses of Charlotte and Jacob Bardsley suggest the ways in which the formal and informal parts of the family economy were gendered. Livestock, small shops, and boarders were women's work, and they were often undervalued. Even when women's labor brought direct economic benefit, it was often invisible, or—especially with boarders and livestock—it was imagined as merely part of the domestic labor that women did anyway.

Yet the disaster exposed the economic value of labor that before had gone unremarked. When women could no longer do their accustomed housework because of explosion-related injuries, their families would either have to keep home another, potentially wage-earning family member, or they would have to hire a domestic servant for cash. Either way, women's domestic labor became financially visible once it could no longer be performed. Mary Green was one such woman. She lived with her second husband, William, a dockhand; their two young children; and her three children from an earlier marriage. When the explosion broke Mary's leg and injured her eye so badly that the doctors removed it, it became suddenly clear how important her domestic labor was. She returned from the hospital after only two weeks to take care of her two young children, a three-month-old boy and a girl who was two and a half. But she was still in no condition to do the housework, so the family kept home Clarence, age fourteen, to do the work his mother could not. Clarence's older brother, Thomas, was an apprentice at a foundry; he too was staying home because of his mother's condition. To help, the relief commission arranged for the family to hire a domestic servant for a month, with the cost borne by a private benefactor. But in April, a visitor from the Relief named G. E. Blakeney found that without Mary's domestic labor, the family was in a deplorable state. When the visitor arrived, Mary was lying on a "rough bed" in the kitchen, the only room in the damaged house where there was a fire. Crucially, in care work, the work and the care were inextricably intertwined; for Mary, the worst part was that she felt she was neglecting her children. She "feels so ill and helpless," Blakeney wrote, and, unable to care for her children and feeling like her illness was causing undue expense, "now she has all gone to pieces." At the next visit, Mary explained that she refused to go to the hospital for treatment because there would be no one to look after the children. In June, still ill, Mary grew increasingly desperate. Blakeney reported that sometimes William or Thomas had to miss work in order to watch over the children. "Woman says she has been a big expense to her husband since the disaster, it has cost so much more to run the house with her not being able to look after things." The disaster laid bare the financial contribution of women's otherwise invisible labor.[34]

In order to handle these sorts of situations, the relief commission created the housework pension. William B. Wallace, a relief commissioner and a Halifax County judge, laid out the policy of what should happen when a woman was so injured that she could not perform her normal domestic labor, thus forcing her family to hire someone to do it for her. If the husband, Wallace wrote in a memorandum for the rehabilitation staff, earned less than twenty dollars per week, "his wife should be regarded as practically a wage-earner, by reason of the fact that she does all the work of the household and indirectly contributes to the maintenance of the family, and therefore there should be some compensation for that partial disability. A small allowance—$3.00 or $4.00 per week should be made in such cases. Care, however, must be taken that the disability should not be treated as a permanent one, & thus the allowance become permanent."[35] Wallace's policy was directed at families in which the father's income was undiminished, so it left considerable leeway in cases where the father had died or become disabled.

Wallace's housework pensions rested on ideas about the family and domestic labor that were gaining currency in progressive North America. Cape Breton's radical labor leader J. B. McLachlan, calling the miner's wife "the greatest financier in the world," announced a "Wage Earner's Contest" in 1917 asking how "wives, mothers, sisters, and sweethearts of men" would support a family on a proposed daily wage of $3.50. Fundamental to McLachlan's question was the understanding that women's hard domestic labor was integral to families' survival.[36] Recognition of the monetary value of women's domestic labor played a role in the movement in Alberta to grant them rights to family property. Feminists argued that if husbands owned all marriage property, women would not see the financial benefit of their labor.[37] The mother's pension movement similarly included ideas about the material worth of women's domestic and reproductive labor, in that instance on behalf of the nation. Some mother's allowance advocates and administrators suggested that women be paid for their housework; others thought the state should compensate mothers for the earnings they would have earned working elsewhere. One social-work manual called recipients "employes of the state" and their money "wages."[38] The relief commission drew on these ideas, but it was unusual and perhaps unique to create pensions that treated domestic labor on a similar basis to formal, wage labor.

The housework pension was very unevenly applied—only twenty-six families in my sample received one, and Mary Green did not—but it explicitly acknowledged the expenses that families like the Greens had to bear when women were unable to do their customary work. As with other forms of relief, despite this acknowledgment, families often had to work to wrest a pension out of the relief commission. John and Mary Christian and their five children provide an example.[39] John was a laborer at the Graving Dock, where he made twenty dollars in an average week. William, twenty-one, also worked in the dockyard, and he paid his parents five dollars for weekly board. His brother Harold, seventeen, was an apprentice plumber, which earned him a paltry three dollars each week. The youngest children, Lillian, Bertram, and Jack, went to school. Meanwhile, Mary kept house. The day of the explosion, John had the misfortune to be on the wrecking tug *Stella Maris*, which tried to tow the burning *Mont Blanc* away from the pier. The captain and twenty of its crew died; John was among only five survivors.[40] His injuries were considerable: his kidneys were "torn," a rib fractured, his back was almost broken, and he had to be strapped into his hospital bed for several days. Although he left the hospital after only six weeks, a visitor reported that his "face is all marked up with powder. Looks wretched." Doctors predicted that he would not be able to work until April. Meanwhile, John's formal work for cash was not the Christians' only loss. The worst was the death of their youngest son, but their financial and material losses mounted. Their house was destroyed, and William could not find new work until the end of January. At the end of January, soon after John's hospital discharge, the family left St. Mary's shelter and moved to a new house on Clifton Street. Lillian was still attending school at St. Mary's—St. Joseph's School had been destroyed in the explosion—which was more than a mile away from the new house. This meant she needed to pay to take a streetcar to school, an additional burden. On top of all this, Mary had a major cut in her hand—a piece of glass remained embedded in it at least until the summer—and she suffered worsening pain and weakness in her hand and arm.

Mary Christian, of course, did not have a formal supervisor who could order her to come into work or send her home if she was unable to do the task assigned. Moreover, if she did not clean, cook, and do laundry, either someone else had to do the work or the family would be dirty and hungry. In January and February, she tried to work, but she often found it impossible; she had to send laundry out and hire a paid domestic worker for other tasks. By the end of February, her hand appeared to be improving, and she went back to her normal duties. In June, her hand got worse again, and she needed again to hire someone else. But hiring someone required access to cash. When the Christians hired someone the first time, the cash came from John's workman's compensation, which provided $21.50 each week. When he went back to work, he earned $20 a week. That even this slight decrease in wages made it impossible to pay a domestic indicates the precariousness of the family's finances. Yet it took much of the summer for the family to get cash to pay someone to do the domestic work. "What I want is the money that should have been given me to pay a woman for working for me since the disaster and until my hand is well again which I hope will be soon," Mary Christian wrote to George Cutten, the head of the Relief's rehabilitation division, in July. "Other people have been getting money to pay for having work done and they did not have as much cause for asking as I have." Considering that her injuries were to her arm and hand and that she still had a piece of glass embedded in her hand, one can imagine the pain and difficulty involved in even writing the letter. She closed with a reminder that there was more to her domestic work than simple labor. "I have been suffering ever since and I lost my child through the disaster. That is the greatest loss of all to me, a loss which no one can repay." Her reminder to Cutten, even while asking for material aid, that her grief was more than material, was a common technique. In this case, it was successful. Christian was eventually ruled 50 percent disabled, and she received eighteen dollars cash to pay for domestic labor for each of the three summer months. The relief commission apparently believed her ordinary, monthly labor to be worth thirty-six dollars, although there is no document to suggest that their valuation was anything but arbitrary.

The relief commission could be capricious. Because women's caring and domestic labor was invisible and illegible, there were not hard rules for what counted and what deserved a pension. The Wasson family—John and Bessie and their children Ida, Lillian, and John—relied on the labor of all family members. John, who had emigrated from Ireland as a young man thirty-five years earlier, worked as a laborer at the tar factory.[41] Bessie, born a Newfoundlander, kept house. Ida, twenty, worked in a cannery labeling cans; John, seventeen, was an apprentice electrician. Lillian, twenty-five, worked alongside her married sister, Annie Rogers, at Moirs chocolate factory. As with many other families, this arrangement was upended by the explosion, which killed Bessie, wrecked the house, and destroyed all its contents with the exception of two chairs. Ida was severely burned when a stove fell on her, and her arm was hurt by collapsing timbers. With Bessie dead and Annie married, only Lillian was left to care for her younger sister. In April, John described the situation in a handwritten letter: "One off [sic] my daughters was burnt very badly in the explosion and she is not able to do anything with her arm yet the other one has to stay home from work to look after her ever since before the explosion they both worked as I am only a poor man I cannot afford to keep both will the relief be able to pay the board for the girl that has to stay come to look after the girl that was burnt." Thanks to Ida's injury and Lillian's shift to domestic labor, the family went from having four formal wage earners to having only two, of whom one was an apprentice and thus earned very little. Yet the relief commission was unsympathetic. Social workers suggested that Ida was malingering, and they refused to pay Lillian for her time caring for her sister.

Lillian, especially, understood the monetary value of her labor, both to herself and to the commission. "Of course if I did not stay at home with her the relief would have had to pay some one to look after her," she wrote in early July, "and off [sic] course I could not afford to lose all that work for nothing." Later that month, having received yet another denial, she wrote again. In her second letter, she tried several arguments. First, she repeated the argument that had she not cared for her sister, the commission would have had to pay someone else to have done the same work: "The relief should have sent word to me at the first start and I could have went to work and they would have to hire some person to look after my sister." She then explicitly referred to her caring labor as work: "How would you or any off [sic] the relief like to work 6 months for nothing[?]" Finally, she insinuated greed and corruption on the part of relief workers and appealed to the rights of donors to have their money distributed fairly: "I suppose the relief are trying to stop the money from the people who suffered through the disaster but it is nothing out off [sic] their pocket because the money was sent to Halifax for us people." The relief commission, however, was unmoved and still refused to pay her. In her letters, Lillian Wasson argued that the informal work that she did to care for her family was indeed work, and that she deserved to be compensated for it. In doing so, she was adopting wholeheartedly the logic that the relief commission applied haphazardly. She sought compensation and recognition for the labor she and other Halifax women had long performed informally.

. . .

Haligonians' social relationships extended far beyond the city. In some ways the city recapitulated Canada's status as both a recipient and donor country for migration.[42] Halifax received migrants from the Maritimes, from Britain, and from Newfoundland, and it sent migrants to central and western Canada and to the United States. All this migration built transregional and transnational networks and communities on which Haligonians relied in the explosion's aftermath. These connections are most visible in figure 15, a map created from data contained in a report from the Halifax Information Bureau. The ad hoc information bureau fielded 6,214 inquiries by telegram and mail, people from afar asking whether their friends or relatives had survived, and whether they needed help.[43] (This total excluded inquiries from Newfoundland, which appear to have been handled by a separate office, staffed by Newfoundlanders; it also excludes those from soldiers abroad in Europe.[44])

The map's most striking feature is the density of inquiries from the United States. Even so, it hides the true extent of Halifax's ties to New England because it displays the towns from which people inquired, not the number of inquiries. The number of towns near Boston from which at least one person made an inquiry must stand in for the large number of inquiries received from Boston proper.

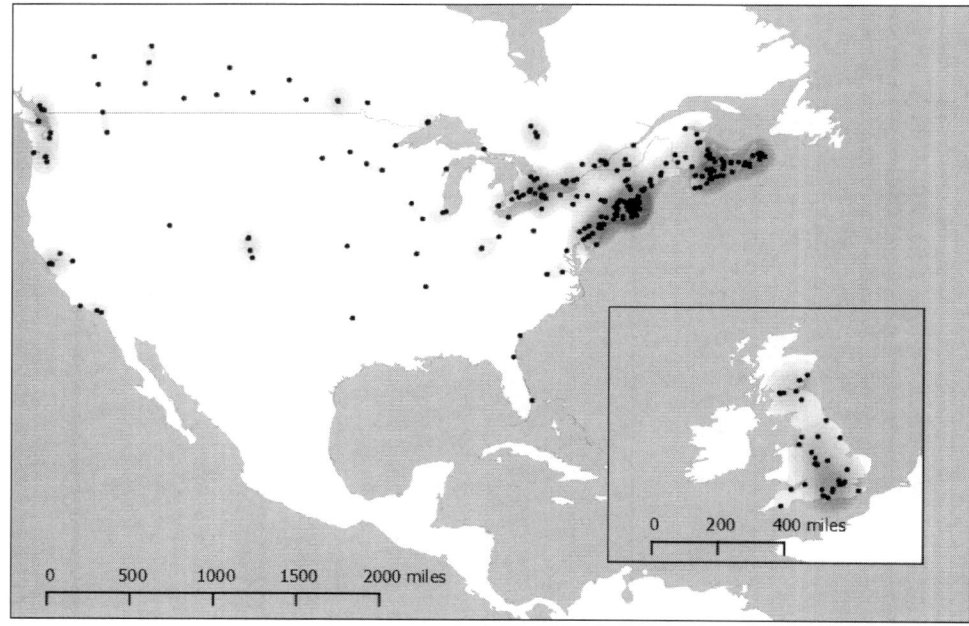

FIGURE 15. The Halifax diaspora, as shown by inquiries to the Halifax Information Bureau. Map by Andrew Rothwell based on data in letter, J. Stredder to T. MacIlreith, n.d., item 94.1e, HRCC. A version of this map appeared in Jacob A. C. Remes, "'Committed as Near Neighbors': The Halifax Explosion and Border-Crossing People and Ideas," *American Review of Canadian Studies* 45, no. 1 (Spring 2015): 30.

Indeed, so many Bostonians feared for their relatives' safety that the city established an information bureau in City Hall to agglomerate inquires.[45] The map shows graphically and geographically the deep connections Halifax maintained to the United States, and that these connections were often deeper than those to other parts of Canada. This observation is reiterated in my sample of 739 Halifax families. Of these, forty-one cases had some explicit connection to the United States: roughly thirty-one of these mentioned family in the United States, twenty went to the United States, and four had family members in the United States who came back to Halifax to help after the disaster. In contrast, only ten families appear to have left Halifax for what was called "Upper Canada"—that is, any point west of New Brunswick. (Again, these are minimums; other families likely had unrecorded connections or migrations.)

The affective ties made apparent by the inquiries were recapitulated by donation patterns. The map shows three regions with particularly strong connections to Halifax: New England, Britain, and rural Nova Scotia. We can add to this Newfoundland, then still a British colony separate from Canada and excluded from the map. We can contrast aid offered from each of these four regions with the $100,000 sent by the Ontario government.[46] Although Ontario's total contribution would be larger if we included private contributions and those from city treasuries, the comparison shows the relative importance of social, cultural, and especially migratory connections rather than the legal and national ties that ostensibly bound Nova Scotia with Upper Canada. Newfoundland, for instance, had deep cultural and migratory connections to Halifax. Although it had less than a tenth the population and much less money than Ontario, its government donated $50,000, the St. John's city government pledged $30,000, and the St. John's Board of Trade gave a further $10,000.[47]

Haligonians relied directly on outlying parts of Nova Scotia for shelter and other material aid; 112 families in my sample went to the country for shelter, where they depended on their social networks outside of Halifax. Bertha Ryan went to stay with her uncle for about a week after the explosion. It was an obvious thing to do, she said sixty-eight years later, "because we belonged to the country, and that is where we—and went back home. That's about all there was to it."[48] Her description of belonging to the country, even though she lived in the city and had a community there—she worshipped at St. Joseph's Church, so she at least had her coparishioners—is telling, for it suggests the deep connection between Halifax and the surrounding rural areas. Rural Nova Scotians had long worked seasonally in industrial cities, and city dwellers retained connections to the countryside.[49] Urbanization was not unidirectional or permanent; families that had moved to Halifax from rural areas retained important ties to their ancestral homes. Yet migrants had come to the city for a reason, and many preferred it there, or they feared what life would be like if they returned to the country. Partly to be helpful, and partly because scarce housing was needed for "able-bodied men and their families," the Relief sent a seventy-one-year-old widow named Marie Walker to her niece in Enfield, about twenty miles away. A month later, she was back in Halifax, demanding help finding a new home. "Says she has always had her own home, and cannot think of going to live with relatives in the country, altho they would no doubt be willing to keep her," wrote May Reid, a social worker from Winnipeg. "Claims she has always made a living in Halifax and wants to continue living here. She is feeling much better and anxious to get to work."[50] As Walker discovered, reserving housing for "able-bodied men" with families privileged the rights of married men to independence over those of adult women. It also suggested that men's formal labor was more important to families and to society—ideologically if not economically—than women's informal work. Reid discovered, though, recipients like Walker had the ability to reject their social workers' authority, in this instance by returning to Halifax.

Britain, linked to Halifax through ties of empire, wartime alliance, and immigration, granted $5 million (£1 million) to the relief fund in the hopes, in Prime Minister Borden's phrase, that it would "greatly strengthen Imperial sentiment."[51] As Borden's comments suggest, donations were inherently political because they enacted claims about community. The British donation, coming as it did immediately before an election that hinged on imperial sentiment, was an intentional assertion of imperial obligation. But these donations could be turned around, and claims about them could be deployed against the government. In April, West Indian medical students at Queen's University, protesting racist mistreatment, referenced that their colonies had "rather liberally" donated to Halifax. That their islands' governments had given to the Canadian relief fund, they argued, entitled them to rights from the Canadian government.[52]

While the British government and West Indian students mobilized contributions for unrelated political purposes, Nova Scotians in Massachusetts did so for more personal purposes. The relief offered by Massachusetts remains one of the most famous stories to come out of the explosion; most memorable was the relief train that departed Boston immediately and arrived in Halifax bearing A. C. Ratshesky, John Moors, and the first relief supplies to arrive from outside the province. Later, Massachusetts raised about $700,000 in cash, plus in-kind donations of clothing, building supplies, and the like. Unlike the British government and most private donors, Massachusetts retained control over its contributions. It formed the Massachusetts-Halifax Relief Committee, which focused originally on sending supplies, first on the train and then on two steamships dispatched from Boston. Later, flush with cash, it shifted to providing furniture to explosion survivors. Through discounts and a remission of the import duty, the committee estimated that it provided about $600,000 worth of aid while only actually spending about $200,000. This left more than $260,000 to spend. After consulting with an expert from the Rockefeller Foundation, the committee decided to establish a public-health program to help alleviate the long-term effects of the explosion.[53]

The initial appeal sent from Boston to local committees throughout the state emphasized Massachusetts' obligation to Nova Scotia because of proximity. Henry Endicott, the Massachusetts-Halifax Relief Committee's chair, promised that "we are all committed as near neighbors of the stricken city."[54] In 1919, the committee's official historian phrased the commitment as related to World War I. The gifts, George Lyman wrote, were "in behalf of a friendly neighbor, now our Ally in the Great War."[55] The references to being "near neighbors" hid the more complex, familial relationship between Massachusetts and Nova Scotia. In 1915, the commonwealth census counted 79,115 residents who had been born in Nova Scotia.[56] These migrants were the latest of several generations of mass migration that bound the Maritimes and New England together economically and culturally.[57] Although some Massachusetts residents may have donated funds out of neighborliness, wartime alliance, or even disinterested altruism, an unknown but large number of donors came from or had family in Nova Scotia.

Many of the Massachusetts residents involved in Halifax relief had direct ties to the city. Thomas Wilson, for instance, worked for the Boston City Building Department and was detailed to the city's information bureau, forwarding to Halifax requests for information about Bostonians' relatives. Included among them was a personal inquiry about a handful of his own cousins.[58] Among the Massachusetts doctors who went to help was a Dr. Fraser, described as "a reputable physician of Lynn [who] formerly lived in Halifax." His personal connection to his former hometown "was one of the reasons why he was selected to go down."[59] The same was true of other Americans who came to Halifax. Gertrude Dobson was a Chicago social worker who came to Halifax with the American Red Cross. Her supervisor wrote in January, "I think as she is a Canadian from New Brunswick she ought to be induced to stay permanently."[60] Personal connections to Canada, the Maritimes, and Halifax encouraged people in the United States to donate money or volunteer their time.

Individual sufferers and survivors in Halifax knew they were enmeshed in a transnational network, and they and their relatives in Massachusetts used it to their advantage. Having friends and relatives in New England gave Haligonians more power in their occasional battles with the relief authorities. In early January, the Massachusetts committee selected five men and two women from Halifax to oversee their operations. The Halifax committee, as it was called, was in charge of adjudicating applications for furniture, including deciding on the amount and quality to be given in each case.[61] In practice, the Halifax committee usually relied on the advice of the relief commission's rehabilitation committee and its social workers in deciding who deserved what furniture. This structure meant that the power structures built into the rest of the Halifax relief work was reproduced in the Massachusetts aid. The presence of the Boston committee and the fact that Nova Scotians in Massachusetts had donated to the fund, however, altered survivors' sense of their rights. They or their Massachusetts relatives could write to authorities demanding that the Halifax sufferer be given their fair share of what the Massachusetts relative had donated.

Most of the time, the Boston committee simply forwarded the complaint back to Halifax and ratified whatever decision the Halifax committee made. The explosion destroyed W. T. Murphy's house, and he had requested furnishings for his new kitchen, dining room, and three bedrooms, including a refrigerator. In May, he wrote to Boston, complaining he had still not received anything. The vice chairman of the Boston committee, James J. Phelan, simply forwarded the letter to the Halifax committee and told Murphy he would have to take up the case there.[62] Yet sometimes, by writing to higher-ups, aggrieved Haligonians could be taken more seriously. Charles Tanner wrote to Massachusetts governor Samuel McCall to complain that he had been mistreated and had not received his share of the relief aid donated by Massachusetts. McCall forwarded the letter to A. C. Ratshesky, who in turn gave it to the chairman of the Halifax committee, G. Fred Pearson. Pearson thought that Tanner's complaint was groundless. "I have investigated this case, and while the facts are generally as stated, Mr. Tanner was blind before the disaster," he wrote to Ratshesky. Given that, Tanner was not due a pension, although "as a matter of fact, he was allowed the sum of $10.00 a week for a number of weeks in order to rest." Because Ratshesky had taken a personal interest in the matter, Pearson promised to "look into the case and see what we can do in the way of furniture or other relief. I am writing Mr. Tanner today to come in to see me."[63] Though there does not appear in the archives a single case of the Boston committee backing the complaint of a sufferer against the Halifax committee, these complaints were still taken seriously.[64] Moreover, the very fact that a Canadian in Halifax like Tanner sought help from an ostensibly foreign politician demonstrates the political community created by donations.

This new political community worked in the other direction, too, when Massachusetts donors wrote to Halifax authorities to demand better treatment of their friends and relatives. Levinia MacKenzie was an old woman who had come to Halifax three years before from rural Moser River to live with her adopted daughter, Rose Bedgood. After the disaster, MacKenzie appeared to have lost her memory, and Bessie Egan, a semiprivate police matron tasked with handling social service cases, took her to the City Home.[65] A month later, MacKenzie felt trapped. Rose, whose husband was overseas in the army and whose own daughter had died in the explosion, was no longer in any position to help. Rose was, the Home matron "judged from . . . appearances and reports concerning her . . . not [of] very good character"; the Relief's head social worker said Rose was "behaving badly and neglecting her child and mother." MacKenzie still had a sister in Mosers River, but she reported that Rose had refused to write to the sister for help, apparently for fear that they would both end up back in the country. She also had sisters in Boston, but she could not remember their addresses. Faced with this situation, Egan had "decided quite firmly" that the best place for MacKenzie was in the City Home.

By June, MacKenzie had apparently found a way to make contact with her relatives in Massachusetts, and one of them relied on her authority as a donor to write an angry letter to the relief commission. "At the time of the terrible explosion in Halifax, this town as well as the whole of Mass contributed money for the relief of those who lost everything in the explosion. I contributed my share as I knew of some people living there, in particular an old lady Mrs Levinia MacKenzie, who lived with her daughter Mrs Rose Bedgood, this old lady lost every thing and we found that that she was placed in the City pour [sic] house for the time being," wrote a furious Mrs. A. E. Anderson of Wakefield. "What was this money give to your Ass. [association?] for, to put people in the poor house?" Anderson suggested that MacKenzie be sent back to Moser River, where, Anderson claimed, she owned a furnished home. "She writes me she will die of a broken heart if she has to stay there any longer," she wrote of the City Home. Ultimately, Anderson's appeal had the same effect as the letters to the Massachusetts-Halifax Relief Committee: the matter was looked into a second time; George Cutten took a personal interest in the case; and Anderson received a strongly worded defense of the relief commission's actions. Nonetheless, we can see in Levinia MacKenzie's case how the existence of the diaspora and its donations to the Massachusetts fund gave extra support and political power to their suffering relatives in Halifax.

Anderson was not the only friend in Massachusetts who feared a Haligonian would wind up the City Home, the humiliating and horrendous institution for the city's poorest—the old, the infirm, or the insane who had no family or community to take care of them. In many respects Sarah Henry, a sixty-three-year-old charwoman—that is, unskilled domestic worker—was like other poor Halifax women. She owned her house and scratched out a living from wages, the support of her family, and some charity. In one key way—her race—she was unusual, being one of among the eight hundred or so African Canadians in the city.[66] In September 1915, a daughter, then living in Cambridge, Massachusetts, with her husband and children, had twice tried to move back to Halifax to take her sister's place caring for their aged mother. Both times, however, the family was stopped at the border. (Henry's daughter, though born Canadian, had lost her nationality when she married her American husband.) Although the Immigration Act of 1910 did not explicitly forbid black migration into Canada, by 1912, what started as merely customary discrimination hardened into something very near an official policy banning their entry.[67] In the case of Henry's daughter and her family, Canadian immigration inspectors claimed, no doubt influenced by their skin color, that they would become public charges. This was a self-fulfilling prophecy, since when the family returned to Cambridge they found the husband's job gone, and they were obliged to rely on charity for several months. With her daughter now on charity, Henry too lost her independence and came to rely on the North Baptist Church.

If things were bad after the daughter's stymied return to Canada in 1915, they got even worse after the explosion. All Henry's unofficial supports were gone: her house, which at its best had been derided by the Canadian immigration inspector as a "very old wooden house of no value," had been destroyed, and the church was overwhelmed. This left Henry dependent on the state. Her house gone, she moved first to the shelter at St. Paul's Anglican Parish and then to the one at St. Mary's Catholic Cathedral. By the beginning of January, eager again for independence from the state, she arranged to move to Cambridge to live with her daughter. Once again, however, Henry was thwarted by the border: American immigration inspectors excluded her and forbade her to try again for a year.

The Americans' decision left Henry stuck in Halifax. With nowhere else to go, she was among the last people to leave the shelter at St. Mary's. To get her out, the relief commission arranged for her to board with Mary Francis. The Francis household was "a colored family in an ordinary poor condition; she has only one day's work a week but had no more before explosion."[68] This last phrase—"had no more before explosion"—was key, since the Relief saw its role as restoring people to the class and social position they were in before the explosion. So in April 1918, the relief commission declared: "From now on Mrs. Henry must live as she did before the explosion, which was chiefly on the generosity of her friends." Cutting her off left Henry at risk of winding up in the City Home. It was only Henry's daughter, still stuck in Cambridge, who had the power to keep her out of there. She continued to advocate for her mother to come to Massachusetts, where she could look after her grandchildren while the daughter worked, but US immigration authorities continued to block her way. When friends in Halifax found her mother a place to live, she agreed to send five dollars a month to help support her. This arrangement meant decreased independence: not only was Henry dependent on her daughter, but the money was routed through the relief commission, which supervised Henry and her expenses. Nonetheless, this money from Massachusetts was the one instrument of the Henry family's power; the daughter was explicit that she sent it to keep her mother out of the City Home.

These explicit cross-border negotiations were the unofficial version of the official thanks that Halifax and Nova Scotia officials gave to their Massachusetts counterparts. Halifax authorities named temporary housing after Governor McCall and designated streets in the complex in honor of Massachusetts, Rhode Island, Maine, and Massachusetts-Halifax Relief Committee chair Henry Endicott, a symbolic recognition of the political community McCall and Endicott

Halifax Families and the Work of Relief 127

FIGURE 16. Massachusetts governor Samuel McCall visits the Halifax apartment complex named for him, 8 November 1918. Note the Massachusetts Avenue street sign and the Massachusetts and US flags on the car. In front are Halifax children living in the apartment complex. The adults, left to right, were E. F. Horrigan, Governor McCall, G. Fred Pearson (chair of the Massachusetts-Halifax Relief Committee, Halifax committee), Captain Hathaway (McCall's aide), Ralph P. Bell (Halifax Relief Commission secretary), and George B. Cutten (director of rehabilitation, Halifax Relief Commission). Behind them, the chauffeur, G. Landry, sat in the car. Photograph from Special Collections, State Library of Massachusetts.

had inadvertently fostered (figure 16). Indeed, the responses penned in Boston and Halifax make clear that authorities recognized the power of the diaspora and its financial contributions. In one case, for instance, Fred Pearson wrote preemptively to James Phelan about a woman named Adelaid Simmons. Pearson wanted Phelan to know about the situation in case she "or any of her friends in Massachusetts call[ed] on you. We frequently receive letters from people to whom we refused to make additional gifts, stating that friends of theirs in Massachusetts have contributed to the fund and it occurred to me Mrs. Simmond's [sic] might be one of those and therefore it might be [of] service to have a statement from me on your records if any of her friends call."[69] The rhetorical strategy of mentioning "friends of theirs in Massachusetts" appears to have been successful in at least getting a second look. Even when it was not, the fact that Haligonians and their Massachusetts advocates thought to make these appeals signals that they understood their relationships as creating a new, transnational form of political power. On both sides of the border, they were building new ideas of what government owed citizens.

⋯

Rita Mariggi, a barely literate Italian immigrant, lived with her husband Cesare in a slummy, four-room apartment on the slopes of Ward 3.[70] According to neighborhood gossip, she "was of good character, only gay and fond of a good time"; he was a gambler and spent his money—about fifteen dollars a week, although stevedores' wages were always uncertain—on other women. Cesare went to work early on December 6 unloading the SS *Picton* at the sugar refinery dock. It took four days to find his body, leaving nineteen-year-old Rita to fend for herself and their three children, including five-week-old Margaret. The apartment was a mess, with only two rooms habitable and even those with rotten floorboards, ruined plaster walls, and a pervasive stench from the basement. When the Relief found her two weeks after the explosion—with two sick older children and an infant, she had been unable to go looking for them—Rita asked to be moved to St. Mary's Shelter, but the next day she changed her mind and asked for help moving to stay with her mother, who had offered to take in the family "for the present anyway." By the spring, she still had not moved, even in the face of increasing harassment from the landlord, who wanted to demolish the building.

Despite this misery and the apparent reluctance of her mother to help, Mariggi had both formal and informal networks to help her survive. Her story shows how Haligonians—even those with the least social capital—worked to balance the formal aid of the state and the informal aid of their communities. As a widowed working-class woman, she needed to work especially hard to maintain her autonomy and independence. In early January, she expected a payment from her husband's union (though there is no record that she ever received it). A few weeks later, she received twenty dollars from an anonymous "Italian friend," who promised to continue helping his compatriots "from time to time." Neither of these was enough to support her and her children, however. In April, Mariggi put her children in an orphanage and went to work as a domestic for the woman she had worked for before her marriage. A month later, the formal apparatus of the orphanage had also failed her. She took her children back, complaining that they had been badly treated and not looked after properly; in particular, the infant had lost weight. She moved into a boarding house, which let her keep the children with her as long as she did some household chores. A month after that, though, the she was forced to leave again, since her children and those of the landlord fought constantly. Mariggi moved in with her mother, but her mother also ran a boardinghouse and was afraid of it developing a bad reputation if her daughter and her children stayed there. Besides, it was almost unbearably crowded there, with nine sleeping in a room, three to a bed. Unable to care for the infant under these new circumstances, Mariggi turned to a childless couple, the Carmonics, whom she knew through her late husband, and gave them her baby. Peter Carmonic, known around the docks as Peter the Greek, ran a "questionable lodging house" for which the police were seeking a legal pretext to shut down. According to a different police report, though, the Carmonic house was simply a sailors' boardinghouse; dirty—like all such places, said the police chief—but reputable. Whatever the repute of the Carmonics' house, turning to them was perhaps the most desperate of Mariggi's attempts to find support within her community, and she was uncertain about it. At least three times she went to the Carmonics to take Margaret back, only to return her when she realized again that she could not take care of the baby.

The survival strategies Mariggi cobbled together with the help, albeit sometimes grudging, of her mother, the Carmonics, and her old employer were important, but they were fragile. As a widow with children, Mariggi was particularly vulnerable. Three children were a burden to feed, clothe, and house. Her gender gave authorities license to police her behavior and put her at the mercy of gossips like the woman who talked about her late husband's gambling. Mariggi knew her vulnerability, and she tried hard to keep Margaret's adoption a secret. She was right to be worried: once the social workers found out, they tried to put a stop to it. After Mariggi was forced to make the adoption official by going to court, Jane Wisdom, the supervising social worker, worked with William Wallace, the judge and a relief commissioner, to block the adoption. Wisdom and her colleague Ernest Blois, the superintendent of delinquent and neglected children, preferred that Mariggi work within formal systems rather than her informal support networks. Blois told her "that Judge [Wallace] would not allow the baby to remain at Peter Carmonic's and that she must take it home and care for it or place it in a Home." As a Catholic, the obvious orphanage for Margaret was the Home of the Guardian Angel—the same one in which Margaret had failed to thrive. "Mrs. Marigge said she would not put it in the Home as they did not care for it," Blois wrote. The Protestant orphanage refused to take the baby because her parents were Catholic. The Catholic clergy, though, were little better. Mrs.

Carmonic refused to go to church since she was convinced that the priest disliked her because, as Blois paraphrased, "she has a big house and no help." Whatever the reason, she was certainly right that the clergyman disliked her. The priest wrote a note of "emphatic protest" against the adoption, on the grounds that because Mrs. Carmonic "does not practice her religion herself she can hardly be expected to guarantee a religious upbringing to her adopted child." He too wanted Margaret sent to the Home of the Guardian Angel "until Mrs. Carmonic proves she is a Catholic by doing her duty."

The priest appeared mollified when Mrs. Carmonic signed a contract agreeing that if she did not become a good Catholic within two years, she would give up the baby, and the case mostly disappeared from the desks of Blois, Wallace, and Wisdom. But Rita Mariggi continued for much of the rest of her life to try to balance the demands of the state's welfare system and those of her informal network. In 1920, Mariggi had an illegitimate child, and the nurse who delivered her baby informed both the relief commission and Bessie Egan, the police matron, even though, the nurse said, Mariggi was "very anxious that H.R.Comm. should not hear of this." She was right to be afraid; the commission cut off her funds until she agreed to put the new baby into the Home of the Guardian Angel. Already, both of her older boys were being cared for at St. Joseph's Orphanage. Two years after that, she married a man named John Dill in a Presbyterian church. This meant that the Relief would no longer pay her widow's pension, but the marriage presumably provided a different kind of security, one without the intrusive gaze of the state. In the next dozen years, Mariggi bore seven more children, and she continued to balance Dill's increasing abuse with her desire to stay free of governmental intrusion in her life. In 1930, for instance, "sick and tired of getting a bawling out" from her husband, she put her eldest son—crippled in an accident three years earlier—into the City Home. She almost immediately regretted it, though, and wrote to the Relief begging help to get him out again. "There was all kinds of men, colored and white, crazy and sick, scrunching, hollering, cursing, swearing, ripping and tearing," she wrote. "He doesnt want to stay there. There was no church on Sunday. Joe said 'For God's sake take me out of here.'" Four years after that, in the depths of the Depression, Dill demanded that the second son, a star student, leave school and either work or live at the Citadel Relief Camp, the humiliating last resort of the city's poorest men.[71] In the ensuing fight, Dill struck Mariggi on the side of her head. Mariggi performed a complicated balancing act between him and the state, prosecuting him for his assault before, in the words of the newspaper, "she became soft-hearted and decided she did not want to prosecute" and warned her husband that should he hit her again, he would find himself back in court. Nearly two decades after the explosion, Mariggi was still trying to maintain her independence from the state while using it to get the support she needed.

Mariggi's was an extreme case, but it highlights the task that faced Haligonians after the explosion. Like Rita Mariggi, like the Hilcheys, like Sarah Henry and her daughter, they wanted—needed, in many cases—the money and other material aid the Relief offered. Yet they rejected, or tried to reject, the new bureaucratic machine that offered it. They preferred instead the reciprocal solidarity of people they knew. Unlike the cold, technocratic aid of the Relief, survivors like Sarah Ellen Powell received both material and emotional help from people they knew. To balance these two—the formal and informal, the hierarchical and reciprocal, the charity and solidarity—Haligonians had to use a variety of techniques, themselves both formal and informal. In addition to doing so as individuals and families, they did so, as we will see in the next chapter, in institutions like churches and unions.

5

"A DESIRABLE MEASURE OF RESPONSIBILITY"

HALIFAX'S CHURCHES AND UNIONS RESPOND TO THE PROGRESSIVE STATE

At an emergency morgue in the Chebucto Street School in Halifax's West End, bodies started to pile up within hours of the explosion. Families clamored at the door, anxious to find their dead and barely held back by a small contingent of police. The men who extracted the bodies from the wreckage sometimes knew who they were and labeled them, but sometimes they were unfamiliar strangers, and sometimes the tags they used to mark the known corpses fell off in transit. When an unmarked body arrived at the school, men there—mostly soldiers, working with little food or rest—catalogued the contents of their pockets and tried to identify them.[1]

For the unknown dead, membership in organizations could literally become their only connection to others, the only thing that enabled them to become known. One father recognized his son from the Catholic prayer leaflets in his pockets. In another case, a teamster was only identified because of "several receipts for Foresters dues, discovered in his clothing." For others, these objects were not enough, and memberships became their only identity. Body number 480, male, age uncertain, wore nondescript clothes and carried four keys and a penknife. The head had come off, so there was no face to identify. The only identifying features were a "crucifix and Roman Catholic emblem." In death the man retained nothing of his identity except his Catholicism.[2]

On Monday, December 17, in the yard of the Chebucto Street School, the city held a funeral—two, really—for the unidentified dead. John Hanlon Mitchell, Archibald MacMechan's undergraduate assistant, was disdainful of the spectacle.

"It was a gathering of sight seers rather than of mourners," he wrote: mothers brought their babies, children played, young people flirted. Survivors who had not yet identified their dead mourned privately or invisibly, or they stayed away in the desperate hope that their loved ones would appear later. In the central enclosure, officiants, officials, "soldiers, clergymen, Salvation Armyist, and gloating Newspaper-women dripping with platitudes, were mixed in an incoherent mass." First came the Protestant service. "Representatives of the different denominations ascended the dais, while those who had not been chosen thrust themselves as far into the foreground as possible," Mitchell wrote. "A hymn was sung, but the singing lacked strength and conviction. Then the service commenced, each minister taking a share. Most noticeable was a tendency to hurry over the proceedings." Accompanied by a military band, the crowd sang two hymns and listened to a sermon by Anglican archbishop Clarendon Lamb Worrell. "The Bishop's address abounded in typical ecclesiastical bromides, but even he seemed to realize that now words counted for little." After the Protestants had finished, it was the Catholics' turn, and the two priests from St. Joseph's Parish in Richmond conducted their own "special service." The Catholics and Protestants together sang "God Save the King," and then they divided into two corteges. The bodies like number 480 that were identifiably Catholic went to Mount Olivet Cemetery; everyone else was claimed by the Protestants for Fairview Cemetery. The corteges left from two different gates and marched separately to the two different cemeteries. Mitchell plainly thought the pomp and sectarianism were a distraction. "Ironically enough," he noted, "owing to a blunder, only two of those over whom they [the St. Joseph's clergy] declaimed were placed in the hallowed ground of a R.C. [Roman Catholic] cemetery."[3]

Such errors notwithstanding, even the citywide day of mourning held on New Year's Day was denominationally divided. The idea had come from Archbishop Worrell on behalf of the committee of Protestant clergy he chaired, but he consulted also with his Catholic counterpart, Archbishop Edward McCarthy. Because there was no building large enough for a single service, Worrell suggested to the mayor that each denomination have one central service all at the same time as each other. Worshippers would stay at their familiar churches and pray beside their familiar coreligionists. In that way, churches would remain separate, their congregations and liturgies distinct, but they would metaphorically come together for a single purpose. Like the funeral for the unidentified dead, the event was simultaneously ecumenical and denominational, and socially, the city remained divided on sectarian lines.[4]

In death and in prayer, Haligonians were sorted by organizational membership—in churches, in clubs, and in other formal institutions. So too were they sorted in applying for and receiving disaster relief. When they applied for aid from the Halifax and Dartmouth Relief Committees and their successor, the Halifax Relief Commission, they were asked, among other things, what organizations they belonged to, including their church, their union, and their fraternal societies. Especially regarding religion, this was considered basic demographic information, but it was also a way that the relief authorities could learn which organizations knew about each family. The relief commission represented a technocratic, interventionist, progressive state that challenged and sometimes wanted to appropriate the knowledge and power of unions and churches. This chapter explores how those two institutions responded to that challenge. In short, the chapter asks how the explosion changed Halifax's churches and unions, and how membership in a church or union altered the individuals' and families' experiences of the disaster.

The fundamental principle of the Halifax Relief Commission was that the public good could best be determined and enacted by disinterested, outside experts, with the advice of local experts if need be but without formal consultation with or deference to the people it claimed to help. Its staff was comprised of professional experts: social workers, a university president, and expert builders and planners. The commission itself was made up of two Nova Scotia judges and an Ontario merchant and former mayor.[5] It was thus an almost perfect embodiment of a certain Progressive ideal of active intervention on behalf of the people, in which expertise was substituted for democracy. The absence of working-class voices within the relief process was not accidental; it was fundamental to the progressive, technocratic project of efficient and benevolent state action. In response, churches and unions—the strongest and most important sites of voluntary association—were forced to adapt. Neither became locations of mutual aid or self-help; rather, they stood between their members and the authorities, sometimes helping and sometimes hurting. Civil society did not act as an alternative to the state; instead it mediated, translated, and buffered survivors and the relief authorities.

For many workers, particularly building tradesmen, the relief commission was simultaneously employer, landlord, relief agent, and claims court. Its tremendous legal power was ideologically wrapped in the authority of a wartime government and disinterested expertise. The Halifax labor movement's ideology of laborism was no match for the commission's legal and ideological power. *Laborism*, variously described by historians as a "diffuse, unsystematic ideology" and a "vague political philosophy," was the ideological and political child of Canadian craft unionism. It was not a specific or unified political platform, but it was associated with union rights, the reform of the political system to allow more working-class participation, and, most important, a "fair deal" for all. Laborists believed in state intervention, and they sometimes allied with Progressive reformers, but their vision of a democratic political system in which labor parties spoke for workers put them in opposition to technocratic politics.[6]

Laborism was a reformist ideology in that its adherents believed that gradualist reforms were needed for liberal, parliamentary democracy to restore its promise, rather than a full reimagination of political economy. Moderation—the belief that existing society needed to be made more fair to everyone—was a fundamental tenet. "We believe that the capitalists have rights that are sacred and must be respected. We believe that the submerged tenth, the floating laborers, have needs that must be granted," wrote the editors of the official organ of the Halifax-based Canadian Brotherhood of Railroad Employees a few months after the explosion. Skilled tradesmen could satisfy both: "We have adopted a middle, a moderate, stand that offers justice to both parties and industrial peace as a result."[7] This chapter shows how, after the disaster, unions' laborist ideology constrained their usefulness to their own members. Faced with a state that did not respect tradesmen's skilled labor and an economy in which they were increasingly replaceable, laborism proved unequal to the task of taking on the technocratic authority and governmental power of the relief commission. The explosion forced Halifax's unions to become more inclusive and move beyond their base of white, male, skilled and semiskilled craftsmen, but they were still unable to retain their power.

Meanwhile, explosion survivors wanted pastoral care and a Christian burial for their dead loved ones, yet they accepted—indeed demanded—them not only from ordained ministers of their own denomination, but also by anyone nearby who seemed able. In the hospitals and the devastated area, as we saw in chapter 1, an ad hoc ecumenism developed in which Protestant ministers prayed for and with Catholics. "Well, it is all the same now," a Presbyterian minister recalled a Catholic woman telling him. "Would you kindly say a prayer for me?"[8] Likewise, laypeople took on clerical functions. The Christian Church was a small North End Protestant congregation; nineteen members died, more were injured, and no family was spared loss. In the absence of their pastor, a layman named L. A. Miles led a burial service. One of Miles's coparishioners told Archibald MacMechan that after the service, a soldier asked Miles if he was a minister. "He was not but the soldier had just prepared graves for his wife and children but wanted some Christian burial. Mr. [Miles] consented to help him, and the solider buried his own family, then proceeded to his duty."[9] When it came to such basics as a prayer in a hospital or over a grave, the formalities of ordination and denomination became suddenly less important. Yet clergymen were acknowledged as experts and authorities by their parishioners and by the relief committee, even as both groups of laypeople sometimes ignored or undermined this authority. Social workers looked to priests and ministers to translate and vouch for their parishioners, but this apparent deference masked that clergymen's power was passing into the hands of lay professional experts. Religion at once remained a deeply important part of survivors' emotional and spiritual lives and became institutionally less significant.

Haligonians' experiences of the explosion—what they lost, whether they had to flee their ruined homes, whether a family member died—were determined mostly by the geographic location of their homes and workplaces. People who lived and worked close to the site where the *Imo* and *Mont Blanc* collided, like the longshoremen working at the sugar refinery pier or the housewives making breakfast on Richmond's slope, were at more risk of injury and death. The very existence of a devastated *area*—a ubiquitous phrase in contemporary descriptions—suggested that damage to the built and psychic environment was spatially bounded. The importance of home and its location in the experience of the explosion was recapitulated in the choices made by survivors. Explosion survivors preferred the company, support, and aid of their families and neighbors; that is, those with whom they most shared their experiences.

There was no residential segregation by religion: Catholics lived next to— sometimes even in the same building as—Anglicans and Methodists and Presbyterians. This residential integration was in marked contrast to Halifax's apparent political and institutional segregation. Each party, for instance, followed a "time-honored rule of Halifax politics" to nominate one Catholic and one Protestant for the two members the city returned to the federal Parliament. Schools and orphanages were also segregated by religion, the former in an unusual arrangement in which they were jointly overseen by a secular government board.[10] Yet these divided institutions were next to each other, creating a literal and figurative neighborliness. The intermixing was symbolized by adjacent blocks of Russell and Kaye Streets about two blocks up from the harbor, where St. Joseph's Catholic, St. Mark's Anglican, and Kaye Street Methodist churches all stood next to each other. The Catholic elementary school, attached to St. Joseph's, stood across the street from a Protestant school so new it had not yet opened.

These Richmond churches were most affected by the explosion. In 1920, a city official estimated that St. Joseph's lost 404 parishioners, the largest number. Two hundred members of St. Mark's died. Kaye Street Methodist lost 167. Grove, a Presbyterian church at the other end of Richmond, lost 170 members.[11] Two days after the explosion, the *Evening Mail* did not yet know the fate of Grove or Kaye Street, but it described St. Mark's as "a parish wiped out," with the rector living in its basement. He reported that all his parishioners who lived north of Russell Street and half of those who lived south of it perished.[12] Down the street, St. Joseph's lost its convent, parish hall, and school. The explosion caused the walls and roof of the main church building to collapse, "and hundreds of tons of brick and slate crashed into the cellar," according to another newspaper article.[13]

Into this city of destroyed homes and churches came representatives of a new profession: trained social workers from the United States and central and western Canada. These women and their male supervisors brought with them a system of organization, an ideology of relief, and a way of knowing that relied on professional expertise. Although the greatest experts were themselves, they acknowledged that they often lacked the local knowledge they needed to judge applicants' need and worthiness. For that they turned to other, local experts, people who knew, or claimed to know, the applicant and who could assess them. Most often, official relief turned to pastors for this local knowledge and authority.

Thus while people's needs had to do with their location and mirrored those of their neighbors, their access to aid was mediated by their denomination and was thus potentially different from their neighbors. This sectarian sorting was imposed by relief authorities. Right after the explosion, Haligonians had flocked to sites of succor, places that in ordinary times they were accustomed to finding help. Though these spaces included religious sites, there was nothing unique about churches and convents; rather, they constituted only a handful of choices among many, including such secular and materialist locations as doctors' homes, drugstores, and barbershops. Even the apparently religious spaces at church-hosted shelters were used without reference to denomination; not only did Catholics stay at St. Paul's and Protestants stay at St. Mary's, but the Knights of Columbus hired a Protestant schoolteacher for the benefit of non-Catholics staying at its shelter.[14] The relief authorities' reliance on priests and ministers was imposed and was not a mere replication of patterns survivors created themselves.

It was the local Halifax Relief Committee that initially formalized a system in which ministers literally signed off on their parishioners' needs. It also used pastors as a conduit of information from the committee to survivors. This early system—established in December, as the city was still scrambling to create a sustainable relief bureaucracy—established clergymen as buffers between the relief authorities and their flocks, passing information in both directions. About a week after the explosion, Archbishop Worrell's committee of Protestant clergy asked the relief committee for formal "access" to the executive committee and its subordinate committees.[15] In response, Dougald MacGillivray, then the chair of the rehabilitation subcommittee, established a system to give clergymen official power within the process. Under this system, survivors received aid based on application forms signed and countersigned by ministers. Worrell's committee selected denominational chairmen, who vouched for their fellow ministers in their denomination; to signal their approval, they would sign out the applications to their colleagues. These other ministers then filled out the forms and countersigned them, and the applicant presented the twice-signed form at the relief depot. Alternatively, the signatures could be obtained in the opposite order; a minister could make a list of supplies and then have it signed by his denominational chairman. This system helped MacGillivray because relief workers had only to learn a limited number of trusted signatures—those of the denominational chairmen. "Moreover," MacGillivray explained to Worrell, "we feel it will serve to place a desirable measure of responsibility on your several chairmen and they must impress upon their brethern [sic] that the granting of orders should be done with all possible care."[16] Catholics were not part of Worrell's committee, but their priests were already organized hierarchically, and it seems they were handled similarly. MacGillivray's scheme meant relief applicants depended on the authority and discursive power of their priests and ministers. When clergy signed and countersigned requests for aid, they were dealing in information—what their parishioners needed—and wrapping it in two levels of authority.

Ministers acted as conduits of information in the other direction, too. A week after MacGillivray formalized his signature system, he asked the city's clergy to announce at their services that anyone who wanted aid needed to register officially.[17] In this instance, MacGillivray arranged to have his information wrapped in the clergy's authority for the purpose of convincing laypeople. Both lay survivors and the relief authorities benefited from this arrangement because the information they wanted to pass along was given greater authority and value by being endorsed by clergymen. In exchange, the clergy benefited by expanding the realm of their authority from the spiritual further into the temporal. Parishioners who might otherwise turn exclusively to secular institutions or to their families had to accept their minister's involvement in order to obtain his signature; likewise, the secular relief authorities had to grant "access" to the clerical committee.

Religious networks also passed information farther afield. The archdeacon of St. Paul's Anglican Church, for instance, received word of parishioners who had been taken to Truro and relayed the news to their friends and to the newspaper.[18] When churches or individuals from outside of Halifax wanted to raise money or collect donated items for survivors, they often turned to clergy through their denominational networks. Sometimes, the money they collected or donated themselves was intended for their coreligionists. The Catholic archdiocese, for instance, received money from around Nova Scotia, Canada, and the United States in various amounts—ranging from five or ten dollars to donations in the hundreds—from priests and bishops. The largest among these was $2,000 sent from Toronto archbishop Neil McNeil, himself originally Nova Scotian.[19] The rector at St. George, an Anglican church in Ward 5, collected and distributed $184, mostly from people who had a personal connection to him.[20] When a minister in Yarmouth asked the children at his church to donate toys "for the unfortunate children of Halifax," he wrote to a religious colleague to ask where they should be sent. His correspondent functioned only as source of information, directing the minister to send the toys to Ernest Blois, the city official in charge of children's welfare.[21]

In none of these cases did the churches function as agents of significant aid or solidarity. These were not the faithful helping coreligionists; even the money received by the Catholic archdiocese seems to have been designated to rebuild St. Joseph's Church, not relieve the parishioners, and the paltry sum collected at St. George came not because the donors were fellow Anglicans but because they were friends of the priest. Rather, in these examples church officials were conduits for information. Clergymen acted as advocates and gatekeepers, not organizers.

The federalized Halifax Relief Commission that replaced the local Halifax and Dartmouth Relief Committees increasingly privileged the expertise of social workers over the local knowledge of clergymen, but it still relied on priests and ministers for judgments. Most of the time, ministers acted as relatively neutral translators and buffers, endorsing the requests made by their parishioners. Sometimes, though, ministers helped their parishioners by encouraging them to request more aid than they had or by otherwise advocating for them. There were also instances of ministers standing in the way of parishioners, for instance by telling relief authorities that an applicant did not need as much aid as he had requested or by disclosing that a family had been poor before the disaster and thus did not deserve disaster-related aid. Church authority appears explicitly in 104 of the total 739 files I sampled from the Halifax Relief Commission's pension files. Of these, 56 (54 percent) were instances of clergy simply vouching for the requests made by laypeople.

In other instances, clergymen took a more active role. Mary Carr was the forty-five-year-old mother of seven children, ranging in age from seven to twenty-four. Her watchmaker husband had died in July 1914, and the family appears to have lived off the earnings of three sons: William, twenty-four, a dry-dockworker; Edward, twenty-one, a sailor; and Alfred, sixteen, an electrician's apprentice. Archibald, twenty-three, worked in Boston. When Mary died in the explosion, it left her family in disarray. The youngest son, Georgie, age seven, was dead. William and Edward were both injured. Archibald quickly came from Boston, but with the family divided among several relatives for shelter, it was hard for the relief commission to get definite information. "Visitor could not get a very straight story from Mrs. Downey," complained a relief investigator of one of the sheltering relatives. There were questions of insurance, about possible migration to the United States, and about when Edward and William could go back to work. Information was scarce, and grief, confusion, and adolescence made the surviving children hard to pin down. Teenaged daughter Gertrude, complained social worker Dorothy Judah, was "usual[ly] very non-committal and hard to get information from." In January, Alfred brought in Father Charles McManus, the family's priest, to help explain the family's situation to the social workers. Although the family's situation still took months more to be settled, it was when McManus came that the relief authorities first began to comprehend. As a translator, he was able to command the respect of the professionals in the relief office even while seeming to understand the specific and confusing information about the family. McManus's involvement was not definitive, and the social workers and the family continued to make decisions on their own, but he was an important early buffer between them.[22]

For the Carrs, the confusion was over who would or could do what, and the priest offered explanation. For Joseph Scallion, a sporadically employed longshoreman, his wife Bridget, and their six children, the problem was trust, and the priest provided credibility. Ordinarily, the eight Scallions lived in four rented rooms, but the explosion rendered two of the rooms uninhabitable. "Family large, surroundings wretched," their investigator wrote. When the Scallions requested new clothing for the children, a social worker suspected that they were just a poor family who saw the relief as a way to get new clothes, and she ordered a home inspection before any aid was disbursed. The family's priest vouched for them and got them a little clothing. As with the Carrs, relief authorities continued to make their own decisions. In the Scallions' case, they acknowledged the priest's opinion and appear to have given the family some of what they asked for. However, the decision remained in the social workers' hands, and they did not grow particularly generous with the Scallions on the priest's say-so.[23]

The Relief's social workers wanted the local expertise that clergymen had, but that knowledge was only one of several things they took into account. The expert knowledge of social workers trumped a priest's opinion. Henry Ward Cunningham, an Anglican priest, highly recommended the Penney family to the relief commission "and advises giving clothing, says that family were hard hit." Yet the authorities remained skeptical. They described only minor damage to the house and suggested that the husband, a laborer, do the repairs himself because he was out of work. In March the Penneys were still living in the basement, which signals that the upstairs remained uninhabitable. Social worker Judah advised that they be given coal, since it was so cold and the house was not yet repaired, "but also advised them that family was practically on the same basis as before explosion which was in all like[li]hood inadequate." Cunningham's say-so was not good enough, and despite his recommendation, Judah made her own determination that the family deserved little aid.[24]

One reason social workers were willing to disregard their clergy experts was that the two groups determined the worthiness of applicants on different scales and based on different evidence. The Penney family was poor before the explosion and Cunningham probably long thought they were appropriate recipients of charity, or at least with all the money flowing into Halifax he thought they should receive some of it. Judah, on the other hand, saw her job as restoring people to the point that they could build back their lives to where they had been before the explosion. People with inadequate livelihoods ironically needed less help.

Nonetheless, without much authority of their own, laypeople continued to rely on their clergymen as translators and brokers. L. J. Donaldson, the pastor at Trinity Anglican Church, sent a note to the relief committee asking that it send blankets and children's clothing to Annie and Frank Wildsmith, noting, "This is to certify that Mrs. Wildsmith is well known to me. Family consists of Mr Mrs Wildsmith + 3 children. They have had a lot of sickness all through the year. And now the loss + cold through the explosion has been very serious for them." To Donaldson, their prior misfortune was a further reason to give them aid, but it may have had the opposite effect. When no help arrived, Annie tried again. Since Frank still had his job, Annie had to stay at home with the children, so she wrote a letter asking for an investigator to visit her. Her plea for consideration rested on her pastor's authority. "You can kindly refer My case to Clergyman Mr Donaldson Trinity Church Brunswick Street, as I received a letter from Him to get an order, & being as it has not been sent to Me, as promised, I am begging to think My case has been overlooked, or the letter has been mislaid," she wrote.[25] Annie Wildsmith recognized that Donaldson had real power, even when his recommendations were not accepted or were ignored by the relief authorities.

When a child or other dependent person needed to be placed with a family or in an institution, clergymen's opinions were particularly sought out. Martha Phelan was a widow and caregiver for her son Michael, who was described as "mentally defective." Martha and Michael had come to the attention of Father McManus a few years earlier when they moved to his parish. Martha was a "feeble old lad[y]" and although Michael occasionally worked and brought home a few pennies, the family was never self-sufficient. Instead, they spent the winters on public charity at the City Home, and in the summers they lived in a shack in Richmond, where they were supported by neighbors. The explosion killed Martha and injured Michael, who was left with no one to look after him; Martha was "said to have relatives in the States but [their] addresses [were] unknown." From the start, relief authorities turned to McManus both as a source of information—he was the one who told them the family's story—and as an authority over where Michael should live. McManus thought that the "City Home [was the] only place for him." Three weeks later, in late January, Michael was finally in good enough shape physically to be discharged from the hospital, and the Relief began the process of moving him to the City Home. However, unspecified "others" objected to the plan, and the Relief became unsure. Faced with this uncertainty, a social worker again contacted McManus, who reiterated his opinion that Michael should live in the City Home. As the Phelans' pastor, the relief authorities deferred to McManus's expertise on who should care for him permanently, even when they did not always agree with his opinions about charity.[26]

Moral guidance and judgment were the special province of clergymen, and this may have been why they were given deference when it came to caring for dependents. When a family was in good stead with their pastor, this could be a good thing; when it was not, it could cause problems. To those on the margins of society like Rita Mariggi, clergymen were just one of many officials, ranging from judges to social workers, who sought to limit their choices and their independence. In Mariggi's case, a Catholic priest worked to block the adoption of her child by a friendly family; he deemed none of the adults adequately pious and preferred for the infant to be raised in the Catholic orphanage. In another case, Mary Swan's Anglican pastor admitted that he "ha[d] no knowledge of her character beyond the fact that he consider[ed] her irrational and irresponsible," but he was not shy about telling a relief investigator that she was unfit to raise her children or passing on a rumor that those children were illegitimate.[27] Social workers gave clergymen deference, but they also came to the same conclusions independently. Neither the priests nor the social workers trusted Mariggi or Swan, and so agreement was easy.

The story of the Carson orphans illustrates the pitfalls of clerical authority—both for the state and for aid recipients.[28] When the explosion happened, Kathleen, age eight, was just beginning the day at St. Joseph's, her parochial school. Her brother Johnnie, age fourteen, was at Victoria General Hospital convalescing from what may have been either kidney or leg surgery. The explosion killed both their parents and their three other siblings and destroyed their home. Kathleen quickly ended up at the Catholic St. Theresa's Home, probably taken there by Father McManus, who may have found her at the school. As the hospital filled up with explosion victims, Johnnie went to stay with his maternal aunt, Gertrude Gaudet. The Gaudets, too, had lost their home, and they were staying in a small flat. The three rooms there housed Mrs. Gaudet, her two small children, and her mother Margaret Major. (Joseph Gaudet was in the navy and was frequently absent.) Because a female lodger took up one of the rooms for herself, with Johnnie there, it meant two adults and three children in two rooms. In addition to the overcrowding, there was a question of money. The grandmother had lived with the Carsons and worked as a charwoman before the explosion, but the physical and emotional strain of the disaster had left her unable to work. The Gaudets agreed to give her shelter, but they complained that they could ill afford her food and clothes. Major's third daughter, referred to only as Mrs. Kaill, either lived in the United States or had gone there immediately after the explosion. "The housing conditions are most undesirable—very congested with poor ventilation and lighting," an investigator sniffed after visiting the Gaudets. "Family impressed visitor as having a very low standard of living. Mrs. Gaudet is easy going and good natured but evidently careless and untidy." In addition, no one seemed to be changing Johnnie's bandages, so the relief authorities soon bundled him back to the hospital where he could be properly nursed.

With the Carson parents dead, Kathleen and Johnnie had a new group of adults to give them official and unofficial supervision. Their grandmother Margaret Major had lived with them before the explosion; she too survived and also went to live with the Gaudets. Gertrude Gaudet, their aunt, also sought to provide continuity within the family, and she housed Johnnie informally and tried to adopt formally both children. In addition, however, the family now had to accept official supervision from the relief commission's children's committee. This meant investigation and criticism of how they lived. "According to opinions expressed by people who knew Mrs. Major and her daughter Mrs. Carson they may possibl[y] have been morons," wrote a social worker. Another investigator wrote of an apparent consensus among those who knew the family that "Mr. & Mrs. Carson were a very low type. The children had absolutely no training whatever, Mrs. Carson, incompetent and her husband drank." Major was illiterate, and investigators were concerned that Kathleen apparently knew neither her letters nor her colors—although it is conceivable that she was just a shy eight-year-old unable or unwilling to perform in front of judgmental strangers.

To help them judge the family, the children's committee workers consulted with clergymen. The Carsons had been raising their children as Catholics, sending them to Catholic school and having them attend church there. Charles McManus, the priest at St. Joseph's, took a special interest in Kathleen. He placed her at St. Theresa's, and in January he personally took her to the relief office to arrange for her to get clothing. He also wanted to oversee Johnnie's placement when he came out of the hospital. But it was not clear that he was the right priest to be handling the family. Henry Ward Cunningham, an Anglican, also knew the family. Margaret Major apparently attended St. Paul's, and Cunningham claimed that the children had been baptized Anglican, with their Catholic father's consent. (Cunningham was rector at St. George's, and why he, rather than the archdeacon of St. Paul's, was involved is unclear.) McManus disputed Cunningham's claim to the family—after all, the children and both parents had always gone to his church—and he produced Catholic baptismal papers for all the children save Johnnie, who it eventually emerged had been baptized Methodist. Cunningham shifted his argument and claimed that the other children had also been baptized Methodist, but that the records had been lost in the explosion. He also claimed jurisdiction because the surviving family, Gaudet and Major, were both Protestant. The relief commission dismissed Cunningham's baptism claims as improbable, since it was unlikely that all the kids would have been baptized twice, and in any case, regardless of his baptism, Johnnie had attended Catholic church. The relief authorities took on the right to adjudicate between the Catholics and Anglicans, and in this case McManus won the right to have authority over the family.

In any case, McManus and Cunningham both agreed that neither the Gaudets nor Major were suitable guardians for the children, especially Johnnie, who was crippled and apparently had severe behavioral problems. When he was finally discharged from the hospital in late February, he was brought to a detention home "until a suitable boarding home can be found for him." Unsurprisingly, this upset the boy. "Johnnie said he had expected his aunt, Mrs. Gaudet, to come for him that day and seemed very disappointed when he found he was to go with" the escort from the relief commission. Johnnie only spent a month at the detention home, though, before he had to be brought back to the hospital. Then, one Sunday in early May, he ran away from the hospital. "He said he was dying to see his grandmother and this was his reason for running away," reported the visitor. His family took him back to the hospital to continue his medical care, but Johnnie pined for them. "The child has a great fear of being sent back to the Detention Home," wrote a visitor. "He wants to go to his grandmother's home and told me he was writing to her all the time." (If so, this would contradict the claims that Major was illiterate.) For their part, Gaudet and Major were eager to have both children come and live with them, but the family had to convince both the social workers and the priest. For the social workers, that meant proving that they were not just eager for the monthly pension payments the children carried with them. For the priest, it meant promising that the children would continue to be raised Catholic. In neither case were they entirely successful. Although McManus acknowledged that Johnnie was "devoted" to his aunt, the priest still preferred that the boy be placed with a Catholic family. Meanwhile the Relief feared that the two women seemed overeager for the children. Though the record is unclear, it seems possible that the hospital delayed Johnnie's discharge while the relief commission figured out where he would go.

When Johnnie finally got out of the hospital in July, he got his way by threatening to the doctor that "he would not stay in a Home or Institution as he wanted to go to Mrs. Goody [Gaudet]." Kathleen, on the other hand, was either more docile or happier at her orphanage, and she stayed there under the supervision of the nuns and Father McManus. The priest remained Kathleen's financial guardian at least through 1930, when her orphan's pension ended. Though Johnnie went to his family, his grandmother and aunt remained under close watch by both lay and clerical authorities. In April 1920, the relief commission wanted Johnnie to appear before its medical board, and Major and Gaudet became "very incensed at the mere idea." A visitor in July 1921 noted, "Mrs. Gaudet states that John is well cared for, but other reprots [sic] state not." In December that year, after an altercation in which the boy hit his grandmother with a flatiron, the family inquired about having him sent to the Industrial School, but they soon changed their mind. Instead they put him to work at a shoe-shine parlor and poolroom on Barrington Street. This time it was the social worker who became incensed. "[I] have told her she must remove him from this at once," wrote M. Lockward, "and that if he wants to work and she can arrange for him to learn a trade, such as plumbing, that the Commission will consider keeping up the payments until he is seventeen, notwithstanding he has left school." Clergymen remained involved too, even Cunningham, who theoretically had no say over a Catholic boy. In 1922, he spoke to Lockward about Johnnie, bemoaning that "he is not very bright and does not seem inclined to work," in Lockward's paraphrase. "Mr. Cunningham says he does not know what to say about the boy."

Kathleen and Johnnie Carson's story highlights several aspects of clerical involvement in relief, though it is unusual that all these elements appeared at once. Most important, the Carsons' experience shows how religion was not an entirely voluntary community. Neither the children nor the adults in their family chose which clergymen would be involved in their case; when Cunningham and McManus squabbled over them, it was the relief commission that decided. Although eight-year-old Kathleen seems to have thrived under McManus's supervision and attention, a girl that young could not be said to have chosen McManus as her pastor. In contrast, it is clear that Johnnie did choose his aunt and grandmother, but he had to struggle to have that choice respected by the other adults who had power in his life. Moreover, the system in which clergymen were given power over their parishioners depended on a world in which it was clear what denomination each person belonged to. This was not the case for this family. The dead mother and her sister were Protestants who married Catholic men, and the confusion over Johnnie's baptism suggests that denomination was not particularly important to the family. McManus and Cunningham cannot be said to have been enacting the Carson parents' religious beliefs; rather, they sought to make choices on behalf of the dead. In this case and others, the clergymen's decisions were not necessarily based on what the families wanted. That is, they were not simply agents or advocates for their parishioners. Rather, they judged them and their characters and made recommendations and requests to the relief commission based on their own judgments. All this adds up to the fact that churches were not an organic expression of their communities' desires or organizing. Clergy did not speak for their congregations when they conferred with relief authorities. Instead they became another level of authority, often imposed from outside by state officials.

In the Carsons' case, Cunningham and McManus battled about who would have authority over the family, each brandishing baptismal certificates and making legalistic arguments about church memberships. In other cases, it was explosion survivors who cast doubt on their own church memberships. In at least sixteen cases in my sample, a family received aid from a church not its own. This indicates that it was the position of clergyman that was important to the relief authorities, rather than the congregation. It also suggests the problem of relying on clergy for this job when not everyone was affiliated.

Cunningham bemoaned the way sufferers had used him. "I think we did a fair share of relief in those awful days immediately following the disaster," he told his parishioners a few months later. "I only wish that all those whose cards I signed and to whom I gave orders, were in attendance on us as they claimed us as their 'spiritual Pastor and Master.'"[29] Cunningham had long been worried about declining attendance, so it frustrated him to vouch for people who did not darken the church's door except in times of crisis. His erstwhile parishioners wanted him for his temporal power vis-à-vis the relief commission; they were not interested in him beyond that.

Not all clergymen objected to this arrangement the way Cunningham did. Hilda Keddy, a twenty-two-year-old from Cow Bay who worked as a salesgirl at a shoe store, boarded with Mrs. Burrus C. MacLeod. Keddy lost nearly everything in the explosion, and flying glass cut her hand, arm, and face, making it hard to return to work. MacLeod took a "motherly interest in the girl" and told a relief investigator that "she would rather keep her in town with her, than let her go home, as she can get more attention in Halifax. . . . [She also] says she likes to have girl staying with her, and will do all she can for her." Part of doing all she could was to advocate on Keddy's behalf. Soon after the explosion, MacLeod went to a bank building on Agricola Street in the North End where volunteers helped survivors register for aid. Among them was F. E. Barrett, the minister at Robie Street Methodist Church. MacLeod included Keddy on the request she filled out for herself. Barrett did not know either woman, but he vouched for them anyway, attesting that Keddy lost her "boots, clothes, coat, etc." When Keddy had not yet received what she needed by the beginning of January, MacLeod telephoned Barrett, asking that he intervene on the girl's behalf. Barrett wrote a letter, and three days later Keddy had a coat.[30] Neither Keddy nor MacLeod attended Barrett's church, and they were not attracted by any spiritual authority he may have possessed. Rather, like those who claimed Cunningham as "spiritual pastor," only to disappear again, they recognized that as a minister, he had the temporal power to move the relief authorities to provide quick aid.

Clergymen's authority rested both on their status as moral leaders and educators—thus their importance in adoption cases—and because they were trusted professionals who were thought to know their parishioners' circumstances. But the secular importance of churches only went so far; in particular, it did not extend beyond the clergy. In the weeks and months following the explosion, churches were not sites of resistance or organization; people did not visibly unite with their coparishioners qua coparishioners to demand more relief, although they may have done so invisibly. If relief applicants shared advice and complaints about the authorities, if they exchanged information about jobs and housing, if they found ways of supporting each other mutually, these conversations were necessarily informal, and they rarely survive in the archival record.

One reason churches did not serve as a space for lay organization is that they literally had no space. Kaye Street Methodist was so badly damaged that it could not even hold a vestry meeting to decide how it should proceed, and the other Richmond churches were in no better shape.[31] St. Paul's, in downtown Halifax, bragged that it was the only church in the city to hold services on Sunday, December 9.[32] Other churches had to find new spaces in which to pray. Catholic St. Joseph's held mass at St. Mary's Army and Navy Club and at Mrs. Hurley's Hospital.[33] Anglicans and Presbyterians in Dartmouth's North End held joint services in space borrowed from a Presbyterian church elsewhere in town.[34] Grove Presbyterian in Richmond met with another Presbyterian church until it could decide what to do about rebuilding.[35] Churches were so dispersed and disrupted that they had to take out advertisements in the newspapers to let parishioners know where to go.[36]

Yet parishioners did share information and sometimes helped each other. Members of the Christian Church, for instance, remained sufficiently well connected that a parishioner was able to tell Archibald MacMechan the number of casualties among its members.[37] Moreover, there is a small handful of examples of relief work being organized congregationally or denominationally. An advertisement in the newspaper, for instance, asked Methodists affected in any way by the explosion to come to the Robie Street parsonage to register and ask for help. While there are no records of the Methodists' activities, the advertisement suggests that there was at least some denominational relief work.[38] At St. George's, the rector noted the "splendid work of those who came at the call and in home and hospital visited many of the St. George's people who had suffered, and arranged for their outfit in clothing, and reported cases to the rector."[39] These glimpses in the archives may hint at a larger pattern, but they are so rare that they are more likely to be exceptions or, at best, a common but small phenomenon. They were short-lived, occurring only in the explosion's immediate aftermath and not sustained for longer. Moreover, they appear to have happened mostly at churches marginally affected by the explosion, not the congregations in Richmond or Dartmouth's North End that themselves were physically destroyed and had the greatest number of casualties.

Some churches did significant relief work, but as charity not directed at their own members. St. Paul's Anglican and St. Mary's Catholic Cathedral both opened as shelters. St. Paul's December magazine bragged that in the first month of the shelter, more than 350 individuals were cared for and 10,000 meals served. The church installed baths and laundry facilities, and it hosted desks for the clothing, food, and information committees. "Over a thousand relief orders [were] written for members of many congregations," the magazine said; the relief work, it emphasized, was not solely directed at the 300 families in the church who were affected but rather at everyone, regardless of religion.[40] Indeed, St. Mary's and St. Paul's acted as secular, state shelters, not as denominational or religious ones. Moreover, although the church was proud that "nearly everyone has been busy in some place," it emphasized that the relief work was done by "members of the church individually," not as St. Paul's members.[41] They had been busy, it seems, in their capacities as family members, neighbors, or friends.

If church efforts were seldom about coparishioners helping their fellows, neither were they about reforming the relief committee, resolving conflicts, or finding the ways people could best get aid. The only political activity churches engaged in was to sign onto a Commercial Club resolution demanding that the Dominion government provide full restitution for property lost or destroyed in the explosion. But this was certainly not specifically church oriented; the campaign included nearly every club or organization in the city, regardless of purpose or class, ranging from the Catholic Archdiocese to the Orange Order and from the Trades and Labour Council to the Dartmouth Board of Trade. Even scrupulously nonpartisan groups like the Masons signed on.[42] From the churches' perspective, the petition was less about parishioners organizing politically for better relief and more about churches wanting money to rebuild their own physical facilities. The entire Catholic episcopacy of the Maritimes wrote a resolution demanding that the federal government "should assume full responsibility for the loss of property in Churches, Schools, and other public Institutions of all denominations in Halifax and the neighboring districts" as soon as possible.[43] Likewise, at a December meeting with cabinet ministers to discuss federal reparations, Anglican archbishop Clarendon Worrell insisted that churches not be overlooked. "The schools and the churches were most important buildings in any community," the newspaper reported him as saying. Worrell emphasized the churches' losses as buildings and institutions, not communities. "Some had lost not only their buildings but largely their flocks; the life lost could never be made good." The loss of its congregations, he implied, increased, rather than decreased, a church's worthiness.[44]

The emphasis on full reparations indicates one reason that churches were not a more important site of lay organization: they were busy. As a Methodist observer observed in 1925, "Everything was unsettled, old ruts disappeared and conditions were such now that necessary adjustments long delayed were possible."[45] All this jolting out of ruts was distracting. After the explosion, the organizing that happened in churches was about the churches—where congregations should meet and in what form. Like other buildings in Halifax, churches had severe physical damage, ranging from the leveling of churches in the North Ends of Dartmouth and Halifax to mere broken windows farther away. An early tally estimated that all told, churches suffered $1.1 million in damages; the relief commission distributed a bit more than $800,000 for repairs. For some congregations, the damage was total and the amount needed was extremely high; the amounts the relief commission gave them were unlikely to cover their full losses.[46]

The money churches needed for repairs was the least of it; even when they were paid it still took much time, effort, and energy to repair or rebuild. Emmanuel Church, an Anglican mission in Dartmouth, "resemble[d] a toppled card house," in J. H. Mitchell's words. The Relief granted the church $12,500 in 1918, but it was not restored until 1920. The displacement and disorientation meant it was hard to focus on rebuilding the church, and the difficulty was compounded by the fact that the pastor left his position to teach at King's College in Windsor.[47] The older and larger of Halifax's two synagogues, the Baron de Hirsch Society, was destroyed in the explosion, but the Relief gave the congregation only $7,625 to rebuild. This was far from adequate, so for the next several years the congregation met in various rented spaces. Meanwhile they looked to raise $30,000 more, an enormous sum for a community of fewer than one hundred families who had just finished raising $12,000 for war relief. Yet by June 1920, when they broke ground at their new location on Robie Street, they had raised two-thirds of their goal. The pressure to rebuild must have taken up much of the congregation's energy and attention.[48]

Although all churches in Halifax and Dartmouth expended considerable energy on rebuilding, for none was it more important than Grove Presbyterian and Kaye Street Methodist, both of which were leveled by the explosion. By 1917, these churches were several generations old, having been founded initially as missions to the railwaymen and other skilled workers who lived in Richmond. Grove, which started as a Sunday school in the train station in 1860, built a hall in 1863 and the church itself in 1872; it was renovated in 1910 and again in 1916. Kaye Street, too, had begun as a mission. It opened its church building in 1868, welcomed its first ordained minister in 1870, and built additions in 1915 and 1916. All these renovations and additions suggested growth, prosperity, and missionary success. The explosion tested all this. "Where [Richmond's] Churches and Halls, Manse and Parsonage had been were now piles of wreckage or smouldering ash heaps," wrote Charles J. Crowdis, the minister at Grove.[49] When the dust settled, the ash heaps extinguished, and the wreckage cleared, these two churches had merged, creating one of the few local union churches in eastern Canada. United Memorial Church presaged the full, national union of the Presbyterian, Methodist, and Congregationalist churches into the United Church of Canada by several years. How they merged illuminates some of the ways church members used and understood their religious communities. In the explosion's shadow, they created new bonds of solidarity, using the patterns and structures available to them from the national church union movement.

Canadian Protestants had been seriously discussing the national interdenominational union of their churches since 1902. The suggestion came after each denomination had itself come together through a series of intradenominational unions, and it was based on a theological commitment to Christian unity and, in the words of historian John Webster Grant, on an increasing "sense of kinship" based on evangelism, the social gospel, and commitment to education.[50] Official negotiations started in 1904, and four years after that a committee had hammered out the Basis of Union, a formal document describing the theological and ecclesiastical form that a united church would take. Within a few years, Methodists and Congregationalists had committed to union and Baptists and Anglicans had definitively taken themselves out. Presbyterians, on the other hand, dickered and debated for more than a decade. Although a continuous majority favored union, a powerful and vocal minority opposed and delayed it. In 1925, the unionists finally won, and the federal Parliament, accompanied by provincial legislatures, passed enabling legislation to create the United Church of Canada.[51] Each local Presbyterian church voted on whether to join the United Church or the continuing Presbyterian church; church property stayed with the winners of the vote.[52]

During the thirteen years between when the Methodists formally accepted the Basis of Union in 1912 and the adoption of the United Church of Canada Act in 1925, church union happened locally. Local Methodist and Presbyterian officials, spurred by a shortage of ministers, agreed to divvy up mission territory, mostly in the West. Picking up the momentum, individual congregations, again mostly in the West, also merged. Though local church unions were often based on temporal and practical need, they were also, in Grant's words, "formal anticipations of the coming union," a way to enact ecumenism and church union before the denominations were able to muster the political consensus to do so. That is, although they may have been initially motivated by financial or other temporal considerations, the men and women who enacted them were performing what they understood to be an important religious act.[53]

For the communities that made up Grove and Kaye Street churches, the explosion was first and foremost a personal and human tragedy. One hundred seventy members of Grove died. Kaye Street counted 167 dead members, including the wife and son of its minister, William J. W. Swetnam. Their survivors "must bear in their bodies or on their hearts the tokens of suffering and sorrow," Grove's minister Crowdis wrote to potential donors a few years later.[54] Homelessness, too, wore on them. Of the 180 families Swetnam tried to visit after the explosion, only fourteen were still in their own houses.[55] This dispersal of the community signals the way the explosion was not just a personal tragedy but one experienced socially. The explosion and its effects were, Crowdis wrote, "a common sorrow and a common concern," and they called out for a common solution.[56]

In early January, the general superintendent of the Canadian Methodist Church convened a meeting of the Methodist clergy from Halifax, Dartmouth, and neighboring Woodlawn. "A most significant feature of the evening," wrote the *Herald*, "was the reception of a deputation from the Presbyterian denomination," represented by ministers at St. Matthew's and St. Andrew's. "This delegation expressed the willingness and desire of the Presbyterians of the city to co-operate with the Methodists in the matter of re-adjustment and overlapping in the work of reconstruction and rebuilding of churches. It is hoped that the different denominations will work together in this spirit of Christian unity, so that the city may be properly churched and yet not overchurched."[57] The next week, when lay representatives from Presbyterian churches from around the city met to discuss their rebuilding efforts, they appointed a committee to confer with the Methodists on reorganizing and potentially merging churches and raising funds from their parent denominations.[58] The joint committee included both Swetnam and Crowdis and was supported by grants of $1,500 each from the Methodist and Presbyterian Churches.[59] It took only a week to agree formally to rebuild Kaye Street and Grove together in a single building. The joined congregations met for the first time in February and elected Crowdis chair. On March 17, the combined church opened in a building made of tar paper, the first church to reopen in Richmond.[60] This Tar Paper Church, as it was called, could hold four hundred worshippers, lasted two years, and cost a mere $4,800 to build.[61]

Officially, the Tar Paper Church remained two congregations, operating under the name Grove-Kaye Church. Crowdis explained the setup: "The members of the respective churches retain their prior relation, the [Presbyterian] Missionary, [Methodist] Connectional and like funds are divided equally, the [Presbyterian] Session cares for the spiritual interest while matters of business are entrusted to a [Methodist] Quarterly Board. All property is held jointly by the [Presbyterian] Maritime Synod and the [Methodist] Nova Scotia Conference."[62] Soon after the combined church opened, Swetnam and Crowdis both resigned to mark a new beginning. Swetnam, who as a Methodist minister was used to itinerancy, was transferred to a church in Bridgetown, Nova Scotia. Crowdis consented to being rehired. Although the congregations remained technically separate entities until national church union in 1925, as things got back to normal in Richmond, the churches continued to shift from mere cooperation to real union, a process that, in the words of the church's historians in 1975, happened "almost unconsciously." In 1920, in order to meet the needs of the growing population of the North End, they built a larger, permanent building and at the same time changed the name to United Memorial Church to signal the union's permanence.[63]

Like the larger church union, United Memorial was both a practical and religious concern. It "came about of necessity, and not of choice," wrote R. W. Ross, a Methodist minister who sat on a joint committee that oversaw the church on behalf of the two denominations; as in Western local unions it was a way to cheaply provide pastoral services to a largely depopulated area. Yet the merger was also clearly a religious and spiritual enterprise, a way of rebuilding a stronger community in Christ. Ross drew a parallel between the way explosion survivors rebuilt their homes and the way they rebuilt their churches. "The people lost everything, and were scattered amongst friends; but they were longing to return to their ruined homes and rebuild some sort of temporary dwelling. What about a church dwelling for these unfortunate people?" Richmond Methodists and Presbyterians were, he wrote, "brethren in adversity. The denominational walls were down, and out of sight."[64] By destroying the physical structures of their churches, the explosion forced these Haligonians to reconsider their churches' theological and ecclesiastical structures. Crowdis, raising money in 1920, emphasized how new and unprecedented was the project. "It must be remembered that this [the initial union] meant in a time of sore stress the formulating and accepting of terms of union by individual congregations, the approval of these by the parent denominations and the erection of the church," he wrote. "Remember, too, that in the East these churches were pioneers in the matter of union. There was no precedent nor was any inquired after."[65] Yet for all the lack of precedent, the merger of Grove and Kaye Street took the Basis of Union as its template and included only the two denominations that were most associated with the national church union movement. Using the national Basis of Union meant the two congregations did not have to innovate or negotiate the ecclesiastical or theological form their new church would take. The Tar Paper congregation created new ways of doing things, but it could only go so far. Local people, both clergy and laypeople, united their two churches for locally meaningful reasons, and they understood their local union church to have a purpose distinct from the national project. Yet they also clearly understood their work as part of a national project.

The unified church was a way of comprehending, memorializing, and living through the horror of the disaster. The *Memorial* in United Memorial Church was an explicit reference to the explosion, and, as the name indicates, the congregation was dedicated to the memory of those killed in the explosion. When the church's spokespeople told the story of the church, it was impossible to start anywhere but the disaster.[66] The memory of dead family members, neighbors, and fellow parishioners lived on in the community created in the unified church, and the religious ritual that took place there was a way of remembering and memorializing them. Though Grove and Kaye Street churches do not appear to have been sites in which people organized to recover materially, the project of rebuilding them was one of the ways parishioners emotionally and spiritually recovered.

The church's memorial purpose was literally built into it. Surviving family members presented the pulpit, chairs, communion table, stained-glass windows, baptismal font, and other church furnishings in memory of lost relatives. A plaque recorded the names of each dead member of both original congregations. Most of all, the church's memory was embodied in its ten bells, presented in 1921 by Barbara Orr, then a young woman. Orr's entire immediate family—both parents and five siblings—had died, and she went to live with an aunt and uncle. They provided money to build a tower, and Barbara paid for ten bells out of her inheritance. As if to emphasize the way lay members of the church had built it and invested it with meaning, Orr played the bells at United Memorial's dedication. Structural weakness in the tower forced the church to stop playing the bells in the 1970s. But in the early 1980s, the province built a bell tower in Fort Needham Park to serve as Halifax's primary civil explosion memorial. Orr's bells were placed inside with four new ones and thus retained their original memorial purpose. To signal the continuity between the original church tower and the new secular memorial, Orr, by then an old woman, played the carillon as she had at the initial dedication ceremony. The plaque on the new tower reminded visitors that the bells originally came from United Memorial, and it retained the list of Orr's dead relatives. It also included a rededication to all of Halifax and Dartmouth's dead, injured, and survivors. Now outside the church, the bells' meaning shifted slightly to become more inclusive, memorializing all the explosion's dead. The transfer of the bells from the united church to the Needham bell tower, and the way that even in their move from a sacred to a secular space they retained and broadened their meaning, highlighted how after the disaster, parishioners came together and built new communities where before differences had divided them.[67]

・・・

At least for Haligonians, the relief commission was the apex of the expansion of the state that had characterized the decade and especially the period of the Great War. The state, through the commission, was developer, city-planning authority, landlord, employer, court, and relief agent. The Relief embodied the central irony of Progressive-Era state intervention: although the government claimed to act on behalf of "the people" and although relief workers were motivated by a desire to rescue and relieve their clients, workers and especially union leaders experienced state expansion as exclusion. The growing, technocratic state devalued and rejected workers' knowledge and power. Historian Bryan Palmer writes that new state initiatives left organized labor "handicapped by its inability to grasp the extent to which the state, as a powerful national force, was engaged in a constant project of regulation and containment."[68] In Halifax, the relief commission forced unions to grasp the extent of the state's growth, but it did them little good. Clergymen were able to take advantage of social workers' need for information from a trusted professional; unions and their leaders were unable to do the same.

Laborism left Halifax's unions ideologically unprepared for this newly interventionist state that simultaneously claimed to speak for the unions' constituency—the homeless workers who lived in Richmond—while denying them a voice of their own. Unions are nearly absent in the records of the rehabilitation section of the relief commission, suggesting that the Relief did not recognize unions' knowledge of and interest in their members. In its capacity of rebuilding the North End, the commission also refused to negotiate with building-trades unions. Its extraordinary power forced Halifax's unions to reconsider their roles. Though unions sought to be included in the relief apparatus and very occasionally tried to help members or members' widows in the relief process, they did not act as sites of spontaneous organization after the disaster. People turned to their families, not their unions or other organizations, for informal relief, and for formal relief they relied on the government. Reacting to their exclusion from the relief process, unions reimagined how they would build and exercise power in Halifax. By the end of the decade, Halifax's unions demanded a political solution to the city's ever-worsening housing shortage, they founded a new newspaper to advocate for themselves, and in 1920 Halifax's shipbuilders engaged the country's largest strike that year while contesting a provincial election in coalition with the United Farmers.

The Halifax Trades and Labour Council sought several times to play a role in the relief process, and relief authorities even invited its representatives to participate. Each invitation, however, was withdrawn, and labor never gained an official role in the distribution of aid. Instead, relief authorities relied on the professional expertise of social workers and the moral expertise of clergymen, relegating class representatives to at best a minor role. This emphasis on professional expertise at the expense of representation was new. When Nova Scotia adopted its Workmen's Compensation Act in 1915, the board it created included a labor representative: John Joy, the "grand old man" of Halifax organized labor. But when, in order to avoid bankrupting the entire workers' compensation system, the Relief took over all claims arising from the explosion, it did not include Joy or any other labor representative.[69] The transfer meant a shift in power away from labor and toward technocratic experts. That the Workers' Compensation Board included a labor representative and the relief commission did not reflected a shift from 1915 to 1918.

On December 21, the provincial premier called a meeting to appoint an official commission to handle relief work, with the intent of replacing the ad hoc relief committee. The premier's meeting included himself, the mayor, and members of the Board of Trade. This was hardly a representative group, and they sought to expand the circle by inviting a variety of men to a second meeting a week later. Among the invitees were the French and American consuls, the chairman of the school board, the secretaries of the Rotary and Commercial Clubs, and a group of insurance companies. They also included Aaron R. Mosher, the president of the Canadian Brotherhood of Railroad Employees, and Ralph H. Eisnor, the president of the local Trades and Labour Council and the carpenters' local.[70] However, that second meeting was postponed and eventually canceled altogether, effectively excluding Mosher and Eisnor from any deliberations.[71] Around the same time, the relief committee's rehabilitation department was suffering severe criticism for seeming to harass aid applicants with endless investigations. Among the reforms promised by the department's new director was that he would bring on a "labor man" to help. The Trades and Labour Council nominated Eisnor, perhaps the most visible labor leader in the city. Eisnor was never installed, however, and soon after the relief committee was replaced with the federally constituted Halifax Relief Commission.[72] Its members included two judges and an Ontario businessman. They, in turn, hired to lead the rehabilitation committee a university president, who was advised by professional social workers. This time the labor movement was not even invited. In fact, there was no notion of representation, and the commission relied instead on disinterested expertise.

The relief coming into Halifax from Massachusetts seemed to provide organized labor another opportunity to influence how aid was given out. The Massachusetts-Halifax Relief Committee was organized as part of the Massachusetts Committee for Public Safety, which had labor mediation in its purview and included labor representatives. Canadian labor leaders wrote to the Massachusetts committee, seeking some input in its activities. The grand chief of the Brotherhood of Locomotive Engineers, for instance, feared that there was "more or less quarreling and friction over the giving of this relief, some claiming that others got more than their share," and he sought to quell this trouble. There is no record in the Massachusetts committee's archives, however, that labor leaders were ever given any formal role.[73]

Labor leaders were not much more successful when they sought to intervene on behalf of individual members. The Brotherhood of Locomotive Engineers, for instance, wrote to commission secretary Ralph Bell about cases of four of its members. Any influence the union was able to exert on the process was subtle, and the letter to Bell was phrased as a request for information, rather than as explicit advocacy. Bell passed the request for information to George B. Cutten, the academic who headed the rehabilitation department. The union's letter was probably akin to the complaints forwarded from Boston and discussed in the previous chapter: it brought the matter to Cutten's personal attention, and it forced the social workers to pay extra attention, but the authorities were careful not to do any more than that.[74]

Even this level of careful and subtle advocacy was rare. In my sample of 739 cases in front of the relief commission, only twenty even mentioned that a family member was a member of a union, and even in these files the union was often mentioned only in the context of an insurance payment.[75] Given the number of railroad workers, longshoremen, and building tradesmen who lived in Richmond, it is inconceivable that this low number represents actual union membership. Rather, it must reflect that unions were unimportant to the relief commission. A comparison with church membership is instructive. No more than 102 of 739 files failed to list the family's religion, and as we have seen, churches were built into the infrastructure of relief through the reliance on clergymen's presumed local knowledge. In contrast, files almost never included information about union membership because relief officials deemed it irrelevant. Although unions and their leaders would likely have had similar—perhaps better—knowledge of members' financial situations, their expertise was not similarly used. When Mary Murphy wrote to the Relief to complain that she had not received any compensation for the death of her son, who supported her, she emphasized that he was "a union man." When social worker Jane Wisdom responded, however, she demanded information about Murphy's other children—who might also support her—and a reference from a clergyman. To Murphy, her son's union membership apparently signified something, perhaps greater respectability or higher wages. To Wisdom, however, it was irrelevant.[76]

Rita Mariggi's husband, Cesare, was a stevedore who belonged to the Halifax Longshoremen's Association (HLA) before he died in the explosion; so too was John Carson, the father of the orphaned Carson siblings. HLA officials spent three months "crawl[ing] over smoking ruins and the shattered debris" looking for their members.[77] In January, as Mariggi's saga with the relief commission was starting, she expected to receive a payment from her husband's union. Assuming that it arrived, that payment was the only thing the HLA did for her—at least that came to the attention of the relief commission. It did not, for instance, come to her defense when the commission tried to force her to put her children in an orphanage or prevented her from giving her baby to a friend.[78] In the Carsons' case, the union did attempt to get involved, but only very modestly. The union president, Michael D. Coolen, wrote to Cutten to find out whether the commission was paying money to orphaned Kathleen and Johnnie; Cutten responded that they were getting the standard pension for orphans and that was the end of the exchange.[79] In both cases the HLA appeared briefly, but in neither case did it make an effort to defend the interests of the family from the prying of the relief commission or the church.

Insufficient archives survive to assess the extent to which unions may have supported their members outside the relief commission's gaze. The carpenters' local held a meeting about two weeks after the explosion "for roll call, to see what members are missing or incapacitated."[80] No record, however, remains of what if anything the union decided to do for members who were incapacitated. Halifax-based Division Number 14 of the Canadian Brotherhood of Railroad Employees held its "first regular meeting" after the explosion at the end of February, where it inducted new members and elected a new executive. The wording—from a report to the national union's monthly magazine—suggests that the union held at least one special meeting, presumably to discuss the explosion, but no records of it remain.[81] Even without knowing the details, however, the fact that these unions held meetings suggests that they had some project, however minor, to help their members who were affected by the explosion.

In the aftermath of the explosion, unions remained active on their traditional terrain, at the work site. They found, however, that the explosion had jostled and shifted that terrain, giving employers more power by allowing them to seem as if they were representing the interests of the entire city. In order to win against employers in this new industrial landscape, unions had to organize on different lines, including outside the workplace and away from the point of production. A union movement that had been dominated by skilled and semiskilled men had to build solidarity with women and with the public. This was an acceleration of shift in strategy that had begun before the explosion and that happened in geographic areas unaffected by it. Throughout Canada, unions temporarily organized more inclusively, welcoming the women, workers of color, recent immigrants, and less skilled workers whom they had traditionally excluded.[82] In the aftermath of the explosion and in the shadow of the relief commission, Halifax's labor movement was at the forefront of this trend.

One example of the difficulties unions faced comes from the streetcar railway, where discipline and hiring practices after the explosion sparked a confrontation between workers and management. Management used a perceived manpower shortage—nine car men had died in the explosion, three of them on duty—to hire women for the first time. Industrywide, managers in both Canada and the United States were hiring women to break down gendered union solidarity and drive down wages, and across the continent women often faced harsh opposition from their male colleagues.[83] In Halifax, the introduction of "conductorettes" was seen as specifically related to the explosion. This meant that unlike in Ottawa and Toronto, where male union members were able to build antiwomen coalitions with the riding public, in Halifax passengers welcomed women workers on the basis that men should be doing the more important (and apparently more masculine) work of rebuilding. The explosion also led to fights over discipline: Workers were incensed when the company fired a twenty-eight-year veteran named James Adams for missing two days of work after the explosion. Adams's comrades retaliated by refusing to keep to the timetables, instead bunching the cars in packs. They also filed an official complaint with the city government against their women colleagues, alleging that they were too weak to operate the brakes. The day after the government refused to take any action, several workers stayed home and more than a third of the streetcars did not run. Surprisingly, nineteen of twenty-two women conductors staged a sick-out in solidarity with the very workers who were rejecting them.[84]

In the middle of February, the company fired two more veteran motormen. The motormen's union struck, sparking solidarity strikes by workers from the car barns, powerhouse, and machine shop, and even the conductorettes. The suddenness of the strike—even the international union was not notified ahead of time—alienated the public, especially since there was bad weather. The lack of public support was in contrast to a strike in 1913, in which riders had sided with workers. Peter Lambly, in a master's thesis on Canadian streetcar labor history, argues that the difference was partially attributable to the explosion. He speculates that for a public already taxed by food rationing, wartime shortages, inflation, and then the trauma of the explosion, the inconvenience of a streetcar strike was too much to bear. Unfamiliar with and unsympathetic to the internal questions of hiring policy and discipline that had sparked the strike, the public sided against whomever they perceived to be disrupting service. Regardless, after only two days, the company agreed to reinstate the fired workers and give them only a temporary suspension, a union victory, albeit an ambiguous one.[85] When the union sought redress from the government against competition from women, it was rebuffed, but when its organizing reached across craft lines and included women, as it did for the two-day strike, they were significantly more successful. Halifax's streetcar workers' local was unusual because in other North American cities, its brother locals rarely accepted women and then only if organized by external women's rights groups.[86] In Halifax, the explosion appeared to do the work of the Women's Trade Union League.

If on the street railway the question was gender, on the intercity rails the fault line was race. Before the explosion, the Canadian Brotherhood of Railroad Employees (CBRE), a national union based in Halifax, faced a mounting challenge by African Canadian sleeping-car porters to be admitted as full members of the union. In the spring of 1917, sleeping-car porters in Winnipeg had chartered a new union, which they sought to affiliate with either the national Trades and Labour Congress or, when that was denied, with the all-white CBRE. They were joined by the Halifax-based Canadian Grand Trunk Railways Sleeping Car Porters Association, which also sought full membership in the railroad brotherhood. At the September 1918 annual convention of the CBRE, George Fraser, the president of the Halifax porters' union, used the language of laborism to request "a square deal." Faced with opposition from CBRE leader Aaron Mosher, though, Fraser and his members lost, and their union was refused affiliation. Instead, after considerable debate, the white union buried the question in a committee in charge of constitutional revisions.[87]

It appears, however, that Halifax railroad workers, at least the ones with the greatest contact with sleeping-car porters, supported the acceptance of black workers. Fraser implied that his demand for full membership had the support of white fellow workers in Division 36, who, he claimed, were "working hand in hand with us."[88] Fraser's suggestion was supported by what a white member of Halifax's Division 12 had written in April, about five months after the explosion and five months before the convention. His report to CBRE's magazine included a discussion of their black colleagues. "As this Division is composed of the dining car waiters, cooks and pantrymen, we come in touch with the colored porters on the railroad, and lately 'feelers' have been put out to know just how they stand with this organization in having a division of their own within the Brotherhood," wrote the correspondent. Although he admitted the question was "very delicate," he urged that porters be admitted. "As there are a great many of these men on all railroads, in my humble opinion I rather think they would be a source of strength rather than otherwise, and the C. B. of R. E. should be able to embrace all employees, irrespective of creed or race."[89] Whether the explosion was responsible for Division 12's greater tolerance is unclear, of course; because the union was based in Halifax, Mosher's continued opposition makes it evident that the explosion did not have that effect on everyone. But it does fit the pattern of the disaster encouraging more inclusive unions. After more debate at the 1919 convention and continued delay by CBRE's national leaders, black sleeping-car porters were finally allowed into the union in 1920, albeit only in a segregated auxiliary. This partial inclusion greatly delayed full integration, which did not come until 1965. Nonetheless, even limited membership allowed black workers to participate in union governance and grievance boards.[90]

The relief commission fundamentally altered governance and labor relations in Halifax, making worker organization and negotiation considerably more difficult. In addition to administering relief and rehabilitation funds, the commission was in charge of rebuilding Halifax: dispatching workers to fix private homes, building temporary apartment buildings at the Citadel and the Exposition Grounds, and eventually building permanent housing in the new Richmond Heights neighborhood. In one of the federal government's first disaster-related acts, it signaled its plans by hiring Robert Smith Low to be reconstruction manager. Low was a politically connected, native Haligonian contractor nationally famous for having built several military encampments but infamous in labor circles for resisting any demands put to him by unions. His militaristic style of discipline was symbolized by the fact that everyone called him "colonel," a reference to his merely honorary rank in the Canadian Expeditionary Force. Ralph Eisnor complained bitterly that Colonel Low had "been a thorn in our flesh for a year, or more back. This is not the first time we have had trouble with the Colonel."[91]

Halifax building tradesmen were at a particular disadvantage in fighting Low and the government that had hired him because the massive rebuilding project attracted an equally massive influx of construction workers. Historian Suzanne Morton estimates that more than ten thousand construction workers came to Halifax, although high turnover meant that most of them probably did not stay for very long. Competitive pressure combined with political pressure to loosen work rules—not just wages, but also hours and overtime—because of the emergency. Unionized tradesmen had to decide between giving up on rules they had negotiated or potentially being replaced by out-of-town sojourners. Carpenters, who were the most susceptible to deskilling, faced the issue in the first three weeks after the explosion; soon enough plumbers, bricklayers, and plasterers had to make a similar decision. In February 1918, when plumbers tried to demand overtime wages for Sunday work, the commission responded with a public campaign in the newspapers questioning their patriotism and a threat to import two hundred replacements from Quebec. Eventually, the plumbers were forced to agree to forfeit overtime pay until May. Plasterers had agreed from the start to forgo overtime pay, but they too demanded a resumption of union work rules in May and went on strike to enforce the demand. Low and the relief commission were initially unwilling to compromise, but the summer brought a labor shortage generally, and so the commission eventually agreed to raise wages and restore overtime. It was not union power, skill, or organization that won the raises, but instead an unrelated labor shortage.[92]

The Halifax unions' problems with Low and the relief commission echoed the fight between the national Trades and Labour Congress and the Imperial Munitions Board, headed by industrialist Joseph Wesley Flavelle. Like Low, Flavelle was both ideologically opposed to unions and practically committed to eking out the highest possible productivity. Unions, in contrast, demanded the power and prestige that would have come from formal fair-wage clauses in munitions contracts. Although unions pressed their point with both the Canadian and Imperial governments, Flavelle refused to give in, denying unions the power that they thought they had earned through participation in the war machine. Unions' frustration and sense of exclusion that resulted contributed to workers' growing dissatisfaction that would eventually lead to the strike wave of 1918 and especially 1919.[93] Halifax unions were likewise excluded from participating in the postexplosion state—embodied in the relief commission—and were unable to force Low to accept their hard-fought rights. The explosion meant that the situation in Halifax was extreme, but the exclusion of labor from state decisions was a broader phenomenon.

By the time the commission withdrew from direct construction work in January 1919, Halifax's unions had grown and changed. Richmond, the neighborhood over which the relief commission had the most direct power as a developer, contractor, landlord, and court, had been home to the skilled workers who dominated the labor movement. So working-class Haligonians interacted with the commission as more than just an employer. They thus increasingly saw class at the point of reproduction: the home. They saw themselves as disadvantaged by the relief commission's standards of rehabilitation, and they increasingly built organizations that challenged the state as both employer and political entity. This included organizing, both in their own crafts—the carpenters' local grew over 425 percent in the year and half after the explosion—and more broadly. In February 1919, the Building Trades Council played a leading role in organizing the provincial Federation of Labour. In May, the Trades and Labour Council started publishing a newspaper. When that month two thousand building tradesmen struck, they demanded an eight-hour day and a uniform wage across crafts, transforming what had been purely craft unionism into something resembling industrial unionism. By the end of 1919, the Trades and Labour Council counted eight thousand members, making it the fourth-largest labor organization in all of Canada. Even with that expansion, unionized workers realized they needed a broader base politically, so in January 1920 a newly revived Halifax Labor Party welcomed membership from all "workers, whether organized or unorganized, mental or manual regardless of race, sex, creed or vocation," and its platform was endorsed by the Great War Veterans Association. The provincial Independent Labor Party's platform reflected this breadth; it included not only production-related planks about the eight-hour day and reforms to workers' compensation, but also reproduction-related planks on school reform and housing. While in some ways this was a retrenchment of laborism, since it was a recommitment to political action organized by unions, it was also a broadening beyond laborism's usual craft-union base.[94]

Ultimately, this expansion proved unsuccessful. In 1920, the Halifax labor movement simultaneously fought the country's biggest strike of the year (as measured by total person-days lost) and a provincial election. The strike was called by the Marine Trades and Labor Federation, an umbrella organization of skilled trade unions in Halifax's shipyards, and it received strong support from other unions, including the Mine Workers, Street Railwaymen, Typographers, and Sheet Metal Workers, who provided the strike fund. Strikers demanded a forty-four-hour workweek and a raise of almost thirty cents an hour, double pay for overtime, and triple pay for Labor Day and Sundays. But the real importance of the strike was about worker power: Workers demanded control over layoffs and rehiring, an end to physical exams, a grievance system with stewards, the right to regulate apprentices, washrooms at every work site, payment on company time, and five minutes of cleanup time. What was at stake in the strike was not lost on the company. It offered a meager five-cents-per-hour raise and refused to negotiate with the union, effectively making the strike a fight over the right to bargain collectively. Members of other unions recognized that a loss at the Halifax Shipyards, Halifax's largest employer, would cripple the city's entire labor movement and undo the gains they had made over a generation, so they tolerated a low level of violence and considered a general strike.[95]

Four weeks into the strike, Premier George Murray called a provincial election. Unions formed a coalition with rural workers and mounted a joint Farmer-Labor campaign across the province. Although the strike continued throughout the election season, the major issue in the campaign was the reproductive question of housing, not the productive question of labor rights. In the rest of the province, the combined party was quite successful, returning eleven members (out of a total of forty-three) and forming the official opposition. The Richmond Heights polling district was Halifax County's firmest Labor bastion, and Labor candidates won a majority there and at other working-class polling sites. But workers had not built a broad enough coalition to win the election, and they did not win a single seat in Halifax. The broader working-class consciousness that the explosion had fostered was apparently not strong enough two years later to elect provincial politicians. It was also not strong enough to win the strike, which collapsed the same week as the Halifax Labor Party lost the election. The company won entirely, blacklisting the union leadership and effectively ending the closed shop throughout Halifax.[96]

In the aftermath of the explosion and after a year of struggle over organized labor's role in rebuilding Halifax, the city's labor movement seemed to shift its ideology to become more inclusive of other workers and other concerns. Galvanized by the way the explosion's effects, relief and rehabilitation payments, rebuilding, and housing were all classed, Halifax workers tried to build a broader movement and fought for power on the job and in reproductive arenas like housing and schools. They did not, however, fully reject their laborist ideology, and craft-based laborism was no more successful in the shipyards than it was with the relief commission. Later in the decade, the Labor Party collapsed with the Nova Scotia economy, and laborism was replaced by the cross-class, regionalist Maritime Rights Movement.[97]

In the Halifax Relief Commission, with its progressive emphasis on disinterested, professional expertise, both churches and unions faced a challenge. Neither were disinterested, and both were repositories of local knowledge about their members that social workers were disinclined to trust. The difference between church and union was largely what determined their role in the relief efforts and their futures. Churches were led by professional clergy, men in whom social workers had a limited trust to judge their congregants. They were also institutions where members were used to coming for support, help, and comfort. Clergymen were able to become translators and sometimes advocates for their parishioners, and churches became a place where people could come to grips with their grief. Unions, with no professional leadership, attempted to change in the aftermath of the explosion, but they were ultimately unsuccessful. The progressive technocracy represented by the relief commission may have had limited space for churches, but it had no room for laborism.

6

"THE SUFFERINGS OF THIS TIME ARE NOT WORTHY TO BE COMPARED WITH THE GLORY THAT IS TO COME"

SALEM WORKERS BUILD POWER IN THE CHURCH AND FACTORY

At ten in the morning the Sunday three days after the Salem fire, during a lull in the nearly constant rain, a militiaman rang a bell to mark the hour. Father Donat Binette arrived at a jerry-rigged, wooden altar that soldiers had been building since dawn. Behind him was his congregation: three thousand men, women, and children, "nearly all homeless, miserable, wet, cold and hungry," in the words of a sympathetic reporter. "They were a motley assemblage, the men and women and children of the tented colony, all clad in nondescript garments, ill-fitting and bedraggled. Their faces were gaunt and worn from suffering." Facing the altar, Binette read the mass, occasionally having to pause when emotion overcame him. "*Kyrie, eleison. Christe, eleison,*" he read in Greek: "Lord have mercy. Christ have mercy." "*Qui tollis peccáta mundi, miserére nobis,*" he said in Latin: "Thou that takest away the sins of the world, have mercy upon us."

When he finished the hymn "Gloria in excelsis Deo," Binette turned to face his audience for the Collect, an opening prayer. "*Dominus vobiscum,*" he said: "The Lord be with you." He spoke in French to his assembled congregation, counseling them to be like Job, to "be patient and everything would come out all right." He urged them to be thrifty and sanitary, to take whatever jobs were offered them, and, in their new domestic circumstances, to watch out for their daughters' chastity. As he spoke, the clouds grew darker and darker, and soon the driving, pouring rain resumed. Some in the congregation put up umbrellas, including an assistant who held one over Binette's head, but nobody left the mass. Binette told his flock not to kneel, since they would get muddy, and he moved onto Communion. "There was no Sanctus bell, but a bugle call sounded—clear, thrilling, and inspiring and then came the elevation of the Host. For a second, as the priest's hands held up the sacred Host, with his face turned toward the heavens, it grew lighter. The sun showed for the briefest period, and then the torrent fell once more," a Catholic reporter wrote. He was clearly moved by this seeming omen, and so were the others at the mass. "The people with one accord dropped to their knees in the wet grass and prostrated themselves in adoration."[1]

By coincidence, the Bible verses Binette read sounded particularly relevant that Sunday. The reading from the Epistles was about enduring misery while waiting for God to redeem his people. "For I reckon that the sufferings of this time are not worthy to be compared with the glory to come," it began.[2] The Gospel reading that followed told the story of Jesus miraculously filling Simon's nets with fish. Though Simon had labored all night without catching any fish, upon Jesus's command he tried again, and that time "enclosed a very great multitude of fishes," so many that their nets broke and their ships were filled to almost sinking.[3] To those for whom the fire had emptied the nets and ships of a lifetime of labor, the lesson about having faith that God would provide would have resonated clearly.

"Never before, perhaps, was there witnessed such a scene in the onetime Puritan city, or in Massachusetts or New England," wrote the somewhat breathless reporter. Father Binette's service was unusual for its location, for the presence of "many non-Catholics and Jews," for the bugle that replaced the Sanctus bell, and for the fact that it appears to have abbreviated the standard High Mass.[4] But more important by far were the ways in which it served as a bulwark of normality. For the three thousand Catholics who attended mass that Sunday and prostrated themselves in the mud when Binette held aloft the host, religion was a fundamental way to make sense of the world, particularly when their corner of it was in crisis. It represented comfort and continuity, a spiritual and emotional architecture that remained intact even when the physical architecture of their lives and neighborhood was destroyed. It provided familiarity amid the unfamiliar, stability amid chaos. It helped them understand the horror that had befallen them. In bad times as in good, Mass was a way to ask for and receive support from God; it was also a celebration of community and a way to ask for and receive support from priests and fellow laypeople.[5] "St. Joseph's structure is not only destroyed, but the whole parish has been scattered to the winds," the *Salem Evening News* wrote the day before, referring to the French Canadian parish where Binette was the senior assistant priest and to which most of his congregants that day belonged.[6] But parishioners had in fact not scattered. Although only about half of the audience was actually living in Forest River Park, most of the others were still close enough that they came to the park to celebrate Mass in the rain with their friends and neighbors.

Catholics were not the only ones who looked to their church for support and guidance on the Sundays after the fire. As Binette was leading his outdoor mass, Harry Newton, the pastor at Crombie Street Congregationalist Church, titled his sermon "A New and Better Salem" and took for his text a verse from Joel: "And I will restore to you the years that the locust hath eaten, the cankerworm, and the caterpiller, and the palmerworm, my great army which I sent among you." To Newton, the years had been lost not from insect infestation but from bad building regulations. "Here is a great opportunity for a city," he preached, blaming cheap construction for the fire's damage. "We need brick houses and slate roofs. I hope the city will immediately step in to control all future building operations in this city."[7] The next Sunday found another Congregationalist minister, Thomas Langdale, leading services in a building that had replaced one that had burned down not a decade earlier. He preached on a text from the Epistle to the Hebrews that described city foundations "whose builder and maker is God." Langdale celebrated the "fellowship in disaster that makes us all kin," and he looked forward to "the new Salem starting with a degree of orderliness made possible by military control and the absence of saloons." DeWitt Clark, the minister at Tabernacle Congregationalist Church, was similarly upbeat. He noted how many of the sufferers were "strangers [who] have come to us, not of the American type," and like Langdale preached that "in seeking to do acts of kindness and goodness lines have been obliterated and are all one people." Henry Bedinger, the priest at St. Peter's Episcopal Church, concurred, praising the "great spirit of kindness and sympathy" that arose from the disaster and promising "out of it will come something better, something more lasting."[8]

As these sermons made clear, church services were for spiritual solace and materialist mobilization. The Protestant ministers praised the spirit of generosity that they perceived in ruined Salem, but they also spoke about the real and practical reforms needed for the rebuilt city. For the Catholics of St. Joseph's, church performed a similar purpose. Later in the day after the rain-drenched mass, Binette put the altar to a different use: as a platform from which to introduce Governor David Walsh. The governor and lieutenant governor Edward P. Barry were touring the camp when they came across Binette. The priest suggested that the politicians—both Catholic Democrats—speak to "his people," and the three of them clamored up on the makeshift altar to address a crowd, again estimated at three thousand, that quickly gathered. Binette introduced Walsh in French and English, and Walsh promised the support of the entire commonwealth in rebuilding the Franco-American community, including its spiritual home. "You who have lost your Church I urge to stay in your home parish," the governor said, "and I shall confer tomorrow on a plan for the State-wide bazaar to raise funds throughout the Commonwealth for the building of a Church, even a more beautiful edifice."[9] As priest, Binette was the translator, perhaps even the literal translator, of Walsh's civil authority. In return, Walsh acknowledged the primacy of the Church—in particular the ethnic Church—in community life and the French Canadians of Salem as Binette's people. By urging Salemites to stay in their home parishes, Walsh also placed himself on the immigrant side of a dispute that would soon develop over where French Canadians should pray.

On the other side of that dispute was the archdiocesan hierarchy. On orders from the vicar general—the cardinal archbishop, like St. Joseph's parish priest Georges Rainville, was on a visit to Rome—the two hundred parishes of the Boston Archdiocese gave up their scheduled collection for Indian and Negro Missions and instead, on July 5, collected money for Salem. Four days later, the archdiocese delivered $5,000 and retained another $28,000 "until plans for aiding the Catholic families who suffered in the fire have been mapped out." Through the second week of July, the archdiocese had collected a total of $37,637.38, of which $17,000 had been disbursed, mostly through the local parishes: $4,000 to French St. Joseph, $2,230 to Irish St. James, $2,000 to Irish Immaculate Conception, $2,000 to Polish St. John the Baptist, $1,300 to Catholic Charities and $500 to French Ste. Anne. The total sum collected represented a mere four and a half cents per Catholic in the diocese and only $16.62 per displaced Catholic family in Salem.[10] But its collection and distribution through the archbishop's mansion and the withholding of its vast majority meant the archdiocese retained for itself the power to decide how the money would be spent.

The power of archdiocesan officials did not go unchallenged. After the fire, French Canadians continued their longtime battle against ecclesiastical domination by what they called the "Irish" bishops who controlled the Catholic hierarchy. They also fought their bosses, and at the end of the decade they struck to demand recognition of their union at the Naumkeag Steam Cotton Company. By exploring the response of Salem's churches and unions to the fire, we can better understand how working people built power in and through those institutions. This chapter begins with the relationship symbolized by Father Binette's translating for Governor Walsh: the system, much like its later counterpart in Halifax, by which clergymen vouched for their parishioners for civil relief. It then returns to Walsh's exhortation for Catholics to stay in their home parishes through an exploration of St. Joseph's parishioners' struggle to do just that. Finally, it combines church and labor history to develop a new understanding of Franco-American unionism at Salem's largest employer, which burned to the ground in the fire.

· · ·

Governmental relief authorities relied on local professionals to help adjudicate claims. In this way, ethnic and religious leaders took on a role that made them extensions of the state. As the initial crisis after the fire ended, relief authorities switched from providing food directly to providing vouchers. Applicants for aid were required to fill out a form and then take it to their doctor or clergyman for endorsement. "If this application was signed by any physician or clergyman of Salem it was accepted without question and an order for a week's supply of food given," Montayne Perry wrote.[11] Requiring that aid applicants receive the endorsement of a local authority was a form of compromise. On the one hand, it infused the official aid system with local knowledge, translated by a "trustworthy" professional. On the other, it put clergymen in the position of judging their congregants. This meant, Perry worried, that ministers could be scammed by their parishioners. "If a refugee donned ragged attire and showed himself to his clergyman with the appeal 'how can I get work, looking like this?'" she wrote, "it was quite natural that the clergyman should give honest sympathy and ready endorsement to the application for more clothing."[12]

Despite Perry's belief that clergymen were easily bamboozled by parishioners, the clergymen considered themselves discerning. John P. Sullivan, the pastor at Immaculate Conception, an English-language parish, reported that he or one of his curates "personally investigated" each family who received any Catholic aid; he probably used the same investigation methods when he vouched for people to receive official aid. "Ordinarily orders were given on the storekeepers," he wrote to the archbishop, "but in many cases, where the people were trustworthy, and had never before accepted charity, to spare their feelings, checks for small amounts—generally $10—were given."[13] As with the official relief, Sullivan sought to craft a compromise between formal investigation and informal trust of his parishioners.

For the official aid from the relief committee, the physician and clergyman did not have complete authority. After they gave their say-so, professional social workers and relief experts investigated the applicant. This system, apparently designed by Ernest Bicknell, chief executive of the National Red Cross, led to conflict between local worthies and professional social workers.[14] Montayne Perry wrote her book to be a guide for future relief managers, and she devoted a cautionary chapter to conflicts between local do-gooders—she focused on doctors and wealthy philanthropists, but this category included clergymen, too—and professional social workers from Boston. She did not use names, and her chapter is best read as a parable rather than as a literal recounting of events. It featured a doctor outraged at the perceived injustice of denying one of his "charity case" patients the relief she wanted. "'Nice state of affairs,' growled The Doctor. 'Hundreds of thousands of dollars up there, and a poor sick woman can't get enough to keep her alive. I'll see what I can do!'" At his club that evening, The Doctor commiserated with The Wealthy Citizen, who complained: "'They got a lot of philanthropy experts and charity specialists down from Boston. Social workers, they call them. "Workers" is a good name for them. They're working Salem all right—drawing fat salaries for looking wise and riding around in autos to "investigate." Time the whole bunch of them were "investigated" themselves, in my opinion.'"[15] The Food Man, whom the two interrogated the next morning, carefully explained to them why an organized, careful, and bureaucratic system was best. The case that originally riled The Doctor, for instance, turned out to be a scam; the family had not been burned out as the wife claimed, the husband had been fired the week before for drunkenness, and they were attempting to live on the dole after he refused work as a unskilled laborer.[16] The Wealthy Citizen, likewise, learned that the experts were all volunteers whose credentials as corporate efficiency experts he admired and respected. Notably, Perry placed her explanation in the mouth of The Food *Man*, a male executive, not a female social worker. Yet The Wealthy Citizen accepts that even the women, whose ranks included a militant suffragist, were paragons of "justice and fair dealing for all."[17] Perry's lesson for future disasters was that that the local experts' knowledge—corruptible by human sympathy, inexperience, and self-importance—was no match for the objectivity, efficiency, and centralized knowledge of professionals. That so many of these professionals were women required only a little more (male) explanation.

Although professionals had the final authority, clergy were the first judges. The daily newspaper explained the system five days after the fire: "Applicants for aid are referred to one of several committees, the St. James and Immaculate Conception parishes having one table, the Polish church another, while the St. Joseph and St. John's churches also have committees to receive their people, as have the other churches, all nationalities affected having representation." At each table, the applicant would be vetted by at least one representative of the committee representing their community. Those committees, wrote the newspaper, were "made up of about 100 people who have either been engaged in charity work or who are well acquainted with the people here so that comparatively few persons 'get by' who are not worthy."[18]

Those seeking help were not just judged by a communal authority; they also organized within their ethnic and religious communities. The pastor at Polish St. John the Baptist, forty-year-old Father Joseph Czubek, had been born in Ohio to immigrant parents.[19] The day after the fire, he bought a bakers' cart full of food and handed it out to his parishioners as they went off to work.[20] But he did more than charitably hand out loaves to the multitudes. The *Evening News* called him the "leader of his people," and in later years he organized parishioners nonreligiously. In 1919, he apparently demanded from the pulpit that Poles not patronize or rent to Jewish stores and "asked the Congregation to raise their hands in a promise to keep away from and boycott the Jews."[21] Five years earlier, after the fire, his congregation included 250 homeless families.[22] He helped organize a response. Czubek, said the newspaper, "had direct oversight" of a crew of about fifty of his women parishioners sewing new clothing to replace what had burned. Crucially, this work took place in communal space: the parish church, school, and convent.[23]

Italians possessed few local institutions or resources and so had to start from scratch. On June 7, a group of seven Salem Italians met to form an ethnic relief committee. Although organized and headed by local men, at least one Bostonian was also involved: Americo Brogi, a US-born, twenty-eight-year-old Boston telephone clerk, who was state secretary of the Italian Citizens' Progressive League and, according to the *Evening News*, "rendered valuable service at the Italian table."[24] The committee elected grocer Giuseppi D'Iorio chairman and shoe worker Frank Salvo secretary. They appealed to Italians around Massachusetts to donate not only to a special fund for Italians but also to the general relief fund; they also intended to solicit aid from the Italian consul in Boston. Apparently none of the men there were of sufficient stature to watch over the money. John Tivnan, the chair of the Committee of Fourteen, was named the trustee of the funds earmarked for Italians, and they promised to send the rest the state Red Cross.[25]

Four days before the Italian meeting, the newspaper announced the formation of a Hebrew Relief Committee. Like its Italian counterpart, the committee intended not only to help Jews but to aid with the general relief effort. Joseph L. Simon, a real estate agent whose office was in neighboring Beverly—and who would later sound the alarm about Czubek's anti-Jewish boycott—and Max Silverman, a shoe man, were dispatched to New York to recruit and consult with national Jewish leaders.[26] At the end of August, the American Jewish Committee held a special fund of $1,800 to "be distributed under [its] auspices," though it is unclear whether any money had yet been disbursed.[27] Although the New York worthies appear to have done little active fund-raising, Jews from cities nearer to Salem pitched in. On July 12, a Sunday afternoon two and a half weeks after the fire, the Lynn Hebrew Citizens' Club hosted a benefit to which Lynn theaters promised to send "their best talent." Also promised were speeches from the Lynn and Salem mayors; the Lynn congressman; Samuel Bailen, "one of the greatest Jewish orators of this state"; and Abraham Alpert, the editor of the Yiddish-language *Boston Jewish American*.[28]

Salem's French Canadians faced special challenges. Their enclave in the Point had burned, and with it their community buildings, including the parish church. Their experience after the fire is thus worth extra consideration. Until recently, historians have focused on two elements of the Franco-American experience: militancy within the church to defend the ideology of survivance and quiescence on the shop floor. *Survivance*—the idea that it was the sacred mission of the French Canadians to preserve Catholicism in North America and, eventually, to spread it across the continent, and that to preserve their religion and culture in the overlapping seas of Anglophones and Protestants that surrounded them, they had to work hard to preserve their language, their institutions, and their religion, all of which were seen as mutually supportive and constitutive—led Franco-Americans to fight bitterly with the "Irish" church hierarchy for lay control.[29] Given the ecclesiastical battles for lay control, it is ironic that what looks like cross-class ethnic solidarity to some looks to others like a priest-ridden community. Labor historians have often accepted nativist complaints that French Canadians would not organize, or that priests held "persuasive power" over them.[30]

The Salem fire offers a way to reconcile the undeniable importance of the church and other cross-class ethnic organizations with the equally undeniable fact of Franco-American unionization.[31] By understanding the peculiarities of ethnic Catholic political culture, we can better understand the way that lay Catholics practiced politics in their lives outside the church. Catholics like the French Canadians in Salem remained Catholics in all their endeavors, even the ones that were outwardly secular.[32] This does not, however, imply clerical domination; indeed, lay Catholics' battles for power within the church could not help but shape their civil politics. Just as women practiced politics in women's clubs and southern African Americans' political activity was shaped in their organizations, working-class Catholics learned politics within their lay religious organizations.[33] This may be particularly true for non-Anglophone Catholics, whose religious practices helped create distinct political practices. As Jennifer Guglielmo argues in the context of Italian American women, historians must be attuned to the ways different communities engaged differently within the political sphere.[34] An equivalent project for Franco-Americans is needed to understand their class and religious politics and the imbrication of the two. The Salem fire and its aftermath provide a starting point. They demonstrate how Franco-Americans created a subtle political culture that valued building power within institutions, sometimes at the expense of material gain.

There is no question that the local parish was intensely important to French Canadians in Salem. In 1914, St. Joseph's parish was led by Father Georges Alphonse Rainville, a fifty-eight-year-old Quebec native. A merchant's son, Rainville had been born in a village called Saint Marc, about thirty miles northeast of Montreal, and went to seminary in Nicolet and Trois Rivières before his ordination in 1883.[35] He was "born to be a bishop," a French actress wrote after visiting Salem in the late 1910s. "It seemed to me that the crozier, mitre, and amethyst ring would suit him very well. He had a simple good-heartedness, a majestic joviality, that let me speak freely to him, though I could not forget for a minute that he was a parish priest."[36] Another observer noted his "unusual executive ability."[37] Rainville's first post had been in Yamaska, south of Montreal, where he was in charge of building a new rectory. When he moved to Massachusetts, he continued as a "bricks and mortar priest," moving from parish to parish building physical infrastructure. After assisting in Marlborough, he founded parishes in Cochituate, where he built a church, and Brockton, where he built a church, a rectory, a convent, and a school. When Rainville arrived in Salem in 1904, he found a church of long standing—St. Joseph's was founded in 1872—but straining at the walls. He and his parishioners built an enormous and imposing new Romanesque church to hold 1,600 people. The building, complete with two 185-foot towers, cost $120,000; it opened a year before the fire. Rainville had arrived just three years after an assistant priest at St. Joseph's, J. Alfred Peltier, had been dispatched to found a second parish for French Canadians; Peltier's church, in the Castle Hill neighborhood, became Ste. Anne's.[38]

When the fire struck, Rainville was out of reach, on a ship to Europe to visit Rome and other pilgrimage sites. "What an awful blow will be dealt him when finally the tidings of this mighty disaster to his loved ones becomes known to the loving leader of his people!" the *Evening News* wrote. "He does not know that his pride, the new church of St. Joseph is a hollow, silent ruin. He does not know that the great and noble looking statue of St. Joseph Still Stands, but like the carven figurehead on a tombstone." Given Rainville's life literally building parishes, the *Evening News* showed remarkable acuity in focusing on the destruction of the built infrastructure. "He does not know that churches, schools, assembly halls, homes, newspapers, employment and even the rudest shelter are denied to the people he so well loves. And when the frightful knowledge of the things that are comes to him, who can say what the result will be?"[39]

The result, when Rainville learned of his neighborhood's destruction upon his landing in Europe, was that he turned around and came home. He booked passage on the *Lusitania* and sailed to New York immediately.⁴⁰ When he got there, he was greeted by his senior assistant, Donat Binette, and Joseph Côté, a priest based in Amesbury. They took him north to Salem, presumably preparing him on the way for what he would see. He got to Forest River Park at seven thirty in the evening on Wednesday, July 15, nearly three weeks after the fire. "M. le Curé mounted the dais erected some distance from the entrance to the park," the French-language newspaper reported. "Fr. Binette rang the bell. Fr. Rainville was already emotional, and when his flock emerged from the tents where they live to join the crowd already pressed at the foot of the dais, he turned his back and broke into tears." In addition to Rainville's parishioners, he was greeted by two Anglophone leaders, a militia captain and the relief committee chair, who formally offered condolences in the name of the people of Salem. It fell to Father J. B. Parent, the French Canadian pastor in Lynn, to console the distraught Rainville. It took some time for Rainville to find words for his flock, but when he did, he bade them *"Courage."*⁴¹

For the three weeks before Rainville came back, his pastoral responsibilities were taken up by Binette and the junior assistant priests. On Sunday, three days after the fire, they planned hourly masses from six to eleven o'clock in the morning; the six masses were two more than offered on a typical Sunday. Each evening, Binette held an outdoor, public prayer service in the camp. But his work and those of the other priests extended far beyond the ceremonies of sacraments and prayer. Each night after the service, reported the *Evening News*, he gave a "talk on matters that concerned [his congregation's] physical and spiritual well being." Despite sleeping indoors in a borrowed room, Binette ate each meal at the camp, where he "consulted on a thousand matters a day" with laypeople and militia officers.⁴²

The officers and other observers recognized Binette—in the absence of his superior Rainville—as the primary leader and spokesman for Salem's French Canadians. Secular respect for Binette's authority was important for two reasons: first, it encouraged French Canadians to accept religious, rather than secular, leaders and organizations as their representatives. Second, it acknowledged Franco-American priests as the spokespeople of their community, rather than "Irish" bishops. The first community meeting of French Canadian refugees was organized under the auspices of the church. The newspaper called Binette "in charge of the French people at the park."⁴³ Montayne Perry agreed and described the priests' role as both figurative and literal translators. "In winning the confidence and co-operation of the refugees themselves, the assistance of the priests and the seminarians was invaluable," she wrote. "Having the entire confidence of their own people, speaking in the language which they could understand, they explained to them the regulations for health and order and the reasons for them, adjusted difficulties and smoothed away misunderstandings as none but the trusted Fathers could have done."[44] Another observer credited the "vigilance of the military authorities and the influence of the priests of St. Joseph's Parish" for the "excellent general moral condition" and good health at the camp.[45]

Recognition of leadership was more than just symbolic. Most important, perhaps, was that priests were given power to dole out relief. When substations for the clothing and employment committees opened in Forest River Park, the clergy "took a hand," and they signed slips to approve requests for clothing.[46] This meant that they had direct authority over who received aid and who did not, and it forced them to judge who among their parishioners were worthy and who were not. Put another way, it meant that Binette and his colleagues could use their local knowledge to the advantage of their parishioners, whose specific needs outsiders might not have otherwise understood. When Franco-Americans in Manchester, New Hampshire, raised money to help the *sinistrés*, they delivered the money not to the general relief fund but directly to Rainville.[47] For Franco-Americans outside of Salem, at least, the logic adopted by the civil and military authorities in Salem seemed to make sense. Binette and Rainville were translators and leaders, local and ethnic authorities with knowledge of their community and respect from others.

All this responsibility was difficult for Binette to handle. On Tuesday night almost a week after the fire, he collapsed and was taken to the armory relief station. "The army physicians there found he had collapsed from exhaustion," the *Evening News* reported. "His shoes were burned, great white blisters covered his feet and his condition was pitiful. After receiving treatment he was taken to his home in an automobile."[48] His responsibility may also have been difficult for his parishioners to handle. We already saw, in chapter 3, that Binette's flock apparently ignored his instructions to expose any "fakirs" in their midst. Though the priest publicly asked refugees to "be on their guard against th[e] vandals" who would try to take relief supplies they did not deserve, such fraudsters continued to live in the camp, unmolested by their neighbors. The refugees also, as we have seen, rejected the advice he gave in his first sermon to take docilely whatever jobs were offered to them.[49] *Sinistrés* apparently did not follow their priests as blindly as observers imagined.

Moreover, Catholics had competition. Members of the French Evangelical (Baptist) Church presumably cared little about what Binette or Rainville had to say. Quebec-born Oliva Brouillette had founded the church as a missionary of the American Home Baptist Mission Society. Eight years later, in 1911, Brouillette had been so successful his congregation had to start raising money to construct a church building of its own; it finally opened in 1913.[50] While Brouillette was certainly a successful missionary, there had been Francophone Protestants in Quebec since the days of New France, and around the turn of the century there was an active evangelist movement that included Baptists.[51] So it is unclear how many of his congregation had been born Protestant or had converted, and if converts, where they had become Baptists. Regardless, their presence, and especially the congregation's growth, complicates the image of a religiously unified French Canadian community. Brouillette, like other clergy, sat on the Committee of 100 and acted as a clearinghouse of information for his congregants. Those who knew the whereabouts of a missing parishioner, for instance, were requested to contact him.[52] After World War I, Brouillette put his disaster-relief experience from Salem to use in northern France and southern Belgium as director of Baptist relief and mission work. He built six "foyers" that became religious and social centers, distributed farm machinery and schoolbooks, and granted relief to orphans.[53] He was apparently as active for his parishioners as Binette was for his.

Taking into account parishioners' willingness to disregard Binette's instructions about fraudulent relief applicants and the presence of Brouillette's church, it is apparent that the Catholic clergy's hold on Salem's Franco-Americans was not complete. Even to the extent that lay Catholics were loyal to their priest and parish, that loyalty was neither blind nor without agency. Rather, their devotion to their ethnic religious institutions and practices was intertwined with an insistence on lay power and authority within the church. The church in Quebec had focused on resistance to Protestant domination. In contrast, migrants built Franco-American Catholicism largely in opposition to domination from "Irish" bishops. Key Francophone Catholic institutions—ranging from Maine's Cause National, founded to demand a French Canadian bishop, to the fraternal organization Forestiers Franco-Américains, which split from its parent over the right to speak French at meetings—served two purposes. They sought to foster the French language, and thus Catholicism. *Qui perd sa langue, perd sa foi*, it was feared: He who loses his language, loses his faith. But equally, these groups were in themselves a challenge to Irish cultural and episcopal domination.[54]

In Quebec, parishes' temporal affairs were managed by a vestry, called a *fabrique*, made up of elected laymen called *marguilliers*. The *marguilliers* oversaw church finances and land, and they served as the representatives of the church vis-à-vis the civil government. Quebec laymen also controlled the school and hired the sexton. While the purview of lay trustees was by definition temporal, their presence meant that laymen had power and authority built into the ecclesiastical structure.[55] When French Canadians migrated to the United States, they sought to replicate the *fabrique* system. Yet from the start, American bishops undercut or refused this style of governance. In Winooski, Vermont, among the first French Canadian national parishes in New England, the French (i.e., from France) bishop insisted on appointing the trustees, rather than let the pew-rent payers elect them. He also explicitly reserved his own power to hire and fire the priest and to "direc[t] all that regards the spiritual matters of the parish." Moreover, he departed from the Quebec *fabrique* tradition by naming the priest as the vestry's president, secretary, and treasurer; insisting that the bishop be consulted for expenditures of more than $300; granting the priest the authority to name the organist, chanters, verger, sexton, and all other officers; placing the school under direct clerical control; and not enforcing *marguilliers'* traditional term limits.[56] In the late nineteenth century, state legislatures aided the bishops' drive for increased authority: Rhode Island made bishops the presidents of local parish corporations in 1866; and in 1887 Maine established the "corporation sole" system, granting the bishop the title to all church property.[57] The Boston Archdiocese, too, was a corporation sole, meaning that the archbishop and the archdiocese were legally indistinguishable.[58]

To demand distinctively French Canadian devotional practice and, especially, French Canadian priests the language and nationality of their priests, Franco-Americans fought for lay prerogatives in church governance and against episcopal domination.[59] Throughout New England, lay Franco-Americans fought their bishops for control of their parishes. In Woonsocket, Rhode Island, just a few months before the Salem fire, Franco-Americans rejected a French-speaking Belgian priest and mounted a monthlong picket, a rent-strike for their pews, and a boycott of the offertory.[60] The bishops defended their prerogatives to hire and reassign priests, and they refused and resented lay attempts to impinge on those prerogatives.

Cardinal William O'Connell was Boston's archbishop in 1914. He was not, claims biographer James O'Toole, an "unremitting Americanizer"; O'Toole depicts him as more concerned with the difficult logistics and practicalities created by a patchwork of national parishes than with forcing ethnic Catholics to adopt American or Irish traditions.[61] But O'Connell had come to Boston from his prior posting in Portland with a bad reputation among Franco-Americans.[62] There, he had met concerted opposition when he tried to install an Irish priest at a mixed parish where there were six French speakers for every English speaker.[63] And when he arrived in Boston in 1906, O'Connell had walked into a long-running dispute over who would control Ste. Anne's parish in Newton. Franco-Americans there had long demanded a compatriot as their pastor, and they appealed beyond the archbishop to the pope's representative in the United States.[64] To O'Connell as to his adversaries, the battle was over who would have power in the church.

"The church is not a democracy," he warned the archdiocese in 1912. "In it each member has his well-defined place. The great body of the faithful are the followers, the disciples of Christ. They neither teach nor command. . . . In a word, the faithful are to be taught, to be led, to be fed." His ecclesiastical ideology brooked neither dissent nor voice for the laity.[65]

The fire burned down both St. Joseph's Church and the neighborhood it served. As we saw in chapter 3, most parishioners went to neighboring cities, of which only Lynn had a French Canadian parish. In Danvers, Franco-Americans had been long stymied in their quest for a national parish by the local Irish priest. As early as late July 1914, 700 Franco-Americans met to begin a push for their own parish, buoyed by their new, larger numbers. The *sinistrés* displaced by the fire into Danvers, the argument went, should have the same access to French-language, Catholic worship and education in their new home as in their old one. Though O'Connell promised in July to consider the matter, by December the activists had made no headway. By then, however, they had abandoned talk of the refugees and instead focused only on the rights of those who now considered themselves Danvers residents.[66] When a delegation met with Annunciation Church's Irish pastor seeking his help, he responded with indignation and barely contained condescension. "I told them that I was fully able to take care of the French, able to understand their language when spoken quite well, able to hear their confessions in French, able to speak their language quite well, able to give a sermon in French, etc etc.," Francis Maley wrote to O'Connell. At the end of the meeting, he pointed out their small donations to Annunciation—not surprising, since they were probably continuing to give money to St. Joseph's—and asked them "how much consideration they deserved in view of their failure to do their duty toward the support of the Church."[67]

Meanwhile, displaced Salemites in Beverly also demanded their own parish. At a late-July meeting there to plan a campaign for a national parish, the chair was a Salem lay leader, Napoleon Levesque.[68] The second, smaller Francophone parish in Salem, Ste. Anne's, also experienced rapid growth as former St. Joseph's parishioners switched churches. A year after the fire, O'Connell sent a second priest to assist the pastor, and in April 1916 he approved the building of a new, larger parish school.[69]

With no French church in either Danvers or Beverly, Rainville's parishioners from St. Joseph's continued to come to him for pastoral care. Although O'Connell ordinarily saved nearly all of his correspondence, there are no surviving records relating to any Salem parish from 1908 to early 1915, and this makes difficult any attempt to assess relations between Franco-Americans and the archbishop in the year after the fire. In March 1915, though, Francis Maley complained to

O'Connell that Rainville was impinging on his turf. In particular, the priest wrote that Rainville had administered last rites and then buried one of his former parishioners, a woman named Rosanna Ducet who had come to Danvers after being burned out of Salem.[70]

In a letter to the archbishop's secretary, Rainville defended himself, insisting that Ducet should be counted as one of his parishioners. He made three arguments. First, he cited the civil authorities, who continued to grant marriage licenses to people using their burned-out addresses, continued to collect taxes from refugees, and, most important, continued to have refugees (though not, of course, Madame Ducet) vote in Salem. Second, he claimed refugees as his parishioners because they "continue[d] to declare themselves parishioners of St. Joseph" and because they retained membership in St. Joseph's lay religious societies. Third, Rainville made a practical argument: that because many of the *sinistrés* spoke only French, they needed him to understand religion. This third was a recapitulation of the standard Franco-American argument that *qui perd sa langue, perd sa foi*. The first claim, resting on legal and civil definitions of residency, would appeal to the archbishop's sense of ecclesiastical order. Rainville's parishioners were still his because they technically fell under his geographical jurisdiction.

It was the second argument that took up the bulk of Rainville's letter and that was the most controversial. It rested on the implicit insistence that laypeople had the right to pick their priests and their parishes—that is, that it mattered of which church they "declare[d] themselves parishioners." This was the right that Franco-Americans had demanded for decades, when they picketed churches or wrote to the apostolic delegate to demand a French Canadian priest. In the territorial parish system O'Connell and other New England bishops preferred, the laity had no such right; they were to attend the parishes to which they were geographically assigned. In all, Rainville concluded, if he did not have the right to perform sacraments for Ducet, he did not have the right to serve any of his parishioners, since all of them were displaced. It was his right, he insisted, to serve those who wanted to be his parishioners wherever they lived. Nonetheless, he closed on a conciliatory note, humbly asking O'Connell's permission to continue to serve the fire's displaced victims.[71]

When O'Connell's secretary responded, he ignored the more controversial parts of Rainville's letter. The cardinal avoided comment on the implicit claim that parishioners had the right to choose their own priest. Instead, O'Connell relied on the civil authorities. If Ducet was still, according to the civil government, a Salem resident when she died, she had the right to use Rainville as her priest not because she chose him but by virtue of her legal residence. Since "her residence in Danvers was merely temporary and begun with the intention of returning to Salem as soon as conditions caused by the fire were suitably overcome, he [O'Connell] judges that she had a sufficient domicile to have had the Sacraments administered to her from your Parish and to be buried therefrom," the secretary wrote.[72]

Because O'Connell's dispensation relied on a determination that the refugees were only temporarily in Danvers, it expired once they were deemed to have left Salem permanently. It did not take long for this to happen. Just over ten weeks after O'Connell granted his temporary permission, he rescinded it. "Good order demands that you cannot go on indefinitely considering your former parishioners as still belonging to your Parish even though they actually have a residence within the territorial limits of other neighboring Parishes such as Danvers, St. Anne's Salem, etc.," his secretary wrote. O'Connell ordered not only that Rainville give up "duties of caring for their spiritual welfare and the rights of administering the sacraments" for his former parishioners, but that he take responsibility for telling them of the change as well. Moreover, perhaps fearing that Rainville would help organize opposition to the order, O'Connell specified that he expected the priest to "cheerfully comply."[73]

Rainville was not cheerful. "I do not understand this change," he wrote the day he received the letter. "But your eminence may be assured that although in the fire of 25 June 1914 I lost everything, what can give me a little comfort on earth is that I have not lost my spirit of obedience to Authority, and I am always ready to sacrifice myself for the Diocese."[74] Rainville's obeisance had an element of irony, however, and the defeat was not as abject as he made it seem. That month, he and his parishioners began a fund-raising campaign to rebuild the church and the school. Children and other laypeople organized benefit concerts and fairs to raise money. The parish had started in the basement of Immaculate Conception, and it returned to its roots by holding services in the basement of the burned church. The community built a small school over the basement, and it continued raising money to rebuild the church, the rectory, and the school. "All sorts of profit making functions took place," wrote the parish historian in 1972, "and at the bazaars and carnivals, choice French Canadian delicacies were served, with 'cortons' being the favorite."[75] Even with the fund-raising, O'Connell continued to stand in the way. Although in September 1915 he granted permission to rebuild the rectory and convent, in spring 1916 he refused to allow St. Joseph's to build a larger school.[76] Not until the next decade would the parish build a school large enough to suit its needs.[77] Though O'Connell had done his best to make Salem's Catholic survivors go to mixed parishes, the laity and clergy subtly undermined him, seeming to cooperate and show him deference while working to rebuild their parish and its institutions.

Rebuilding the church and the school worked for the *sinistrés* who returned to the Point, but thanks to O'Connell's ruling, those who remained in outlying areas were still without a Francophone priest. The pressure lessened in Danvers when O'Connell replaced an Irish assistant priest with a French Canadian, although Franco-Americans faced continued pressure to assimilate with their English-speaking coparishioners.[78] In Beverly, by the fall of 1916, half of the fire refugees had returned home, but the remaining 490 still needed a church.[79] So Rainville returned to his roots as a builder of parishes, as he had been in Cochituate and Brockton, and he began a mission church.[80] After delegations and mass meetings and considerable lay agitation, in 1917 O'Connell finally allowed Rainville to buy a decrepit old Methodist church near the town line with Danvers, and Rainville forwarded the cardinal a list of three Franco-American priests who could serve there.[81] Naturally Rainville preferred one of his subordinates from Salem, but O'Connell chose an outsider from Lowell.[82] Again we see a back-and-forth struggle for power within the church, with O'Connell and Rainville both contesting and undermining each other's authority.

In the longer term, both parishes thrived. St. Joseph's opened its new school in 1921 and then built a larger one in 1925, and the church remained open and consciously Franco-American until it fell victim to the archdiocese's financial crisis in 2004.[83] St. Alphonsus, as the Beverly church became, also lasted until the same wave of parish closures.[84] But Rainville did not live to see this flourishing. In 1948, the parish historian remembered that he had been "worn out by this work [of rebuilding] and undermined by sickness"; he died in 1920 at age sixty-two.[85] Even in death, Rainville worked to build Franco-American institutions. In his will he gave $3,000 to the Collège de l'Assomption, the new Francophone college in Worcester, to create a scholarship for a St. Joseph's parishioner. The rest of his estate went to the archdiocese, but from the grave he again undermined the hierarchy; he restricted his bequest to the specific purpose of building the parish school for boys that O'Connell had stalled. Donat Binette, his protégé, helped raise even more money for the school, rallying parishioners to donate in Rainville's memory.[86] Binette was even more explicit than his mentor in his defiance of the hierarchy. By 1927, then the pastor at Rainville's old parish in Cochituate, he was a public supporter of the *Sentinellistes*, a Rhode Island group whose opposition to their bishop was so extreme that they were excommunicated.[87]

The *fabrique* system and the battles with the episcopacy created a subtle and nuanced political culture. Thanks to the deliberate work of Quebec bishops, the French Canadian church was ultramontanist, yet it battled O'Connell, the most ultramontane of the American bishops.[88] The movement fought for lay authority, but the power the laity sought was to be subservient to the clerics of their choice. The *fabrique* was not, that is to say, the republican and democratic, almost congregationalist, lay trustee system Jay Dolan valorizes from the early days of the American church.[89] The village parish world French Canadians sought to recreate in industrial New England was arranged with the priest and *marguilliers* at the top of a strict hierarchy.[90] Women were excluded from the *fabrique*, as were those who could not afford to rent pews. Neither anticlerical nor even entirely antiepiscopal, it was a certain version of democracy, in which laypeople demanded to pick their own leaders.

The fight to let laypeople pick their priests and choose their churches—even while remaining loyal to the greater Church—helped to create a political culture that fostered cooperation and compromise amid conflict. The aftermath of the Salem fire was a skirmish in the long-running war between Franco-American priests and laity and the Irish hierarchy. As in Salem, sometimes Franco-Americans won, and sometimes they lost. O'Connell's acquiescence to the Beverly mission was a rare unambiguous ethnic victory. More often, as with the survival of a Francophone St. Joseph's, there was a continual process of contestation. This ethnic conflict required, at least to some extent, a cross-class alliance. Rainville, as the parish priest, is most dominant in the archives, and it was middle-class businessmen and professionals who spoke publicly at meetings. But it was working-class women and men who made and bought the *cortons* and paid to attend the fund-raising dances and fairs. Although the spokespeople of the movement for French Canadian autonomy in the church were an elite group, workers and their families also participated. Together, they created a system in which power was rarely absolute or explicitly contested. Rather, they found advantage here and there, cooperated and compromised where they could, and resisted episcopal domination in subterranean, quiet ways. The fire gives us an opportunity to understand this political culture. But political cultures are never stable, and the fire and the resulting fight over St. Joseph's parish forced Salem's French Canadians to practice, reenact, and recraft theirs.

The fight to save St. Joseph's parish in the wake of the fire can help us make sense of the labor history of the community's largest employer, the Naumkeag Steam Cotton Company, which manufactured Pequot brand sheets and pillowcases. That labor history gives credence to the idea that there was something particular about Franco-American politics in Salem. As Salem's Franco-Americans rebuilt their schools, grew their churches, and watched as their priests publicly battled New England's bishops, the Massachusetts textile industry faced an existential crisis of deindustrialization.[91] Salem's elite eventually turned to historical tourism—creating historic sites in or on the edges of French Canadian and Polish neighborhoods—to help the city's economy and remind residents of the city's long American and protestant history.⁹² Naumkeag workers relied on the pattern they cut during the crisis of the fire to respond to the new economic crisis. They first organized an unusual industrial union and then developed a nationwide reputation for labor-management cooperation.⁹³ Yet cooperation did not mean quiescence. Indeed, Naumkeag workers spent the 1920s building power in the factory and in their union just as they worked to build lay power in their church.

Loom fixers, the most skilled craft in the factory, had organized a union in 1901. It was led by French Canadians, and so it made sense that when Franco-American fraternal organizations needed space for their relief work, they set up shop in Loom Fixers Hall.⁹⁴ Although only the most skilled minority of the 1,500 to 3,000 Naumkeag workers belonged to a union in 1914, much conversation in Loom Fixers Hall that weekend must have revolved around whether any of them would ever get their jobs back.⁹⁵ Naumkeag's owners soon promised to rebuild and "establish a model manufacturing community at the Point, with new-type mills and new-type dwelling houses."⁹⁶ Indeed, they reopened in February 1916 as one of the country's most modern textile mills.⁹⁷ This late capital investment meant workers and owners were on somewhat different ground from their counterparts at textile mills elsewhere in New England.

Almost exactly four years after the fire, in late June 1918, Naumkeag's doffers, roving boys, and oilers—all skilled, male workers—walked off the job to demand a large pay increase. By the next afternoon, most of the rest of the workers in the mill were on strike too. Although management thought that it was the Poles who were the most militant, a majority of the strike committee was French Canadian. So too were those from whom the union received out-of-town assistance, including Horace Riviere, a United Textile Workers (UTW) organizer from Manchester, New Hampshire, and Albert Langevin. Langevin was the union's representative on a state-organized arbitration committee, and he was so hated by mill owners that when the committee visited a Fall River plant to investigate industry-standard conditions, management there refused him admission.⁹⁸ Possibly Franco-Americans' reputation for quiescence hid their involvement in the strike, making it seem as if Poles were the militant workers, despite French Canadians' leadership.

The workers won their monthlong strike. In the context of very high wartime inflation, management had initially offered a 10 percent raise, and the workers walked off demanding 17.5 percent, although the demand later went up to 25 percent. The arbitration board granted a 17 percent wage increase to most workers and made the increase retroactive to the start of the strike. Management complained, "The decision of the state Board of Conciliation and Arbitration was in effect the granting of the demands of the striking operatives" but nonetheless accepted the board's decision. More important, the strike meant unionism had spread beyond the loom fixers. Together, other crafts and unskilled workers asked for and received a charter as Local 33 of the UTW and elected plumber John O'Connell business agent.[99]

Unlike their counterparts in cities like Fall River, who remained suspicious of industrial unionism, Salem workers led by O'Connell spent the next year organizing a broad, industrial union comprising all levels of workers. "O'Connell took advantage of a strike among the unorganized (or almost unorganized) help to form a strong union," one observer recalled fourteen years later. "The union looked to Mr. O'Connell for leadership and they got it."[100] A series of strikes and threatened strikes followed as O'Connell tried to organize the union. In November 1918, apparently emboldened by their success over the summer, the workers tried again, demanding a 25 percent wage increase; they coordinated with unions in New Bedford and Fall River, who each demanded 15 percent raises. Citing slackened business at the end of World War I, the Naumkeag board refused. The skilled loom fixers appear to have been in the lead—or at least the board understood best how to deal with a long-standing craft union—because the board instructed the manager to "inform the Fixers" of their refusal and "to make the same answer to any demands of the other unions in the Mill." During that meeting, the manager at the Danvers Bleachery, a satellite operation, telephoned to say that fifty of his workers had walked off the job, also demanding a pay raise. These workers, too, were refused.[101] It also appears to have been the loom fixers, working with counterparts in Fall River and New Bedford, who called off a threatened strike the next month.[102]

In August 1919, the UTW again demanded a 25 percent raise, union recognition, and a closed shop. After a month of negotiations, union and nonunion workers went on strike together. Having learned the year before that they could lose state arbitration, management avoided it, and the strike dragged on until November. Again, the workers won. Though only the lowest-paid workers received an immediate pay increase, the board of directors agreed to peg wages at 5 percent more than any other New England mill. The company recognized the union and agreed to a dues checkoff, a closed shop, and a grievance-adjudication process. In exchange for all this, however, the company sought to increase productivity by between a quarter and a third.[103]

The 1919 settlement was the turning point. O'Connell had his industrial union—loom fixers retained an independent union for a few years but eventually affiliated with the UTW[104]—and the contract set the tone for the next decade. While other factories closed or moved to the US South, this unusual labor accord kept wages high and Naumkeag in Salem.¹⁰⁵ Throughout the 1920s, the union grew its power in the factory and signed contracts that covered not only wages but also working conditions, job classifications, discipline, and seniority.

The trade-off was formalized and expanded in the 1927 and 1928 contracts. In 1927, the company challenged the accord and began to resist further union demands in wages, conditions, and authority. Workers responded in kind, and Poles, by then about half the workforce, challenged the union's leadership. O'Connell was still the business manager and the officers were still predominantly French Canadians. They also demanded greater militancy in the face of increasing management intransigence. The Poles tried to fire O'Connell and replace him with one of their own, but they were rebuffed by French Canadians, who made up the other half of the workers. (Irish, who were a small proportion of the workforce, also sided with O'Connell.) The union executive responded to the challenge by intensifying its position, exchanging greater productivity for more power to make decisions.¹⁰⁶ The union promised to give two months' notice before a strike; to promote the distribution of Pequot-brand products; and to encourage higher productivity in terms of quality and quantity. In exchange, the company promised high wages, continued employment, good working conditions, and, perhaps most important, monthly meetings between management and union to discuss the factory's operation. In December 1928, workers rejected a management proposal that would have laid off 300 workers and forced the remaining weavers to tend twenty-four looms instead of thirteen. O'Connell worked to stave off a strike by successfully proposing a union-controlled stretch-out. With Morris Llewellyn Cooke, protégé of Frederick Winslow Taylor who spent the 1920s recruiting unions to accept and adopt scientific management, O'Connell developed a system of "joint research." O'Connell's handpicked delegates to a joint union-management committee controlled industrial engineering, and the union nearly took over in sales and marketing.¹⁰⁷

The 1927 and '28 contracts gave the union increasing power over management prerogatives, but in practice, O'Connell became increasingly autocratic. He handpicked the union's members on the joint research committee, and they saw it as a stepping-stone to management and so represented the bosses, not their constituents. He started to refuse even to explain his policies to his members; as he told a researcher a few years later, his language dripping with ethnic contempt, "I didn't bother to report because they are a bunch of ignorant Canucks and Polaks who wouldn't understand anyway." O'Connell was even rumored to own stock in the company.¹⁰⁸ Nevertheless, the initial trade-off of higher productivity in exchange for power for workers seems to have been popular, especially among French Canadian workers. Even Morris Cooke, who considered the union's embrace of scientific management one of his life's major achievements, noted the labor peace that maintained before his arrival.[109] As O'Connell's victory in 1927 indicates, it was only with the support of French Canadians that he held power for so long, and even in his autocratic period he held power with the support of an executive dominated by Franco-Americans.

But the rank-and-file rebelled when O'Connell, a boss in unionist's clothing, took power rather than the workers themselves. The stirrings of dissent that had begun in 1927 and 1928 grew louder in 1930, when O'Connell's stretch-out came to the weaving room, where two-thirds of the workers were women. He agreed to increase the number of loom machines each weaver watched from thirteen to twenty. This meant that 100 to 150 workers would be laid off. Moreover, management proposed that seniority be decided based on the amount of time on the job, rather than the date that a worker had begun to work at Naumkeag. Since women took time off when they had children, under the company's system they would have significantly less seniority than their male colleagues who had been initially hired at the same time. Though women objected to this arrangement, O'Connell and the other men who ran the union agreed with management.[110]

Discontent finally boiled over in 1933, and workers rebelled against the union-management collaboration. With the help of the left-led National Textile Workers Union, they struck, led out by the married women who had suffered most under O'Connell.[111] The head of the strike committee was Wilfred Levesque, "a swarthy, stout French Canadian" who had been the sole dissident on the union's executive board.[112] Despite Levesque's leadership, managers again believed that Poles had started the insurrection, although they acknowledged that French Canadians "were not long in taking part"; a state conciliator, looking back on the strike two years later, remembered "mutual suspicion" between Poles and French Canadians. O'Connell, however, who surely knew best, thought that opposition to him came from both groups. And all agreed that it was married women who rebelled first, objecting to their loss of seniority.[113] The strike ended with Levesque as business agent of the new Independent Sheeting Workers of America, Local 1, and a different outsider of Irish descent, leatherworker James Burke, as secretary. The new union rejected joint research and the stretch-out, but it still demanded the rights its predecessor union had won, including recognition, the ability to collect dues inside the mill, and seniority rights. Management, meanwhile, had tired of any union and sought to deny recognition. In the fall of 1935, workers stuck again. This time, French Canadians and Poles "stood together," said the government conciliator, although "the French were the leaders and the Poles the followers." Having learned from the failure of the autocratic O'Connell, Levesque and Burke insisted on bringing each company proposal to a vote of the full membership.[114]

A generation after the fire, working-class Salemites were, in the factory and neighborhoods rebuilt from the ashes, still battling to maintain power over their own lives and employment. Starting the story with the fire shows the continuity between the union drive of 1918 and 1919 and the militant strikes of 1933 and 1935 and how the experience of one crisis shaped a response to another. It also demonstrates the connection between Franco-American ecclesiastical battles and labor struggles. After the fire, Salem's Franco-Americans and other *sinistrés* experimented with new ways of organizing themselves and their obligations to each other. Parishioners and clergy at St. Joseph's carefully and subtly fought with their archbishop for power within the church. Building on this political culture, they and their coworkers at the Naumkeag Steam Cotton Company worked to build power on the shop floor, even experimenting with scientific management. In the church, workers tried to build a disaster citizenship that emphasized lay power while remaining loyal to the Church. In the union, lay Catholics similarly emphasized worker control while willingly trading productivity. In was only when bishops, managers, and union leaders went too far that workers and the laity rebelled outright.

CONCLUSION
CITIES OF COMRADES

In the summer of 1919, movie star Tom Moore finally got the serious role he had been coveting, and the critics raved it was his best film yet.[1] In *The City of Comrades,* he played Frank Melbury, a scion of a fine Montreal family. Trained as an architect, transplanted to New York, and ruined by drink, he climbs back to respectability with the help of other drunks. These derelicts help each other at a club—the title's City of Comrades—founded on an explicit model of solidarity; refusing charity from outsiders, they offer help, support, and aid as friends to friends. The mutuality of this aid imbues it with spiritual significance. "What I had thought of only as human aid I now perceived to be the celestial bread and wine," Melbury says in the novel on which the film was based. The brotherly love of his fellow man was itself God.[2]

The club's literal and figurative communion works, but despite Melbury's recovery his love interest rejects him, so he follows the path of so many scorned lovers: he enlists in the army. On his way to the Great War, he stops in Halifax just in time for the explosion. "To get the proper realistic effect," wrote a reviewer, "director Harry Beaumont ordered that the set be blown up with dynamite, causing the wreckage to fall on Moore."[3] Melbury is blinded, the love interest nurses him back to health, and through the help of solidarity they live happily ever after.

Basil King, who wrote the movie and the novel it was based on, was born in Prince Edward Island, educated in Nova Scotia, and had been a priest at the Anglican cathedral in Halifax. In 1900, by then only forty-one years old and a rector in Cambridge, Massachusetts, his eyesight failing and his thyroid diseased, King gave up the ministry and began to write best-selling novels. "His prose style is undistinguished; his plots are little more than ingenious mechanisms;

his characters rarely come to life," his biographer Randall Stewart wrote, calling King's prose "ponderously didactic or mawkishly sentimental." He continued: "It is not likely that his novels will be read in the future, except perhaps by students of popular literary taste."[4]

King's original novel was called *The City of Comrades*, but the phrase appears nowhere else in the text. In the book, Melbury never goes to Halifax, and the disaster never appears, but the title nonetheless appears to gesture at the explosion. King probably got the phrase *city of comrades* from Samuel Prince, a fellow Anglican priest from Nova Scotia. Prince had been an assistant priest at St. Paul's Church in Halifax at the time of the explosion, and in his Columbia sociology dissertation, he wrote that the city "gained the appellation City of Comrades."[5] Although Prince's dissertation was not yet published, it seems likely that he and King knew each other through their regional and religious connections, and the phrase *city of comrades* does not appear to have been used in print before.

The City of Comrades contains, for our purposes, three key insights. First is the very existence of everyday solidarity practiced by ordinary people. The film and novel are reminders that although the state and its agents could have difficulty discerning this informal order, it was not invisible. Just as King saw it from afar, so too did some of those who witnessed it in Halifax. For Samuel Prince it inspired a brief meditation on Peter Kropotkin's anarchist tract *Mutual Aid*.[6] Florence Murray, a Halifax medical student and future missionary, saw it in the hospitals there and called it "organization without any organization."[7] Moreover, this solidarity works. Melbury's friend Lovey affirmatively does not want to sober up. He is afraid of the club precisely because he knows solidarity is more powerful than charity. He tries to discourage Melbury from going because he fears its success: "Worse than missions and 'vangelists, they are."[8]

King's second important insight is that this solidarity waits latently. He presages Colin Ward's famous metaphor: "Society which organizes itself without authority is always in existence, like a seed beneath the snow, buried under the weight of the state." For Ward, the state's oppressive snow melts during what Aristide Zolberg calls moments of madness, when everyday rules are suspended.[9] Everyone, from the bums on the street to the denizens of the City of Comrades club to the people of Halifax, stands waiting to help their fellows, friends, and neighbors. Melbury does not inspire anything new; rather, his personal disaster allows him to see what has always surrounded him.

Finally and most important, to King the value of solidarity is not merely material; rather, it is also spiritual and emotional. Frank Melbury's comrades give him food, a bed, clothes, and eventually a job. But they also give him emotional and spiritual support. King depicts both material and emotional aid together as prayer through action. Jointly, they are the species of the communion of man, just as, for an Anglican priest, bread and wine are the two species that comprise the communion of God. The spiritual and material, in other words, are indivisible, and each constitutes the other. This is one of the fundamental points about solidarity: that the solace comes jointly and inseparably from the aid offered and from the relationship with the person or organization who offers it.

...

This book is about two cities of comrades, Salem and Halifax. Both were cities of comrades before their disasters, and the fire and explosion served to make evident what had existed before. Families, neighbors, friends, and coworkers had patterns and traditions of self-help, informal organization, and solidarity that they developed before crisis hit their cities. Those traditions were put to unusual purposes and extreme stress when the disasters happened. They were also challenged by the new agents of the state who were given extraordinary powers in the wake of the disasters. The working-class people who most directly experienced the disasters understood them and their cities starkly differently than the professionalized relief authorities.

The first way that survivors and their relievers differed were in their experiences of order and disorder after each disaster. Relief managers sought to read their destroyed cities using their progressive ideologies of order, which valued the knowledge of experts, officials, and professionals. Their ideology prevented them from understanding or even seeing the spontaneous order that survivors and relief workers built. This spontaneous order was based not on aimless altruism but on the connections, networks, and practices of daily solidarity that existed before. Soldiers in Halifax worked on rescue and salvage not in response to orders but because they had solidarity with each other and with their civilian neighbors; the militiamen in Salem who responded without their officers were similarly motivated. Halifax women who volunteered at hospitals went with their friends; their "organization without any organization" was spontaneous in that it was a new application, but it was based on the relationships they had before the explosion. Survivors in both cities went to places and people they used for support and aid in normal times, including doctor's homes, drugstores, churches, and, most important, their families.

The order that the state imposed was not valueless or neutral; it had real political implications. The specifics of its politics were contingent on the specifics of municipal politics in each city. But they were also general: the politics of official rescue were the politics of the middle class, and so its emphasis on professional knowledge privileged the middle class. In Salem, the relief effort became a tool of Salem's Yankee elite in its political battle against an Irish mayor. In Halifax, the relief committee continued a national trend that chose efficiency, expertise, and state power over democratic participation. The patterns set in the initial relief period held in rebuilding, as when the Halifax Relief Commission worked to dominate building tradesmen and the Salem Rebuilding Commission forbade cheap apartments in the name of safety.

Both for survivors and for the state, aid represented a series of trades. Aid recipients got material resources in exchange for granting the state power over their lives. Of course, the loss of homes, jobs, and possessions in the disasters made this not a free choice. Thus, *sinistrés* lived in Forest River Camp, accepting shelter and food in exchange for a loss of autonomy and privacy. Halifax survivors took cash and other relief but had to accept investigators' snooping and social workers' judgments. Those with more resources—those with greater social capital or economic capital—suffered less intrusion. Poor women like Rita Mariggi faced the most difficulty balancing their personal autonomy with their need for the state's assistance.

For the state, there were also trade-offs. The legibility that social workers and other relief authorities tried to impose on their cities allowed the state to function, but it also permitted new demands. This explains the contradiction we saw in Salem, where the state both wanted people in militarized refugee camps, because that made them easier to control, while at the same time it feared that the camps would become permanent. In Halifax, we see it in the handling of housework pensions and family economies. On one hand, social workers tried to understand these complex, informal economies. This meant imposing legibility and, especially, assigning a monetary value to all labor, including domestic labor. At the same time, the Relief sought to limit the state's financial liability and responsibility. If it had succeeded in making all labor visible, legible, and monetized, it would have quickly gone bankrupt.

Individuals and families were not passive in the face of the state's sometimes contradictory claims. Although poor and working-class people had for years engaged in a complex balancing act and negotiation with the state in order to extract the maximum resources while giving up the least autonomy, the Progressive Era was new in that the state signaled that it would now actively pursue rescue and relief as among its core objectives. As the state sought to do more for its citizens, citizens responded by demanding more while at the same time seeking to maintain the practices of informal solidarity in which they had long engaged.

The Progressive Era saw a change in the way states interacted with and understood their citizens. The technocratic state, with its fetishization of experts, desired to help, regulate, and intervene. With the state expanding, civil society faced new challenges, which different sorts of organizations handled variously. Clergy, with their advanced degrees, professional titles, and well-respected positions, had knowledge and authority that was useful both to the technocrats and to their congregants. Although parishioners continued to build their own meanings into religious ritual and spaces, as they did with United Memorial Church, they used the authority of their pastors to press their material demands for relief. The relief authorities in both cities, likewise, used clergy to give them the local knowledge they needed to judge applicants.

Unions had less ability than churches to adapt to the new technocratic order because the state was more interested in what it presumed to be the disinterested local knowledge of priests and ministers than in the type of knowledge unions could offer. Unions tried various tactics to remain relevant: in Halifax, they broadened their membership and their concerns, most notably to cover reproductive questions of housing and schooling. Ultimately, this was unsuccessful, as the failed 1920 election and strike demonstrated. In Salem, workers at the Naumkeag Steam Cotton Company experimented with collaboration and cooperation with their bosses. In the factory, the union's knowledge was useful to eke out ever greater productivity. This attempt failed too, though, when autocratic union leaders went too far and workers rebelled in 1933.

Though formal and informal organizations evidenced continuity before, during, and after the disasters, they also changed and adapted. The friends who went with each other to volunteer in Halifax hospitals not only relied on their preexisting social networks, but they also renegotiated status, hierarchy, and roles once they were working with strangers in the hospital. Parents, siblings, and grown children who opened their houses to shelter their relatives had to alter their relationships with their family members who came to depend on them. Churches and unions found they had to change and adapt to new conditions, as when the pastor of St. Joseph's Parish in Salem fought to keep his parishioners. Halifax's unions and churches, too, tried to adapt—successfully and unsuccessfully—to their new terrain. Changes were most obvious for the members of what became United Memorial Church, who built a new sacred institution as a memorial to their dead loved ones. The Halifax Relief Commission represented a new type of state intervention, dominated by professionals and their disinterested expertise, and churches, unions, and families had to work with and around it.

At the center of the conflict between these two styles of aid—professional and lay, hierarchical and reciprocal, formal and informal—lay questions of labor. Charity was often a way to discipline workers. In Salem, this was evident in authorities' fear of idle workmen in relief camps; in Halifax it was visible in social workers' careful monitoring of survivors' employment. Reproductive labor was also to be disciplined, again through both military coercion and social workers' more subtle power. Meanwhile, survivors added to their labors a new type of work: the managing of relief authorities and the maximizing of relief aid. Halifax workers grew used to dealing with the relief commission within the family, in the neighborhood, and at the job, and they built a labor politics that included reproductive questions. Likewise, Salem workers learned from the fire the lessons they used in the economic crisis of deindustrialization.

Understanding the Halifax Relief Commission as the apex of the progressive, technocratic state highlights the central irony of progressive governance. The middle-class and elite outsiders who tried to help—from May Sexton and Jane Wisdom in Halifax to Frank Graves and Montayne Perry's "Food Man" in Salem—were sincere and heartfelt in their desire to relieve and rescue the disasters' victims. With the possible exceptions of George S. Campbell's smug lack of interest in Halifax and Captain Blanchard's crushing hostility in Salem, the problem was not ill will but structural inability. They thought that their training, experience, and especially their status as outsiders meant that they possessed the dispassionate expertise to most efficiently and effectively dispense relief and rescue the sufferers. Their centralized and hierarchical relief, however, had the effect of disrupting the informal solidarity practiced by the sufferers themselves. It was their very outsiderness that rendered them unable to see or understand the patterns of support and solidarity that they were disrupting.

When Salemites clashed with National Guardsmen over the rules in refugee camps and when Haligonians refused to leave their wrecked houses to stay in an official shelter, survivors appeared to reject the centralized, formalized, and hierarchical relief the state offered them. In most cases this was not explicit, intentional resistance to the state. Rather, by demanding relief on their own terms and by, when they could, choosing solidarity over charity, the *sinistrés* and explosion survivors resisted the ideological, intellectual, and organizational trends of the Progressive Era. In a period of increased middle-class dominance, of the rise of the professional expert, and of the ever-growing emphasis on managerial efficiency, survivors' choices to reject those things constitute resistance to the growing progressive state. Instead, they worked to craft a disaster citizenship, in which they combined the solidarity they enacted with their colleagues with demands for state assistance on their own terms.

Halifax's North End and Salem's Point district were both borderlands neighborhoods. Their obvious differences—one in Canada, one in the United States; one French Canadian, one English Canadian; one a city at war, one in peace; one a city that sent migrants, one that received them—pale in comparison to their similarities. They were linked by people, by ideas, and by region. Halifax's progressive reformers looked toward Boston at least as much as they looked at Toronto or Montreal—as when they started their civic beautification campaign, built supervised playgrounds, or sought to hire a woman police officer—just as Nova Scotia's emigrants picked as their destination the Boston States more than they did Upper Canada. Studying Salem and Halifax together brings to the fore the counterintuitive contours of the Maritimes and Quebec diasporas and the ways Salemites and Haligonians understood and experienced migration and the border. The stories of the Salem fire and Halifax explosion emphasize the importance of studying American and Canadian history together, not only comparatively but as a transnational, North American whole. Crucially, this is as important for American history, often unaware of Canada, as it is for Canadian history, so often hyperaware of the United States.

Ties of literal and figurative kinship bound together Massachusetts and Nova Scotia. These ties encouraged people living in the former to donate money to Halifax relief and sometimes even to travel there to help. More important, their donations allowed Haligonians who felt slighted by the insufficient aid they received from the relief commission to appeal through another channel. Despite the national border that ran between the Maritimes and New England, migration, family ties, and the Massachusetts-Halifax Relief Committee helped to create a political community that existed in both countries. In figure 15 (p. 120), effectively a map of that political community, the density of inquiries from New England, especially in comparison to those from central and western Canada, highlights the importance of understanding the Maritimes in their North American, rather than purely Canadian, context. This does not mean that the Canadian context is unimportant—the religious and political choices made by Haligonians in their unions and churches were particularly Canadian—but it does mean that a national perspective is decidedly incomplete.

If the explosion shows the importance of diaspora to the lives of those still in Halifax, the Salem fire shows the perhaps surprising lack of importance of the broader diaspora to Franco-Americans in Salem. Refugees from the Point for the most part chose to stay in neighboring cities rather than go to larger French Canadian diasporic centers. French Canadians elsewhere showed a corresponding lack of interest, raising little money for their compatriots' benefit and, in the case of the Rimouski newspaper, not even covering the fire. This was not, however, evidence of assimilation, since French Canadians in Salem fought to retain their ethnic parish and crafted a particularly French Canadian industrial union. Paradoxically, although diaspora is by definition a translocal phenomenon, Salem's French Canadians experienced diaspora deeply locally.

. . .

Disasters exposed the tensions of the Progressive Era and the growth of the interventionist state. The state and its actors in the military and in civilian relief bureaucracies sought to impose order on what they imagined to be a chaotic social landscape. This brought them into conflict with the very people whom they sought to help. Working-class people had durable and effective modes of support of their own, and although they wanted the increased material resources that the state brought, they did not want to cede power to the state. The citizenship they built—what I have called *disaster citizenship*—was complex and perhaps contradictory; it was not a program, but rather a set of themes expressed in a series of contestations, negotiations, and compromises. Some of these took place at societal or institutional levels, and some took place individually. Workers in Halifax sought to have a voice in the relief process through their unions, but they were repeatedly rebuffed. In Salem, French Canadians sought to influence municipal government by backing the successful recall of Mayor John F. Hurley. Individuals and families contested the order imposed by the state either literally by demanding more relief or implicitly through the choices they made to rely more on their families than on officials. They made choices on geographical scales that did not easily correspond to governance, by organizing shelters with their neighbors and building political communities across the US-Canada border. The ordered altruism of the initial shock of disaster helped them see connections with new people, and they experimented with organizing in different, more inclusive ways and on broader issues. "The people" for whom reformers claimed to speak had their own alternative modes of support and rescue that they quickly and effectively mobilized in times of crisis, but which remained illegible to elites. The objects of reform responded by creating a citizenship that simultaneously resisted and used the progressive state.

While the Salem and Halifax disasters were particular to their era, they offer lessons for contemporary disaster relief. The history of disaster relief helps us better understand the conflicts and tensions inherent in it. Rescue and relief are unavoidably political—that is, not simply a technical challenge—because they are inherently about the distribution of society's resources and about power. Planners, reformers, and relief professionals should be humble, remembering that the objects of their assistance have local knowledge that is inaccessible to them. Practically, the Salem and Halifax disasters explain why outside and hierarchical relief is sometimes unsuccessful, and why some people may appear to reject the help offered by outsiders. Because of the importance of organic and spontaneous organizations and communities in responding to disasters, the ingredients for healthy and strong communities in ordinary times—high social capital and dense social networks—also make them resilient to disasters. The best disaster policy is to build strong, multilayered communities in which friends, neighbors, and family members look out for each other and have the resources to help in an emergency. This means, among other things, building and protecting organizations, especially unions, that teach and practice solidarity.

The Salem and Halifax disasters represent a critical turning point in the history of urban destruction. In nineteenth-century urban North America, industrial fires had been common. In 1918, a Canadian government researcher counted 528 urban conflagrations worldwide between 1815 and 1915, causing $2 billion in damages. The United States and Canada accounted for 55 percent of these, a total of 290 great fires, costing a total of $1.4 billion.[10] Although the decline of candlelight, wood heat, and volunteer firefighting had, by the mid-nineteenth century, changed the nature of urban fire to make each one a less threatening prospect, urban conflagrations continued.[11] In the forty years prior to Salem's fire, Chicago (1871); Boston (1872); Seattle (1889); St. John's, Newfoundland (1892); Hull and Ottawa (1900); Jacksonville, Florida (1901); Toronto (1904); Baltimore (1904); San Francisco (1906); and Chelsea, Massachusetts (1908), among others, had all suffered major conflagrations. Changes in building and firefighting led to a drastic decline in large-scale urban fires in the twentieth century. If the Salem fire was not the last of its kind, it was among the last.[12] That at most six people died in Salem testifies to the strides made in containing and fighting fires, even when fire departments could not ultimately save property.[13]

In contrast to the industrial accidents of the nineteenth century, the iconic form of urban destruction in the twentieth century was intentional bombing. Understood, as it was, as part of World War I, the Halifax explosion stood at the precipice of a new age of disaster, between the preaerial urban destruction of the American Civil War and the Spanish shelling of Valparaiso (1866) and the explosive horrors of twentieth-century total war. Standing next to Haligonians on that edge were the victims of colonial wars in which European powers had begun dropping bombs from airplanes just before the start of World War I. In the 1930s and '40s, bombers attacked cities not for their military or strategic value, but for the ideological and propaganda value of the wanton death and destruction. First in Chechaouen in Spanish Morocco, then in Guernica, and then, most devastatingly, in Coventry, Dresden, Tokyo, Hiroshima, and Nagasaki, war planes sought not to disrupt the enemy's armies but to destroy their people and their cities. As in Halifax, they seemed to do so in an instant. Area bombing destroyed 39 percent of Germany's total urban area and an astounding 50 percent of Japan's. Neither these statistics nor the equally startling numbers of the dead and bombed out (60,595 dead and 750,000 homeless in the United Kingdom; 550,000 and 7.5 million, respectively, in Germany; and 500,000 and

8.3 million in Japan) adequately conveys the destruction of families, communities, and institutions that came from those urban attacks. The Halifax explosion presaged the technology and ideology of twentieth-century wars, which created a special brand of horror and made their urban destructions starkly different from the industrial fires of the nineteenth century.[14]

If urban destruction in the nineteenth century was largely a result of industrial accidents and that of the twentieth century from war, the twenty-first century will likely be a period of meteorological and seismological disasters. The chronic effects of global warming, especially coastal erosion and rising sea levels, plus other development and construction choices, have left cities less able to withstand extreme storms and floods.[15] Climate change has already brought more frequent and worse extreme-weather events, including floods, heat waves, droughts, and storms.[16] In other words, extreme weather will continue to become more common and worse, and our ability to handle what comes will degrade. As always, the social effects of these "natural" disasters are felt most by the poor, both globally and within developed countries.[17]

Seismological disasters, too, will likely become worse this century. In the twentieth century, urbanization in relatively rich countries led to a decline in the fatality rates of earthquakes, thanks to improved building codes and other infrastructural advantages of cities. But urbanization in poorer countries has not been accompanied by stronger buildings and better infrastructure. To the contrary, poor cities globally are marked by construction that even in ordinary times is shoddy and dangerous and by infrastructure that is inadequate even absent a disaster. The massive growth of cities in poor, seismically active countries, something that happened rapidly in the latter part of the twentieth century and will only accelerate in the twenty-first, will almost certainly lead to an increased rate of fatalities per earthquake. Cities in what geologists call the Alpine/Himalayan/Indonesian collision zone that runs along the southern edge of the Eurasian tectonic plate—a region where, already, 85 percent of all earthquake fatalities occur—are in particular danger. The Port-au-Prince earthquake in January 2010 provided a shocking example of what powerful earthquakes can do to cities in poor countries.[18] Meanwhile, the same erosion that has made coastal regions more susceptible to storms also makes them more susceptible to tsunamis. As with the meteorological disasters that will come as a result of global warming, this century's earthquakes will be worse for poorer countries than for richer ones and worse still for the comparatively poor within each country.

New Orleans's and Port-au-Prince's disasters were precipitated by natural events—Hurricane Katrina and an earthquake. But in both cases, the human systems designed to prevent or relieve disaster failed catastrophically. New Orleans flooded only when the government's levees failed. Haitians continue to suffer from a wholly preventable cholera epidemic brought by United Nations soldiers there to help after the earthquake.[19] These human failures point to the need to better understand not just the resources and benefits states bring to rescue, but also their blind spots and detriments, and what they erase, flatten, and take away.

Our current, neoliberal historical moment often seems to erase politics through appeals to the supposed naturalness of the market and individual competition. "Solutionism," as Evgeny Morozov has called it, seeks technical and engineering solutions to social and political problems; meanwhile, the erosion of the public sphere and welfare state—indeed, the systematic disassembly and ideological devaluation of all forms of social solidarity—have left only military and police institutions and logics to respond to crises. Contemporary technocracy, then, is different from the technocratic ideology that arose in the Progressive Era. Rather than granting power to social workers in disaster's aftermath, neoliberal technocracy gives power to soldiers, businessmen, economists, and programmers. Yet state and official disaster response in both eras have similarities: an emphasis on hierarchy, legibility, and order; a devaluation of local, unofficial, and horizontal responses; an emphasis on the coercive power of the state; and a reliance on engineers' responses and ways of knowing.[20]

Contemporary counter-responses to official disaster relief are likewise different from their Progressive-Era counterparts. Yet the history of technocratic state disaster relief and the response to it helps to clarify the spontaneous organization we saw in New Orleans, Port-au-Prince, New York, and the sites of other contemporary disasters.[21] In New Orleans, we saw a forceful rejection of neoliberal attacks on social solidarity in Common Ground, which started by organizing medical assistance and self-defense against white vigilantism and ended up modeling a new form of urban citizenship.[22] In Port-au-Prince, Haitians reinscribed long-standing practices of mutual aid and solidarity into postdisaster politics.[23] In New York, antiausterity and democracy protesters from the Occupy Wall Street movement remobilized as Occupy Sandy, a radically democratic and strikingly helpful disaster-response organization.[24] These projects do not look like the Progressive-Era disaster citizenship I have described, but they are also new forms of citizenship created in response to new forms of governance. By helping us to see how disasters are moments of empathy and solidarity that can take on political meaning, the stories of how Salemites and Haligonians crafted their disaster citizenship provide, ironically, some hope for our own era of disaster.

NOTES

Abbreviations

ACRC	A. C. Ratshesky Collection, P-586, American Jewish Historical Society, Boston, Mass.
ADR	*Acadian Daily Recorder*
AHQCR	1903 Army Headquarters Central Registry, RG24-C-1-a, Department of Defence fonds, LAC
AMF	Archibald MacMechan fonds, vol. 2124, MG 1, NSA
AMP	Archibald MacMechan papers, MS-2-82, Archives and Special Collections, Killam Library, Dalhousie University
AN	*L'Avenir National*
ARC	American Red Cross, Central Files, 1888–1916, Box 56, RG 200/130/77/1/1, National Archives II, College Park, Md.
CCH	Commercial Club of Halifax fonds, vol. 91, MG 20, NSA
CdS	*Courrier de Salem*
DE	*Daily Echo*
EJM	"Letters from Archbishop" folder, box 4, Archbishop Edward J. McCarthy fonds, Archives of the Halifax Archdiocese, Halifax, Nova Scotia
EM	*Evening Mail*
HEC	Halifax Explosion Collection, MG 27, NSA
HH	*Halifax Herald*
HRCC	Series C, Correspondence, Halifax Relief Commission fonds, MG 36, NSA
HRCP	Series P, Pension Files, Halifax Relief Commission fonds, MG 36, NSA
JKF	Janet Kitz fonds, accession 2007–006, NSA
KOH	Janet Kitz oral histories, Halifax Explosion Memorial Bells Committee Collection, no. FSG 31, MF 298, NSA
LAC	Library and Archives Canada, Ottawa, Ontario
LCWH	Local Council of Women of Halifax fonds, NSA
MC	*Morning Chronicle*

MD6	Records of Military District No. 6, RG 24-C-8, Department of National Defence fonds, LAC
M-HRC	Massachusetts Halifax Relief Committee correspondence and papers, Special Collections Division, Massachusetts State Library, Boston, Mass.
MNGMA	"Massachusetts National Guard—Salem Fire—Salem Mass 1914—Correspondence—Box 1," Massachusetts National Guard Museum and Archives, Worcester, Mass.
NARA I	Archives I, National Archives and Records Administration, Washington, D.C.
NDR	Directors' Records, 1905–1938, vol. AB-3, Naumkeag Steam Cotton Company records, Baker Business Library, Harvard University, Allston, Mass.
NSA	Nova Scotia Archives, Halifax, Nova Scotia
PC	Parish correspondence files, 1907–1940, RGIII.C.3, Archives of the Roman Catholic Archdiocese of Boston, Braintree, Mass.
PEM	Phillips Library, Peabody Essex Museum, Salem, Mass.
RBP	Sir Robert Borden Papers, MG 26, LAC
RCNCR	Royal Canadian Navy First Central Registry System, RG24-D-1-a, Department of National Defence fonds, LAC
RS	A. C. Ratshesky scrapbook, item 3, box 4, ACRC
SEN	*Salem Evening News*
SEO	*Saturday Evening Observer*
SNP	Smith-Nyman Papers, Manuscripts and Archives, Sterling Memorial Library, Yale University, New Haven, Conn.
St.GAP	St. George's Anglican Parish fonds, MG 4, NSA
St.JAB	St. John's Ambulance Brigade (Nova Scotia Council) fonds, 2003-039/003, MG 20, NSA
St.MAC	St. Matthias Anglican Church fonds, MG 4, NSA
St.PP	Archives of St. Paul's Parish, Halifax, Nova Scotia
USJBA	Union St.-Jean-Baptiste d'Amérique Corporate Archives, Woonsocket, R.I.
USJBAS	Union St.-Jean-Baptiste d'Amérique scrapbooks, Institut Français, Emmanuel d'Alzon Library, Assumption College, Worcester, Mass.
YMCAS	Scrapbook collection, Salem YMCA, Salem, Mass.

Introduction

1. Joseph B. Treaster, "At Stadium, a Haven Quickly Becomes an Ordeal," *New York Times*, 1 September 2005; Julie Mason and Michael Hedges, "Washington Sends More Troops, Begs for Patience," *Houston Chronicle*, 2 September 2005.

2. *Washington Post*, 2 September 2005. That day, the *New York Times* used the same words to describe New Orleans on its front page.

3. Larry Bradshaw and Lorrie Beth Slonsky, "The Real Heroes and Sheroes of New Orleans," *Socialist Worker* 9 September 2005, 4–5, available at http://www.socialistworker.org/2005-2/556/556_04_RealHeroes.shtml, accessed 27 August 2012; "After the Flood,"

This American Life, episode 296, 9–11 September 2005, available at http://www.thisamerican-life.org/radio-archives/episode/296/After-the-Flood, accessed 12 December 2013. See W. Joseph Campbell, *Getting It Wrong: Ten of the Greatest Misreported Stories in American Journalism* (Berkeley: University of California Press, 2010), 163–83.

4. Christopher Capozzola, *Uncle Sam Wants You: World War I and the Making of the Modern American Citizen* (New York: Oxford University Press, 2008).

5. Kathy Peiss, *Cheap Amusements: Working Women and Leisure in Turn-of-the-Century New York* (Philadelphia: Temple University Press, 1986), 5, 7–8; Susan A. Glenn, *Daughters of the Shtetl: Life and Labor in the Immigrant Generation* (Ithaca, N.Y.: Cornell University Press, 1990), 3–4; Sarah Deutsch, *Women and the City: Gender, Space, and Power in Boston, 1870–1940* (New York: Oxford University Press, 2000); Carolyn Strange, *Toronto's Girl Problem: The Perils and Pleasures of the City, 1880–1930* (Toronto: University of Toronto Press, 1995).

6. Stuart D. Brandes, *American Welfare Capitalism, 1880–1940* (Chicago: University of Chicago Press, 1976), 1–9; David Montgomery, *The Fall of the House of Labor: The Workplace, the State, and American Labor Activism, 1865–1925* (Cambridge: Cambridge University Press, 1987).

7. Roger Finke and Rodney Stark, *The Churching of America, 1776–2005: Winners and Losers in Our Religious Economy* (New Brunswick, N.J.: Rutgers University Press, 2005), 123–24, 151–55; Jay P. Dolan, *The American Catholic Experience: A History from Colonial Times to the Present* (Notre Dame, Ind.: University of Notre Dame Press, 1992), chap. 6.

8. Bradley R. Rice, "The Galveston Plan of City Government by Commission: The Birth of a Progressive Idea," *Southwestern Historical Quarterly* 78, no. 4 (April 1975): 365–408.

9. Gregory Clancey, "Toward a Spatial History of Emergency: Notes from Singapore," in *Beyond Description: Singapore Space Historicity*, ed. Ryan Bishop, John Phillips, and Wei-Wei Yoo (London: Routledge, 2004), 55–56, n. 36; Minami Orihara and Gregory Clancey, "The Nature of Emergency: The Great Kanto Earthquake and the Crisis of Reason in Late Imperial Japan," *Science in Context* 25, no. 1 (2012): 103–26.

10. John M. Barry, *Rising Tide: The Great Mississippi Flood of 1927 and How It Changed America* (New York: Simon and Schuster, 1997).

11. Michele Landis Dauber, *The Sympathetic State: Disaster Relief and the Origins of the American Welfare State* (Chicago: University of Chicago Press, 2012).

12. For more recent disasters as generative of citizenship claims, see Adriana Petryna, *Life Exposed: Biological Citizenship after Chernobyl* (Princeton, N.J.: Princeton University Press, 2002); Emily Gilbert and Corey Ponder, "Between Tragedy and Farce: 9/11 Compensation and the Value of Life and Death" *Antipode* 46, no. 2 (March 2014): 404–25.

13. Gunther Paul Barth, *City People: The Rise of Modern City Culture in Nineteenth-Century America* (New York: Oxford University Press, 1980), 149; Grace Elizabeth Hale, *Making Whiteness: The Culture of Segregation in the South, 1890–1940* (New York: Vintage Books, 1999), 199–239.

14. Karen Sawislak, *Smoldering City: Chicagoans and the Great Fire, 1871–1874* (Chicago: University of Chicago Press, 1995); Andrea Rees Davies, *Saving San Francisco: Relief and Recovery after the 1906 Disaster* (Philadelphia: Temple University Press, 2012).

15. Dauber, *Sympathetic State*; Gareth Davies, "The Emergence of a National Politics of Disaster, 1865–1900," *Journal of Policy History* 26, no. 3 (2014), 305–26.

16. ADR, 2 January 1918, 2.

17. Craig Heron and Myer Siemiatycki, "The Great War, the State, and Working-Class Canada," in *The Workers' Revolt in Canada, 1917–1925*, ed. Craig Heron (Toronto: University of Toronto Press, 1998), 14–15; J. A. Corry, "The Growth of Government Activities in Canada, 1914–1921," *Canadian Historical Association Annual Report* (1940): 63–73; David Edward Smith, "Emergency Government in Canada," *Canadian Historical Review* 50, no. 4 (December 1969): 429–48.

18. "Certified Copy of a Report of the Committee of the Privy Council, Approved by His Excellency the Governor General on the 9th of March 1918," item 23, AMF.

19. John Griffith Armstrong, *The Halifax Explosion and the Royal Canadian Navy: Inquiry and Intrigue* (Vancouver: University of British Columbia Press, 2002), 191, 196.

20. Morris Mott, "Tobias C. Norris," in *Manitoba Premiers of the 19th and 20th Centuries*, ed. Barry Ferguson and Robert Vardhaugh (Regina, Saskatchewan: Canadian Plains Research Center Press), 145–49; William Lewis Morton, *The Progressive Party in Canada* (Toronto: University of Toronto Press, 1950), 33.

21. Jill L. Grant, Leifka Vissers, and James Haney, "Early Town Planning Legislation in Nova Scotia: The Roles of Local Reformers and International Experts," *Urban History Review* 40, no. 2 (Spring 2012), 3–14; John English, *The Decline of Politics: The Conservatives and the Party System 1901–20*, 2nd ed. (Toronto: University of Toronto Press, 1993).

22. Tamara K. Hareven, "An Ambiguous Alliance: Some Aspects of American Influences on Canadian Social Welfare," *Histoire sociale/Social History* 2 (April 1969): 82–98.

23. Theda Skocpol, *Protecting Soldiers and Mothers: The Political Origins of Social Policy in the United States* (Cambridge, Mass.: Belknap Press, 1992); Linda Gordon, *Pitied but Not Entitled: Single Mothers and the History of Welfare, 1890–1930* (New York: Free Press, 1994), 7; Michael B. Katz, *Shadow of the Poorhouse: A Social History of Welfare in America*, 10th anniv. ed. (New York: Basic Books, 1996), 113.

24. Veronica Strong-Boag, "'Wages for Housework': Mothers' Allowances and the Beginnings of Social Security in Canada," *Journal of Canadian Studies* 14, no. 1 (Spring 1979): 24–34; James Struthers, *No Fault of Their Own: Unemployment and the Canadian Welfare State, 1914–1941* (Toronto: University of Toronto Press, 1983), 12–43; Elisabeth Wallace, "The Origin of the Social Welfare State in Canada, 1867–1900," *Canadian Journal of Economics and Political Science* 16, no. 3 (August 1950): 383–93.

25. Michele Dauber makes a similar argument in more detail. See her *Sympathetic State*, 15.

26. Jay Driskell, *Schooling Jim Crow: The Fight for Atlanta's Booker T. Washington High School and the Roots of Black Protest Politics* (Charlottesville: University of Virginia Press, 2014); Jessica Toft and Laura S. Abrams, "Progressive Maternalists and the Citizenship Status of Low-Income Single Mothers," *Social Science Review* 78, no. 3 (September 2004): 447–65.

27. Katz, *Shadow of the Poorhouse*, ix–x, 154–57; Gordon, *Pitied but Not Entitled*, 76; Davies, "Emergence of a National Politics of Disaster," 319–20.

28. James C. Scott, *Seeing Like a State: How Certain Schemes to Improve the Human Condition Have Failed* (New Haven, Conn.: Yale University Press, 1998), 2–3.

29. Robert C. Post and Nancy L. Rosenblum, "Introduction," in *Civil Society and Government*, ed. Nancy L. Rosenblum and Robert C. Post (Princeton, N.J.: Princeton University Press, 2002), 1, 3; Marian Moser Jones, *The American Red Cross from Clara Barton to the*

New Deal (Baltimore: Johns Hopkins University Press, 2013), 81, 115; Julia F. Irwin, *Making the World Safe: The American Red Cross and a Nation's Humanitarian Awakening* (Oxford: Oxford University Press, 2013), 4–5.

30. Pierre Bourdieu, "The Forms of Capital," in *Handbook of Theory and Research for the Sociology of Education*, ed. J. G. Richardson (New York: Greenwood, 1985), 248; Robert Putnam, *Making Democracy Work: Civic Traditions in Modern Italy* (Princeton, N.J.: Princeton University Press, 1993), 167. For a useful overview of social capital in the disaster studies literature, see Michelle Annette Meyer, "Social Capital and Collective Efficacy for Disaster Resilience: Connecting Individuals with Communities and Vulnerability with Resilience in Hurricane-Prone Communities in Florida" (PhD diss., Colorado State University, 2013), 30–38.

31. Kai T. Erikson, *Everything in Its Path: Destruction of Community in the Buffalo Creek Flood* (New York: Simon and Schuster, 1976).

32. Daniel P. Aldrich, *Building Resilience: Social Capital in Post-Disaster Recovery* (Chicago: University of Chicago Press, 2012).

33. Eric Klinenberg, *Heat Wave: A Social Autopsy of Disaster in Chicago* (Chicago: University of Chicago Press, 2002).

34. Staughton Lynd, "Communal Rights," *Texas Law Review* 62, no. 8 (May 1984): 1423–30; Jacob A. C. Remes, "Solidarity, Citizenship, and the Opportunities of Disasters," in *Labor Rising: The Past and Future of Working People in America*, ed. Daniel Katz and Richard Greenwald (New York: New Press, 2012), 144–45.

35. Daniel T. Rodgers, "In Search of Progressivism," *Reviews in American History* 10, no. 4 (December 1982): 126. The classic text on this aspect of the Progressive Era is Robert H. Wiebe, *The Search for Order, 1877–1920* (New York: Hill and Wang, 1967).

36. Amy Bridges, *Morning Glories: Municipal Reform in the Southwest* (Princeton, N.J.: Princeton University Press, 1999).

37. Montgomery, *Fall of the House of Labor*, 217.

38. Julie Greene, *The Canal Builders: Making America's Empire at the Panama Canal* (New York: Penguin Press, 2009); Jessica B. Teisch, *Engineering Nature: Water, Development, and the Global Spread of American Environmental Expertise* (Chapel Hill: University of North Carolina Press, 2011).

39. John Herd Thompson and Stephen J. Randall, *Canada and the United States: Ambivalent Allies*, 4th ed. (Athens: University of Georgia Press, 2008), 73.

40. Michael Willrich, *City of Courts: Socializing Justice in Progressive Era Chicago* (Cambridge: Cambridge University Press, 2003).

41. Paul Starr, *The Social Transformation of American Medicine* (New York: Basic Books, 1982); Willrich, *City of Courts*, 19.

42. Kathleen D. McCarthy, *Noblesse Oblige: Charity and Cultural Philanthropy in Chicago, 1849–1929* (Chicago: University of Chicago Press, 1982), 126.

43. Barbara Clow, *Negotiating Disease: Power and Cancer Care, 1900–1950* (Montreal and Kingston: McGill–Queen's University Press, 2001), esp. chap. 4.

44. Evelyn Savidge Sterne, *Ballots and Bibles: Ethnic Politics and the Catholic Church in Providence* (Ithaca, N.Y.: Cornell University Press, 2003); Evelyn Higginbotham, *Righteous Discontent: The Women's Movement in the Black Baptist Church, 1880–1920* (Cambridge, Mass.: Harvard University Press, 1994).

45. Robin D. G. Kelley, *Race Rebels: Culture, Politics, and the Black Working Class* (New York: Free Press, 1996).

46. Daniel T. Rodgers, *Atlantic Crossings: Social Politics in a Progressive Age* (Cambridge, Mass.: Belknap Press, 1998).

47. Morton, *Progressive Party in Canada*, 30–31; John H. Thompson, "American Muckrakers and Western Canadian Reformers," *Journal of Popular Culture* 4, no. 4 (Spring 1971): 1060–70.

48. On the role of humanitarian aid in American foreign relations, see Irwin, *Making the World Safe*.

49. Richard Plender, *International Migration Law*, 2nd rev. ed. (Dordrecht: Martinus Nijhoff, 1988), 66–78.

50. Thomas A. Klug, "The Immigration and Naturalization Service (INS) and the Making of a Border-Crossing Culture on the U.S.-Canada Border, 1891–1941," *American Review of Canadian Studies* 40, no. 3 (2010): 395–415; Jacob Remes, "Movable Type: Toronto's Transnational Printers, 1867–1872," in *Workers across the Americas: The Transnational Turn in Labor History*, ed. Leon Fink (New York: Oxford University Press, 2011), 402; Mae M. Ngai, *Impossible Subjects: Illegal Aliens and the Making of Modern America* (Princeton, N.J.: Princeton University Press, 2004), 64–67; Bruno Ramirez, *Crossing the 49th Parallel: Migration from Canada to the United States, 1900–1930* (Ithaca, N.Y.: Cornell University Press, 2001), 36–40; Sarah-Jane Mathieu, *North of the Color Line: Migration and Black Resistance in Canada, 1870–1955* (Chapel Hill: University of North Carolina Press, 2010), 27–33; Hugh J. M. Jackson, *The Voyage of the* Komagata Maru: *The Sikh Challenge to Canada's Colour Bar*, expanded ed. (Vancouver: University of British Columbia Press, 2014).

51. John F. Moors to Mr. [August] Cunningham, 3 September 1914, "File 835.08—MASSACHUSETTS, Salem Fire—Reports, 6/25/1914," ARC.

52. French Canadian descendants, as historian Mark Richard calls them, had many changing names for themselves when they were in the United States. In the time I write about, they, or at least their ethnic spokespeople, usually called themselves *Franco-Américains*. Sources produced by Anglophone Americans sometimes used *Franco-American*, sometimes *French Canadian*, and sometimes simply *French*. I refer to them interchangeably as *Franco-Americans* and *French Canadians*. See Richard, *Loyal but French: The Negotiation of Identity by French-Canadian Descendants in the United States* (East Lansing: Michigan State University Press, 2008). Unless otherwise noted, all translations from French-language sources are my own.

53. The US Census Bureau estimated 46,994 residents slightly more than a month before the fire; SEN, 11 May 1914, 1. The Canadian Census found 46,616 Haligonians in 1911 and 58,372 in 1921; it is likely that Halifax's wartime population was at the higher end of the range. See Canada, *Sixth Census of Canada, 1921*, vol. 1, *Population* (Ottawa: F. A. Acland as King's Printer, 1924), 17.

54. Telegram, Frank C. Jordan to G. H. Murray, 27 December 1917, item 95; Murray to Secretary of State of Ohio, 10 January 1918, item 163; George F. Howard to Murray, 29 December 1917, item 118; and Albert P. Langley to Murray, 26 December 1917, item 94; all on reel 15,123, HEC. A few months later, Halifax officials again sought guidance from Massachusetts about how to establish a relief commission. See telegram, Robert T.

MacIlreith to A. C. Ratshesky, 9 February 1918, item 113.14, and MacIlreith to Ratshesky 18 February 1918, item 113.16; both in HRCC.

55. Frederick W. Jenkins to G. H. Murray, 27 December 1917, item 98, HEC.

56. John F. Moors, "Review of *Disasters and the American Red Cross in Disaster Relief*, by J. Byron Deacon," *The Survey* 39, no. 27 (26 January 1918): 472.

57. Janney Byron Deacon, *Disasters and the American Red Cross in Disaster Relief* (New York: Russell Sage Foundation, 1918).

58. On the book's cover page, Deacon is identified as on a leave of absence as the general secretary of the Philadelphia Society for Organizing Charity and the division director of civilian relief of Pennsylvania. A book review identified him as the Red Cross's assistant director general of civilian relief. By 1919 he was listed as the director general. See "Review of *Disasters*," *Journal of the Association of Collegiate Alumnae* 4 (April 1918): 551–52; and J. Byron Deacon, "The Future of the Red Cross Home Service," *Proceedings of the National Conference of Social Work* 46 (1920): 365–71.

59. On the shift in the Red Cross, see Jones, *American Red Cross*.

60. Deacon, *Disasters*, 13, 40, 26.

61. Ibid., 196.

62. Ibid., 93–94, 181–85.

63. Ibid., 127, 48.

64. Ibid., 60, 137.

65. Ibid., 84.

66. Ibid., 63–64.

67. Ibid., 175.

68. Ibid., 176–77.

69. Ibid., 181.

70. On Deacon's influence in Halifax, see Russell R. Dynes and E. L. Quarantelli, "The Place of the Explosion in the History of Disaster Research: The Work of Samuel H. Prince," in *Ground Zero: A Reassessment of the 1917 Explosion in Halifax Harbour*, ed. Alan Ruffman and Colin D. Howell (Halifax: Nimbus/The Gorsebrook Research Institute for Atlantic Canada Studies at Saint Mary's University, 1994), 63. Reviews or notices included those in *The Survey*, which offered to sell the book to readers; *Journal of the Association of Collegiate Alumnae*; *National Municipal Review* 7 (March 1918): 194; *South Atlantic Quarterly* 17 (April 1918): 182; *The Bookman* 46 (February 1918): 735; *New Republic*, 16 March 1918, 214; and Ethel Bird, "Review of *Disasters and the American Red Cross in Disaster Relief*, by J. Byron Deacon," *American Journal of Sociology* 24 (July 1918): 112–13.

71. Deacon, *Disasters*, 19; Jones, *American Red Cross*, 135–36.

72. Jones, *American Red Cross*, 116–75.

73. Jacob A. C. Remes, "'Committed as Near Neighbors': The Halifax Explosion and Border-Crossing People and Ideas," *American Review of Canadian Studies* 45, no. 1 (Spring 2015): 26–43; Michelle Hébert Boyd, *Enriched by Catastrophe: Social Work and Social Conflict after the Halifax Explosion* (Halifax, Nova Scotia: Fernwood, 2007); John C. Weaver, "Reconstruction of the Richmond District in Halifax: A Canadian Episode in Public Housing and Town Planning, 1918–1921," *Plan Canada*, March 1976, 36–38.

74. Clipping from *Boston Post*, 9 December 1917, RS; Jones, *American Red Cross* 116–22.

75. "Moors, John Farwell," *National Cyclopaedia of American Biography* 41:382–83 (Clif-

ton, N.J.: James T. White, 1891–); "John Farwell Moors, '83, Elected Fellow," *Harvard Alumni Bulletin* 20, no. 16 (17 January 1918): 280–81; Edward J. Kopf, "The Intimate City: A Study in Urban Social Order: Chelsea, Massachusetts, 1906–1915" (PhD diss., Brandeis University, 1974), 61–62, 66.

76. SEN, 3 July 1914, 9; clipping from *Christian Science Monitor,* 14 December 1917, RS; "List of Committees," undated, MNGMA; *Salem Register,* 17 July 1914, 1.

77. SEN, 26 June 1914, 5.

78. Joseph Scanlon, "Rescuers or Troublemakers? The Massachusetts Response to the 1917 Halifax Catastrophe," *Journal of the American Society of Professional Planners* (1999): 55–69; Janet F. Kitz, *Shattered City: The Halifax Explosion and the Road to Recovery* (Halifax, Nova Scotia: Nimbus, 1989), 84–87; Blair Beed, *1917 Halifax Explosion and American Response* (Halifax, Nova Scotia: Dtours Visitors and Convention Service, 1998), 18–30, 65–74.

79. On the coat, see [Ratshesky's brother] to Ratshesky, 11 December 1917, in RS.

80. Clipping from *Boston Globe,* 9 December 1917, RS.

81. Personal narrative of Mr. Dougal[d] MacGillivray, item 196, AMF; MC, 19 December 1917, 5. Christian Lantz's Christian name was sometimes incorrectly given as Christopher.

82. Clipping from *Boston Globe,* 23 December 1917, YMCAS.

83. CdS, 24 July 1914, 1.

84. Clipping from *Boston Globe,* 23 December 1917, YMCAS.

85. Unlabeled clipping, 13 December 1917; unlabeled clipping, 14 December 1917; both in YMCAS.

86. Newspaper advertisement enclosed with Arthur N. Phippen to Henry B. Endicott, 12 December 1917, reel 12, M-HRC.

87. Naomi Klein, *The Shock Doctrine: The Rise of Disaster Capitalism* (New York: Picador, 2007).

88. Kenneth Maxwell, *Pombal: Paradox of the Enlightenment* (Cambridge: Cambridge University Press, 1995), 17–18, 24–32; Alan Wolfe, *The Future of Liberalism* (New York: Knopf, 2009), 217–19, 223–25.

89. Rebecca Solnit, *A Paradise Built in Hell: The Extraordinary Communities That Arise in Disasters* (New York: Viking, 2009), 107.

Chapter 1. "Organization without Any Organization"

1. John Griffith Armstrong, *The Halifax Explosion and the Royal Canadian Navy: Inquiry and Intrigue* (Vancouver: University of British Columbia Press, 2002), 29–40.

2. Jay White, "Exploding Myths: The Halifax Harbour Explosion in Historical Context," in *Ground Zero: A Reassessment of the 1917 Explosion in Halifax Harbour,* ed. Alan Ruffman and Colin D. Howell (Halifax: Nimbus/The Gorsebrook Research Institute for Atlantic Canada Studies at Saint Mary's University, 1994), 251–274.

3. Hugh MacLennan, *Barometer Rising* (1941; repr. Toronto: McClelland and Stewart, 1989), 171.

4. Personal narrative of Miss Florence J. Murray, item 192, AMF.

5. Personal testimony of Mrs. F. H. Sexton, item 224, AMF.

6. Michelle Hébert Boyd, *Enriched by Catastrophe: Social Work and Social Conflict after the Halifax Explosion* (Halifax, Nova Scotia: Fernwood, 2007), 67. Boyd's numbers, being the most recent published estimates, rely on the painstaking work of the Nova Scotia Archives to identify every person who died from the explosion. However, the exact numbers are beside the point.

7. MacLennan, *Barometer Rising*, 4.

8. David Sutherland, "Halifax Harbour, December 6, 1917: Setting the Scene," in Ruffman and Howell, *Ground Zero*, 4.

9. Michael Bourdreau, *City of Order: Crime and Society in Halifax, 1918–35* (Vancouver: University of British Columbia Press, 2012), 18.

10. Henry Roper, "The Halifax Board of Control: The Failure of Municipal Reform, 1906–1919," *Acadiensis*, 14 no. 2 (1985): 46–95; Andrew Nicholson, "Dreaming of the 'Perfect City': The Halifax Civic Improvement League 1905–1949" (MA thesis, Saint Mary's University, 2000); D. A. Sutherland, "The Personnel and Policies of the Halifax Board of Trade, 1890–1914," in *The Enterprising Canadians: Entrepreneurs and Economic Development in Eastern Canada, 1820–1914*, ed. Lewis R. Fischer and Eric W. Sager (St. John's: Maritime History Group, Memorial University of Newfoundland, 1979), 205–29.

11. "Work Done by the Halifax Local Council of Women during the Presidency of Dr. Agnes Dennis, C.B.E. M.A.—1905—1920," Scrapbook, reel 20,204, LCWH.

12. Nicholson, "Dreaming of the 'Perfect City'"; Ernest R. Forbes, "Battles in Another War: Edith Archibald and the Halifax Feminist Movement," in *Challenging the Regional Stereotype: Essays on the 20th Century Maritimes* (Fredericton, New Brunswick: Acadiensis Press 1989), 69, 75–76.

13. "Work Done by the Halifax Local Council of Women during the Presidency of Dr. Agnes Dennis, C.B.E. M.A.—1905—1920," Scrapbook, reel 20,204, LCWH.

14. Nicholson, "Dreaming of the 'Perfect City,'" 52, 12, 43.

15. John English, *The Decline of Politics: The Conservatives and the Party System, 1902–20*, 2nd ed. (Toronto: University of Toronto Press, 1993).

16. Henry Roper, "Archibald MacMechan and the Writing of 'The Halifax Disaster,'" in Ruffman and Howell, *Ground Zero*, 85–99; Janet E. Baker, *Archibald MacMechan: Canadian Man of Letters* (Lockeport, Nova Scotia: Roseway, 2000); Archibald MacMechan, *The Halifax Explosion: December 6, 1917*, ed. Graham Metson (Toronto: McGraw-Hill Ryerson, 1978).

17. Early draft by MacMechan labeled "The Halifax Disaster," item 271, AMF.

18. Personal narrative by Lt. Ray Colwell and Lt. Arthur, item 133; personal narrative of Engr. Cmdr. Richard Howley, R.N., item 158; personal narrative of Capt. Lorne Allen, item 113; all in AMF.

19. Personal narrative of Dr. M. J. Burris, item 122, AMF.

20. Personal narrative of H. D. Brunt, item 119, AMF.

21. Personal narrative of Arthur S. Frye, item 147, AMF.

22. Lt. Eric Grant to MacMechan, 29 December 1917, item 151, AMF.

23. J. E. Furness to MacMechan, 10 April 1918, and enclosed narrative, items 54 and 55, AMF. Confusingly, Furness reported eighty casualties, but his subtotals added up to eighty-one.

24. On the longshoremen's union, see Catherine Ann Waite, "The Longshoremen of Halifax, 1900–1930; Their Living and Working Conditions" (MA thesis, Dalhousie University, 1977).

25. Lt. Eric Grant to MacMechan, 29 December 1917, item 151, AMF.

26. See reports from various officers, mostly lieutenants, items 260–69, AMF.

27. Personal narrative of Col. Ralph B. Simmon[d]s, item 225, AMF.

28. Personal narrative of C. Sutherland, item 231, AMF.

29. Personal narrative of Mrs. J. G. MacDougall, item 195, AMF.

30. Personal narrative of Col. Ralph B. Simmon[d]s, item 225, AMF.

31. Personal narrative of Lt. O. B. Jones, item 44, AMF.

32. Personal narrative of [Lt.] Col. Phinney, item 215, AMF.

33. "My Experience in the Halifax Disaster," catalogued as "Account Related to the Halifax Explosion," MG 30-E90, LAC.

34. Janet F. Kitz, *Shattered City: The Halifax Explosion and the Road to Recovery* (Halifax, Nova Scotia: Nimbus, 1989), 61–62.

35. Personal narrative of Mrs. Annie Anderson, item 114, AMF.

36. Personal narrative of Dr. W. W. Woodbury, item 238; Timeline of first day, by J. C. (Jim) Reid, to Mrs. J. G. MacDougall, item 195d-e; both in AMF.

37. "Victorian Order of Nurses," item 25, AMF.

38. Personal narrative of Mrs. H. Bryant, item 121, AMF.

39. Report, [R. V. Harris] to Assistant Commissioner, St. John's Ambulance Brigade Overseas in the Dominion of Canada, folder 1, St.JAB.

40. *Dalhousie Gazette*, 29 January 1918, 1, 2; "Dalhousie's Part in the Relief Work," item 26, AMF.

41. Jean L. Ross diary, reel 15,127, item 1, vol. 9, HEC.

42. *Dalhousie Gazette*, 29 January 1918, 1.

43. Personal testimony of Miss Josephine Creighton [Crichton], item 135, AMF; Jean L. Ross diary, reel 15,127, item 1, vol. 9, HEC.

44. Individual testimony of Mrs. John F. Stairs, item 48, AMF.

45. Clara Smith to "Katherine," 2 January 191[8], item 38, AMF. Emphasis in original.

46. Personal narrative of Rev. D. G. Cock, item 132, AMF.

47. Personal narrative of Warrena Maddin, item 180; Clara Smith to "Katherine," 2 January 191[8], item 38; both in AMF.

48. "Intermediate Medical Report re Halifax Disaster, Dec. 6, 1917," MG 27 (copy also in file 71-26-99-3, "Medical Services, Halifax Disaster," box 6359, AHQCR).

49. Personal testimony of Miss Josephine Creighton [Crichton], item 135, AMF.

50. Personal narrative of Marjorie Moir, item 188, AMF.

51. "Notes of Medical Relief Committee of Halifax Disaster," item 10, AMF.

52. Personal narrative of Miss Velma Moore, item 47, AMF.

53. Personal narrative of Christine MacKinnon, item 200, AMF. MacMechan's notes are unclear, but MacKinnon may have been a nursing student.

54. Personal narrative of Mrs. H. Bryant, item 121, AMF.

55. Jean L. Ross diary, reel 15,127, item 1, vol. 9, HEC. Emphasis in original.

56. Personal narrative of Miss Florence J. Murray, item 192, AMF; Ruth Compton

Brouwer, "'Home Lessons, Foreign Tests': The Background and First Missionary Term of Florence Murray, Maritime Doctor in Korea," *Journal of the CHA*, n.s., 6 (1996): 103–28.

57. Personal testimony of Rev. Hugh Upham, item 234, AMF.

58. Statement re Explosion at Halifax, J. P. D. Llwyd, Dean of Nova Scotia, item 168, AMF. Double underlining in original.

59. HH, 11 December 1917, 1; MC, 12 December 1917, 2; MC, 21 December 1917, 8; DE, 11 December 1917, 5.

60. J. H. Mitchell, description of burial service for unidentified dead, item 282, AMF.

61. "Notes of Medical Relief Committee of Halifax Disaster," item 10, AMF.

62. Document written by Miss Emily Brown, Providence, R.I., item 118, AMF.

63. Personal testimony of Miss Constance Bell, item 117, AMF.

64. David C. Miller, *Introduction to Collective Behavior and Collective Action*, 2nd ed. (Prospect Heights, Ill.: Waveland, 2000), 289.

65. Memorandum [of conversation with Henry S. Colwell], item 270; early draft by MacMechan labeled "The Halifax Disaster," item 271; both AMF; *Prominent People of the Maritime Provinces* (Montreal: Canadian Publicity, 1922), 39.

66. "Minutes of Morning Meeting," 6 December 1917, item 3–3c; W. A. Duff to Mac-Mechan, 5 February 1918, item 51; memorandum [of conversation with Henry S. Colwell], item 270; all in AMF.

67. MacMechan, *Halifax Explosion*, 51; "Minutes of Morning Meeting," 6 December 1917, item 3–3c, AMF; *Prominent People*, 88, 79–80.

68. "Minutes of Morning Meeting," 6 December 1917, item 3–3c, AMF.

69. Early draft by MacMechan, labeled "The Halifax Disaster," item 271, AMF.

70. "Minutes of Afternoon Meeting," 6 December 1917, item 3d-3e, AMF; [Bertram M. Chambers], "Halifax Explosion," *Naval Review* 7, no. 1 (1920): 451. Notably, Chambers had only been in Halifax for two weeks, and civilians at the meeting did not recognize him.

71. "The Fire Department," notes by J. H. Mitchell based on conversation with Controller Hines, item 20, AMF.

72. Memorandum [of conversation with Henry S. Colwell], item 270, AMF.

73. Personal narrative of Mrs. H. Bryant, item 121, AMF.

74. Personal narrative of Mr. George S. Campbell, item 124, AMF; Sutherland, "Personnel and Policies of the Halifax Board of Trade," 213.

75. Personal narrative of Mr. Dougal[d] MacGillivray, item 196, AMF.

76. Personal narrative of Edmund A. Sanders, item 222, AMF.

77. "Experiences of a Relief Worker," item 182a, AMF.

78. "Minutes of Morning Meeting," 7 December 1917, item 3f-3j, AMF.

79. Forbes, "Battles in Another War."

80. Undated (but after 1935) biographical sketch, presumably written by Agnes Dennis, in her scrapbook, reel 10,219, Agnes Dennis fonds, NSA; "Work Done by the Halifax Local Council of Women during the Presidency of Dr. Agnes Dennis, C.B.E. M.A.—1905—1920," Scrapbook, reel 20,204, LCWH; William March, *Red Line: The Chronicle-Herald and the Mail-Star, 1875–1954* (Halifax, Nova Scotia: Chebucto Agencies, 1986), 361–62.

81. Personal narrative of Mrs. Clara MacIntosh, item 198, AMF.

82. Boyd, *Enriched by Catastrophe*, 56; personal testimony of Miss Jane Wisdom, item

237, AMF; Suzanne Morton, *Wisdom, Justice and Charity: Canadian Social Welfare through the Life of Jane B. Wisdom, 1884–1975* (Toronto: University of Toronto Press, 2014), 118; Daniel J. Walkowitz, *Working with Class: Social Workers and the Politics of Middle-Class Identity* (Chapel Hill: University of North Carolina Press, 1999), 57–85.

83. "Minutes of Morning Meeting," 7 December 1917, item 3f-3j, AMF.
84. Personal testimony of Miss Jane Wisdom, item 237, AMF.
85. "Minutes of Morning Meeting," 7 December 1917, item 3f-3j, AMF.
86. MacIntosh, "A Brief Account of Relief Work Undertaken by Mrs. MacIntosh & Assistants at City Hall following Explosion Dec. 6th," item 22; personal testimony of Miss Jane Wisdom, item 237; both in AMF.
87. Personal testimony of John Hanlon Mitchell, item 186, AMF.
88. Lois K. Yorke, "Best, Edna May Williston (Sexton)," in *Dictionary of Canadian Biography*, vol. 15 (Toronto/Quebec: University of Toronto/Université Laval, 2005).
89. Personal testimony of Mrs. F. H. Sexton, item 224; MacIntosh, "A Brief Account of Relief Work Undertaken by Mrs. MacIntosh & Assistants at City Hall following Explosion Dec. 6th," item 22; both in AMF.
90. Personal testimony of Mrs. F. H. Sexton, item 224; MacIntosh, "A Brief Account of Relief Work Undertaken by Mrs. MacIntosh & Assistants at City Hall following Explosion Dec. 6th," item 22; MacIntosh to Charles Copp, 13 December 1917, item 29; all in AMF; Forbes, "Battles in Another War," 73–74, 81; Morton, *Wisdom, Justice and Charity*, 112.
91. "Relief Work in Dartmouth," information from A. C. Johnson, item 27, AMF.
92. Answers to questionnaire requested by W. C. Milner, 12 July 1920, item 7, reel 15,125, MG 27; Memorandum, [Benson] to Secretary, Militia Council, 15 December 1917, file 86-2-1, "Explosion—Reports," box 4548, MD6 (a copy is item 250, AMF).
93. Personal narrative of G. H. Libby, item 165; "Experiences of a Relief Worker," item 182a; both in AMF.
94. C. A. Hayes, "How the Government Railways Responded to the Call," item 6, AMF.
95. Interview of Helen Facey, then Mrs. Cooper, 27 June 1985, interview 13, KOH.
96. Interview of Miss Margaret M. Smith, 31 July 1985, interview 106, KOH.
97. Interview of John J. Flemming and May Flemming, née Gerroir, 28 June 1985, interview 21, KOH.
98. Interview of Nellie Billard, 27 June 1985, interview 18, KOH.
99. Personal narrative of Christine MacKinnon, item 200, AMF.
100. MS narrative of Lt. Rod Macdonald, item 194, AMF.
101. Interview of Margaret Nowlan, 26 June 1985, interview 7, KOH.
102. Personal narrative of Mrs. Henry Dunstan, 12 February 1918, item 141, AMF.
103. Interview of Julia Coleman, née DeYoung, 28 June 1985, interview 19, KOH.
104. "Maggie" to "Precious Cousin" [Mrs. Copp], 19 December 1917, folder 5, "Explosion Letter," vol. 10, HEC.
105. Personal testimony of Mrs. Moore, item 191, AMF. On the tsunami, see Alan Ruffman, David A. Greenberg, and Tad S. Murty, "The Tsunami from the Explosion in Halifax Harbour," in Ruffman and Howell, *Ground Zero*, 327–44.
106. Personal narrative of Harvey Jones, item 161, AMF.
107. Jane Jacobs, *The Death and Life of Great American Cities* (1961; repr. New York: Vintage, 1992), 68–71.

108. Personal narrative of H. D. Brunt, item 119, AMF.
109. Personal testimony of Principal Matheson, item 184; "St. Joseph's Girls' School," item 18; both in AMF.
110. "Report of the Y.M.C.A.," item 45, AMF.
111. Personal testimony of C. Sutherland, item 46f, AMF.
112. Personal narrative of Miss Florence J. Murray, item 192, AMF.
113. Personal narrative of Mrs. Clara MacIntosh, item 198, AMF.
114. Personal narrative of Mrs. Annie Anderson, item 114, AMF.
115. Personal narrative of Dr. M. J. Burris, item 122, AMF.
116. Interview of Catherine MacDonald, née Boudreau, 29 July 1985, interview 99, KOH.
117. "Medical Work in the North End," item 166, AMF; Bridglal Pachai and Henry Bishop, *Historic Black Nova Scotia* (Halifax, Nova Scotia: Nimbus, 2006), 44.
118. A. W. Duffus to Col. W. E. Thompson, 22 January 1918, item 254, AMF.
119. "Notes of Medical Relief Committee of Halifax Disaster," item 10, AMF.
120. Capt. John Flint Cahan, item 123; personal testimony of Miss Josephine Creighton [Crichton], item 135; personal narrative of Miss Florence J. Murray, item 192; all in AMF.
121. Interview of Julia Coleman, née DeYoung, 28 June 1985, interview 19, KOH.
122. Personal narrative of Mr. L. A. Myers [Miles], item 205, AMF.
123. "Report of the Y.M.C.A.," item 45, AMF.
124. MS narrative of Lt. Rod Macdonald, item 194, AMF.
125. Copy of letter, Jean H. Armstrong to Mr. Kent, 30 December 1917, item 274, AMF.
126. Memorandum, [Benson] to Secretary, Militia Council, 15 December 1917, file 86-2-1, "Explosion—Reports," box 4548, MD6 (a copy is item 250, AMF); A. W. Duffus to Col. W. E. Thompson, 22 January 1918, item 254, AMF.
127. Personal narrative of H. D. Brunt, item 119, AMF.
128. MS narrative of Lt. Rod Macdonald, item 194, AMF.
129. Capt. John Flint Cahan, item 123; [Testimony of] Col. W. E. Thompson, item 233; both in AMF.
130. Personal narrative of Warrena Maddin, item 180, AMF.
131. "Morgue at Chebucto School," AMF.
132. [Testimony of] Col. W. E. Thompson, item 233, AMF.
133. MS narrative of Lt. Rod Macdonald, item 194, AMF.

Chapter 2. "A Great Power Had Swept Over It"

1. SEN, 26 June 1914, 1, 5; SEN, 25 June 1914, 1; SEO, 27 June 1914, 1; Bertram E. Ames, "Report No. 150 on Conflagration, Salem, Mass., June 25, 1914," report to Underwriters' Bureau of New England, E S1 F6 1914_{15}, PEM; E. V. French to Willis H. Ropes, 29 July 1914, box 1, E S1 F6 1914_2, PEM; John F. Moors to Mr. [August] Cunningham, 3 September 1914, "File 835.08—MASSACHUSETTS, Salem Fire—Reports, 6/25/1914," ARC. These numbers are significantly higher than those reported by freshly graduated civil engineer Ames: 1,600 buildings destroyed; 1,500 to 1,600 families made homeless, comprising 10,000 people; and 4,000 left jobless.
2. SEO, 27 June 1914, 1.
3. Selskar M. Gunn and Samuel M. Schmidt, "An Investigation of Housing Conditions

in Salem, Mass." (report to the Housing Committee of the Associated Charities and the Committee on Nuisances of the Civic League, [December 1911]), 33, 38, 40–41, 46 (quote on 33).

4. Gunn and Schmidt, "Housing Conditions," 32–33.

5. Ernest P. Bicknell to August Cunningham, 14 August 1914, "File 835—MASSACHUSETTS, Salem Fire 6/25/1914," ARC; Aviva Chomsky, "Salem as Global City, 1850–2004," in *Salem: Place, Myth, and Memory*, ed. Dane Anthony Morrison and Nancy Lusignan Schultz, 219–47; Robert Phippen Donnell, "Locational Response to Catastrophe: The Dynamics of Locational Change in the Shoe and Leather Industry of Salem, Massachusetts, and after the Conflagration of June 25, 1914" (MA thesis, Clark University, 1971).

6. Gunn and Schmidt, "Housing Conditions," 31.

7. S. Cannon, "On parle Français à Salem, Massachusetts," *Etudes canadiennes* 22 (1987): 104–5, 109.

8. Gunn and Schmidt, "Housing Conditions," 31.

9. Although O'Keefe's religion was not discussed in the newspapers, I suggest that he was Protestant because he attended the dedication of a Masonic temple; see the entry for 24 June 1915, *Abstract of the Proceedings of the Most Worshipful Grand Lodge of Free and Accepted Masons of the Commonwealth of Massachusetts for the Year 1915* (Cambridge, Mass.: Caustic-Claflin, 1916), 204. On Hurley and the Irish capture of the police force, see Theodore N. Ferdinand, "Politics, the Police, and Arresting Policies in Salem, Massachusetts, since the Civil War," in *Deviant Behavior and Social Process*, ed. William A. Rushing, 2nd ed. (Chicago: Rand McNally College Publishing, 1975), 151–54; and Joan M. Maloney, "John F. Hurley: Salem's First Hurrah," *Essex Institute Historical Collections* 128 (January 1992): 27–58. On the recall, see "The Recall in Salem, Mass.," *National Municipal Review*, April 1915, 304. On Franco-American distrust of Hurley, especially over police appointments, see CdS, 13 February 1914, 1; CdS, 24 December 1914, 1.

10. CdS, 31 December 1914, 4.

11. John F. Moors to Mr. [August] Cunningham, 3 September 1914, "File 835.08—MASSACHUSETTS, Salem Fire—Reports, 6/25/1914," ARC.

12. Montayne Perry, *The Salem Fire Relief* (Salem, Mass.: Milo A. Newhall, 1915); originally printed in SEN on 28 October 1914, 29 October 1914, 30 October 1914, 31 October 1914, 2 November 1914, 3 November 1914, 4 November 1914, and 5 November 1914. Cooper C. Graham et al., *D. W. Griffith and the Biograph Company* (Metuchen, N.J.: Scarecrow Press, 1985), 125; note attached to clippings of the articles, YMCAS; William E. Chancellor, *Reading and Language Lessons for Evening Schools* (New York: American Book Company, 1912), iii.

13. William B. Emery to Charles H. Cole, 19 October 1914, MNGMA.

14. SEN, 30 June 1914, 4; James J. Flink, *The Car Culture* (Cambridge, Mass.: MIT Press, 1975), 42–66; James J. Flink, *The Automobile Age* (Cambridge, Mass.: MIT Press, 1988), 27–39, 73–80; and Kathleen Franz, *Tinkering: Consumers Reinvent the Early Automobile* (Philadelphia: University of Pennsylvania Press, 2005), 2.

15. SEN, 26 June 1914, 9; R. E. Eagan to Capt. P. B. Chase, 17 July 1914, MNGMA.

16. Postcard in Miscellaneous collection of circulars, etc., box II, Salem Fire Collection, E S1 F6 1914_2, PEM.

17. Edward Dunbar Johnson, "The Salem Fire," after June 1915, E S1 F6 1914_{11}, PEM.

18. SEN, 26 June 1914, 15.
19. Capt. P. B. Chase to Samuel H. Batchelder, 22 July 1914, MNGMA.
20. SEN, 26 June 1914, 5 (the phrase appears twice on the page), 6 (once more).
21. Perry, *Salem Fire Relief*, 5.
22. SEN, 26 June 1914, 9.
23. Ibid., 7.
24. Ibid., 9; Perry, *Salem Fire Relief*, 5–6.
25. Perry, *Salem Fire Relief*, 6.
26. Ibid., 6.
27. SEN, 26 June 1914, 2.
28. Paul Starr, *The Social Transformation of American Medicine* (New York: Basic Books, 1982); Susan Reverby, "A Caring Dilemma: Womanhood and Nursing in Historical Perspective," in *The Sociology of Heath and Illness: Critical Perspectives*, ed. Peter Conrad and Rochelle Kern, 3rd ed. (New York: St. Martin's, 1990), 184–95.
29. Perry, *Salem Fire Relief*, 8.
30. SEN, 26 June 1914, 1.
31. SEO, 27 June 14, 1; Perry, *Salem Fire Relief*, 22; SEN, 26 June 1914, 5.
32. Reprinted in the SEO, 27 June 1914, 8.
33. SEN, 1 July 1914, 10.
34. SEN, 26 June 1914, 15; SEO, 27 June 1914, 1; *Boston Globe*, 22 January 1915, 8.
35. A list of members appeared in the SEN, 26 June 1914, 15. The list appears incomplete, in that it excludes a few people who were subcommittee chairs. With some exceptions, I have determined members' ethnicities based on their last names, but this is an imperfect method.
36. Hurley's defeated opponent was William S. Felton (Maloney, "First Hurrah," 52). The two former mayors were Arthur Howard, who had been mayor in 1910 (1910 census, series T624, roll 587, p. 139), and David M. Little, a Spanish-American War veteran, former customs collector, former naval architect, and now officer of the Merchant's National Bank (Maloney, "First Hurrah," 42; SEO, 25 July 1914, 1). In addition, Margaret and Harriet Rantoul, two unmarried, grown daughters of another of Hurley's vanquished Yankee opponents, served on the committee (1910 census, series T624, roll, 587, p. 104; on the election between Hurley and Robert Rantoul, see Maloney, "First Hurrah," 51).
37. *Columbiad*, August 1914, 16.
38. They were Morris Newmark (1920 census, series T625, roll 697, p. 164), Harry Pitcoff (1910 census, series T624, roll 588, p. 31), and Max Winer (1920 census, series T625, roll 697, p. 119).
39. The following numbers should all be read as minimums because the occupations of nearly a fifth of the members are unknown. Where relevant, I have included wives under their husbands' occupations. Some people, if they held multiple positions, are included in more than one category.
40. John F. Cabeen (SEN, 3 July 1914, 9; 1920 census, series T625, roll 697, p. 127; *Naumkeag Directory*, 221).
41. Labor committee chair William F. Cass, shoes (1910 census, series T624, roll 588, p. 27); housing committee chair John H. Deery, in the shoe industry (1920 census, series T625, roll 697, p. 170; SEO, 27 June 1914, 1); J. Willard Helburn, a leading leather

manufacturer, and his wife, who chaired the clothing committee (SEO, 18 July 1914, 1); food committee chair John E. Spencer, machine manufacturer (Perry, *Salem Fire Relief*, 42); Alvan Thompson, Helburn's vice president (*Naumkeag Directory*, 428, 138); Greeley C. Curtis, a principal in a Marblehead airplane company (*Naumkeag Directory*, 246, 941).

42. Henry M. Batchelder (1920 census, series T625, roll 697, p. 138); Leland H. Cole, who became chair of the joint finance committee (1920 census, series T625, roll 696, p. 164); Harry P. Gifford (1920 census, series T625, roll 969, p. 164); treasurer Josiah Gifford (1920 census, series T625, roll 697, p. 29); rehabilitation committee chair Eugene J. Fabens (SEO, 11 July 1914, 8); David M. Little (SEO, 25 July 1914, 1); William S. Nichols (1920 census, series T62, roll 697, p. 171); James Young Jr. (1920 census, series T625, roll 697, p. 125).

43. Proprietors: Walter K. Bigelow (1920 census, series T625, roll 697, p. 140); transportation committee chair Dan A. Donahue (SEN, 3 July 1914, 9); Stanislas Levesque (1920 census, series T625, roll 697, p. 170); William B. Mansfield (1920 census, series T625, roll 697, p. 111); Pierre Michaud (1910 census, series T624, roll 588, p. 178); Morris Newmark (1920 census, series T625, roll 697, p. 164); Max Winer (1920 census, series T625, roll 697, p. 119). Salesmen, managers, and clerks: George A. Ashton (1920 census, series T625, roll 697, p. 169); John F. Browning (1900 census, series T623, roll 648, p. 116; subsequent censuses show him retired); May Goldsmith (1920 census, series T625, roll 697, p. 201); Martin D. Hoyte (*Naumkeag Directory*, 304); George C. Silsbury (1910 census, series T624, roll 588, p. 181).

44. John A. Bagley (1920 census, series T624, roll 587, p. 239); James Brenan (*Naumkeag Directory*, 213, 311); Frederick W. Broadhead (1920 census, series T625, roll 697, p. 4); Paul N. Chaput (CdS, 17 July 1914, 8); Leland H. Cole (*Naumkeag Directory*, 235, 414); J. Frank Dalton (*Naumkeag Directory*, 247); William S. Felton (*Naumkeag Directory*, 269); Robert M. Martin (*Naumkeag Directory*, 349); Patrick A. McSweeney (1910 census, series T624, roll 587, p. 253); Fred A. Norton (*Naumkeag Directory*, 366); Patrick F. Tierney (1910 census, series T624, roll 588, p. 43); Fred E. Warner (1910 census, series T624, roll 587, p. 248); J. Stoddard Williams (*Naumkeag Directory*, 447); James J. Welch (1910 census, series T624, roll 588, p. 85).

45. Samuel H. Batchelder (1910 census, series T624, roll 588, p. 160); Stephen W. Phillips (1920 census, series T625, roll 696, p. 229); Alden P. White (*Naumkeag Directory*, 445).

46. Unless otherwise noted, affiliations come from a list published in the SEO, 20 June 1914, 4. The Protestants were: Theodore D. Bacon, North (Unitarian); Henry Bedinger, St. Peter's Episcopal; Oliva Brouillette, French [Canadian] Baptist; Frederick W. Buis, First Baptist; John E. Charlton, Wesley Methodist Episcopal; DeWitt C. Clark, Tabernacle Congregational; James P. Frank, Grace (Episcopal); John L. Ivey, Lafayette Street (Methodist); Edward D. Johnson, First Unitarian; Thomas G. Langdale, South Congregational; Alfred Manchester, Second Unitarian; the unaffiliated Robert M. Martin, who was also an insurance agent (*Naumkeag Directory*, 349); Harry L. Newton, Crombie Street (Congregational). Catholics included John J. Cronin, curate at Immaculate Conception (*Naumkeag Directory*, 243); Joseph Czubeck, pastor of St. Joseph's (Polish); Matthew J. Gleason, curate of St. James (*Naumkeag Directory*, 284); Jeremiah J. Herlihy, curate of Immaculate Conception (1920 census, series, T625, roll 696, p. 115); Michael J. McCall,

pastor of Immaculate Conception; Frederick Muldoon, curate at Immaculate Conception (*Naumkeag Directory*, 359); J. Alfred Peltier, pastor of St. Anne's (French); Rosario Richard, former curate at St. Joseph's, now pastor at a French church in Shirley, Massachusetts (SEN, 3 July 1914, 13). Rabbi Hyman Pitkoff may also have been a committee member (*Naumkeag Directory*, 383).

47. Christian Lantz; Ethel Osborne (*Naumkeag Directory*, 372); and Arthur Bodwell (*Naumkeag Directory*, 209).

48. Perry, *Salem Fire Relief*, 23. A list of the Committee of Fourteen appeared in the SEN, 3 July 1914, 9.

49. SEO, 11 July 1914, 8; SEN, 3 July 1914, 9.

50. *New York Times*, 13 April 1931, 19; *Columbiad*, August 1914, 16.

51. *Naumkeag Directory*, 421; SEN, 3 July 1914, 9.

52. census, series T624, roll 581, p. 151; *Boston Globe*, 6 August 1955, 1.

53. SEN, 11 July 1914, 8.

54. J. E. Spencer to C. H. Cole, 3 July 1914, MNGMA; SEO, 27 June 1914, 1; Perry, *Salem Fire Relief*, 42.

55. Susan Traverso, *Welfare Politics in Boston, 1910–1940* (Amherst: University of Massachusetts Press, 2003), 18.

56. Ernest P. Bicknell to August Cunningham, 14 August 1914, "File 835—MASSACHUSETTS, Salem Fire 6/25/1914," ARC.

57. Unsigned memorandum on Salem Relief Committee letterhead, 17 July 1914, "File 835—MASSACHUSETTS, Salem Fire 6/25/1914," ARC.

58. Winthrop D. Lane to Bicknell, 17 July 1914, resending letter of 10 July 1914; Bicknell to Lane, 16 July 1914, both in "File 835—MASSACHUSETTS, Salem Fire 6/25/1914," ARC.

59. Mabel T. Boardman to [Lindley M. Garrison], 23 July 1914, "File 835—MASSACHUSETTS, Salem Fire 6/25/1914," ARC.

60. Boardman to Maj. Gen. George W. Davis, 15 July 1914, "Folder 835.2—MASSACHUSETTS, Salem, Finances & Accounts, 6/25/1914"; Bicknell to Henry Breckinridge, 13 August 1914, "File 835.08—MASSACHUSETTS, Salem Fire—Reports, 6/25/1914"; Boardman to Jerome Green, no date but c. 17 August 1914, "Folder 835.2—MASSACHUSETTS, Salem, Finances & Accounts, 6/25/1914"; all in ARC.

61. SEN, 1 July 1914, 10.

62. Thomas H. O'Connor, *The Boston Irish: A Political History* (Boston: Northeastern University Press, 1995), 214–15; Geoffrey Blodgett, "Yankee Leadership in a Divided City: Boston, 1860–1910," in *Boston, 1700–1980: The Evolution of Urban Politics*, ed. Ronald P. Formisano and Constance K. Burns (Westport, Conn.: Greenwood Press, 1984), 90, 97, 106; Sarah Deutsch, *Women and the City: Gender, Space, and Power in Boston, 1870–1940* (New York: Oxford University Press, 2000), 137–38, 147. Thanks to Danny Scholzman for his help with the history of Massachusetts municipal-commonwealth relations.

63. Edward J. Kopf, "The Intimate City: A Study in Urban Social Order: Chelsea, Massachusetts, 1906–1915" (PhD diss., Brandeis University, 1974), 60–62.

64. Constance K. Burns, "The Irony of Progressive Reform: Boston, 1898–1910," in Formisano and Burns, *Boston, 1700–1980*, 139–44.

65. SEO, 4 July 1914, 1.

66. Ibid., 11 July 1914, 8; SEN, 3 July 1914, 9.
67. SEN, 1 July 1914, 10.
68. *Salem Register,* 17 July 1914, 1; SEO, 18 July 1914, 4; *Christian Science Monitor,* 14 December 1917, RS.
69. Scott Gabriel Knowles, *The Disaster Experts: Mastering Risk in Modern America* (Philadelphia: University of Pennsylvania Press, 2011), 100–108.
70. *Congressional Record* 63rd Cong., 2nd sess., 1914, 51, pt. 12:11800–01; SEN, 3 July 1914, 1, 14.
71. Mabel Boardman to [Lindley M. Garrison], 23 July 1914, "File 835—MASSACHUSETTS, Salem Fire 6/25/1914"; Ernest P. Bicknell to Henry Breckinridge, 13 August 1914, "File 835.08—MASSACHUSETTS, Salem Fire—Reports, 6/25/1914;" Boardman to Jerome Green, c. 17 July 1914, "Folder 835.2—MASSACHUSETTS, Salem, Finances & Accounts, 6/25/1914"; all in ARC.
72. SEN, 2 July 1914, 5; SEN, 6 July 1914, 9.
73. Michele Landis Dauber, *The Sympathetic State: Disaster Relief and the Origins of the American Welfare State* (Chicago: University of Chicago Press, 2013), 45–47.
74. *Congressional Record,* 63rd Cong., 2nd sess., 1914, 51, pt. 12:11918.
75. Ibid., pt. 12:11945–47.
76. Ibid., pt. 12:11945.
77. SEN, 11 July 1914, 7; *Congressional Record*, 63rd Cong., 2nd sess., 1914, 51, pt. 12:11948. Congressmen voting yes came from all parties and regions, but those voting no were nearly all Democrats and more than half were from the South.
78. Garrison to Boardman, 23 July 1914, "File 835—MASSACHUSETTS, Salem Fire 6/25/1914," ARC.
79. SEO, 15 August 1914, 1.
80. Ibid., 28 November 1914, 4.
81. CdS, 17 December 1914, 8. Unfortunately, Lane's records have disappeared.
82. John Lombardi, *Labor's Voice in the Cabinet: A History of the Department of Labor from Its Origin to 1921* (New York: Columbia University Press, 1942), 144–51, 156.
83. Report of T. V. Powderly to [William Wilson], 28 July 1914, reprinted in United States Department of Labor, *Reports of the Department of Labor, 1914* (Washington, D.C.: GPO, 1915), 295–97.
84. Lombardi, *Labor's Voice in the Cabinet*, 151–57; [O. L. Harvey], *The Anvil and the Plow: A History of the United States Department of Labor* (Washington, D.C.: GPO, [1963]), 17–19.
85. SEN, 25 June 1914, 1, 5.
86. Perry, *Salem Fire Relief*, 6.
87. Ibid., 7.
88. George F. Heustis, "The Night of the Salem Fire," *Tenney Service* 1, no. 9 (September 1914): 4.
89. Perry, *Salem Fire Relief*, 10.
90. SEN, 26 June 1914, 6.
91. John Porter Sumner, "Reminiscences of the Salem Fire of June 25, 1914," E S956 1970, PEM.
92. SEN, 26 June 1914, 2.

93. Perry, *Salem Fire Relief*, 13–14.
94. John Porter Sumner, "Reminiscences of the Salem Fire of June 25, 1914," E S956 1970, PEM.
95. Perry, *Salem Fire Relief*, 7.
96. Ibid., 8–9.
97. Ibid., 9–10.
98. SEN, 26 June 1914, 5.
99. Perry, *Salem Fire Relief*, 9.
100. SEN, 2 July 1914, 12.
101. Perry, *Salem Fire Relief*, 13.
102. SEN, 27 June 1914, 4.
103. Ibid., 26 July 1914, 5.
104. Quoted in the SEN, 26 June 1914, 5. See also p. 9, where Cole repeats his threat.
105. Ibid., 26 June 1914, 9.
106. Ibid., 27 June 1914, 4.
107. Ibid., 29 June 1914, 4.
108. Edward Dunbar Johnson, "The Salem Fire," after June 1915, E S1 F6 1914_{11}, PEM.
109. SEN, 26 June 1914, 5. On the growing culture of car tourism and touring, see Franz, *Tinkering*, 3–8.
110. Perry, *Salem Fire Relief*, 19–20. On fund-raising from the tourists, see, e.g., SEN, 3 July 1914, 1.
111. SEO, 4 July 1914, 1.
112. Memorandum, Brig. Gen. Jas. G. White to [Cole], 2 July 1914, MNGMA.
113. SEN, 26 June 1914, 5, 9; SEN, 1 July 1914, 10; SEN, 2 July 1914, 1; AN, 7 July 14, 1.
114. [Emily A. Murphy], "Merchants, Clerks, Citizens, and Soldiers: The Second Corps of Cadets in Salem, Massachusetts," pamphlet (Salem, Mass.: Salem Maritime National Historic Site, National Park Service, US Department of the Interior, [2005?]), 11–12, 14–15; Philip S. Foner, *The Industrial Workers of the World, 1905–1917*, vol. 4 of *History of the Labor Movement in the United States* (New York: International, 1965), 330–31.
115. Edward Dunbar Johnson, "The Salem Fire," after June 1915, E S1 F6 1914_{11}, PEM; 1910 census, series T624, roll 587, p. 221; 1920 census, series T625, roll 697, p. 205.
116. Provisional Troops Special Orders 36, 29 June 1914, MNGMA (copy also in unlabeled accordion folder, Miscellaneous collection of circulars, etc., box I, Salem Fire Collection, E S1 F6 1914_2, PEM).
117. Murphy, "Merchants, Clerks," 12, 15.
118. CdS, 3 July 1914, 4. Beaucage's phrase *"silence farouche sinistre"* was a pun: *sinistre* meant "sinister," whereas a *sinistré* was a victim of the conflagration.
119. Edward Dunbar Johnson, "The Salem Fire," after June 1915, E S1 F6 1914_{11}, PEM.
120. Ibid.
121. *Salem Register*, 3 July 1914, 1; SEN, 29 June 1914, 12.
122. Perry, *Salem Fire Relief*, 11.
123. SEN, 26 June 1914, 6.
124. Provisional Troops Special Orders 6, 27 June 1914, MNGMA.
125. SEN, 3 July 1914, 1.
126. Provisional Troops Special Order 54, 3 July 1914, MNGMA.

127. SEN, 2 July 1914, 5.

128. Ibid., 26 June 1914, 5.

129. Chase to Samuel H. Batchelder, 22 July 1914; Capt. James F. Hickey to [Charles H. Cole], 26 December 1914; Hickey to Cole, 26 January 1915; Cole to Josiah Gifford, 6 October 1914; Batchelder to Bay State Creamery Co., 16 July 1914; B. L. & G. R. Beeman to Cole, 17 October 1914; Maj. William L. Casey to Cole, 19 September 1914; Cole to Casey, 15 September 1914; Capt. Julian I. Chamberlain to Col. Frank A. Graves, 1 July 1914; all in MNGMA.

130. R. E. Eagan to Chase, 17 July 1914, MNGMA.

131. SEO, 4 July 1914, 4.

132. Robert M. Fogelson, *America's Armories: Architecture, Society, and Public Order* (Cambridge, Mass.: Harvard University Press, 1989).

133. SEO, 15 August 1914, 1.

134. Col. Frank A. Graves to Adj. Gen. Charles H. Cole, 1 July 1914, MNGMA.

135. Nayan Shah, *Contagious Divides: Epidemics and Race in San Francisco's Chinatown* (Berkeley: University of California Press, 2001).

136. SEN, 1 July 1914, 11.

137. SEO, 4 July 1914, 4.

138. SEN, 29 June 1914, 1.

139. Perry, *Salem Fire Relief*, 10.

Chapter 3. "It Is Easy Enough to Establish Camps"

1. SEN, 26 June 1914, 2; SEO, 27 June 14, 1.

2. SEN, 26 June 1914, 5.

3. Edward Dunbar Johnson, "The Salem Fire," after June 1915, E S1 F6 1914$_{11}$, PEM; Albert Goodhue Jr., ed., "The Salem Fire," *Essex Institute Historical Collections* 100 (1964): 184.

4. *Pilot,* 4 July 1914, 1.

5. SEN, 27 June 1914, 1. Although Forest River Park was dominated by French Canadians, there were also some Italian families there. As we will see, Jews were sent to Bertram Field. It is unclear where Greeks and Poles stayed.

6. Philip S. Foner, *The Industrial Workers of the World, 1905–1917*, vol. 4 of *History of the Labor Movement in the United States* (New York: International, 1965), 435–49; Carey McWilliams, *Factories in the Field: The Story of Migratory Farm Labor in California* (Boston: Little, Brown, 1939), 164–67; Alexander Keyssar, *Out of Work: The First Century of Unemployment in Massachusetts* (Cambridge: Cambridge University Press, 1986), 235–36.

7. SEN, 29 June 1914, 5.

8. Ibid., 1 July 1914, 10; CdS, 9 July 1914, 6.

9. SEN, 2 July 1914, 12.

10. Daniel Levinson Wilk, "Cliff Dwellers: Modern Service in New York City, 1800–1945" (PhD diss., Duke University, 2005), 292–310; SEN, 29 June 1914, 1.

11. Montayne Perry, *The Salem Fire Relief* (Salem, Mass.: Milo A. Newhall, 1915), 14.

12. SEN, 29 June 1914, 1; [Ernest Bicknell?] to F. J. Mulhall, 30 June 1914, "File 835—MASSACHUSETTS, Salem Fire 6/25/1914," ARC. For reassurances, see SEN, 29 June 1914, 1, 5; SEN, 2 July 1914, 1, 12; SEN, 3 July 1914, 4.

13. SEN, 30 June 1914, 10.
14. Ibid., 10.
15. Ibid., 1 July 1914, 10.
16. Minutes of Health Committee meeting, 29 June 1914, 9:00 AM, "Letters, Reports and Records Relating to the Salem Fire," E S1 F6 1914$_{10}$, PEM.
17. SEN, 3 July 1914, 9.
18. Ibid., 1 July 1914, 6; Col. Frank A. Graves to Adj. Gen. Charles H. Cole, 1 July 1914, MNGMA.
19. SEN, 3 July 1914, 4.
20. Ibid., 6.
21. Ibid., 1 July 1914, 6, 11; Graves to Cole, 1 July 1914, MNGMA.
22. Perry, *Salem Fire Relief*, 17.
23. Edward Dunbar Johnson, "The Salem Fire," after June 1915, E S1 F6 1914$_{11}$, PEM.
24. See, e.g., SEN, 26 June 1914, 5; SEN, 27 June 1914, 1.
25. SEN, 1 July 1914, 11; SEN, 2 July 1914, 9.
26. SEN, 30 June 1914, 1; Spencer to Cole, 30 June 1914, MNGMA.
27. Spencer to Cole, 3 July 1914, MNGMA.
28. For example, see Ernest Bicknell's draft article for a Red Cross magazine, enclosed in a letter, Bicknell to August Cunningham, 14 August 1914, "File 835—MASSACHUSETTS, Salem Fire 6/25/1914," ARC; SEN, 3 July 1914, 3.
29. SEN, 3 July 1914, 9.
30. John M. Barry, *Rising Tide: The Great Mississippi Flood of 1927 and How It Changed America* (New York: Simon and Schuster, 1997), 307–17. For concerns after the San Francisco earthquake and fire, see Marian Moser Jones, *The American Red Cross from Clara Barton to the New Deal* (Baltimore: Johns Hopkins University Press, 2013), 125.
31. Pierre Anctil, "Aspects of Class Ideology in a New England Ethnic Minority: The Franco-Americans of Woonsocket, Rhode Island, 1865–1929" (PhD diss., New School for Social Research, 1980), 69–70.
32. SEN, 29 June 1914, 1.
33. Ibid., 2 July 1914, 13.
34. Card from Employment Bureau, second unlabeled folder, Miscellaneous collection of circulars, etc., 1914, box I, Salem Fire Collection, E S1 F6 1914$_{2}$, PEM.
35. The tour of the camp was reported in SEN, 3 July 1914, 6, and unless otherwise specified, the details and quotes in the next three paragraphs come from that article.
36. The French Local 1210 of the Carpenters' Union had successfully established a fifty-cent hourly wage in 1913. See Memorandum to Charles Johnson Jr., "Local Union No. 1210, Salem, Mass.," 18 October 1954, Local Histories folder, United Brotherhood of Carpenters and Joiners of America, Massachusetts State Council Records, 1892–1980, MS col. 015, W. E. B. DuBois Library, University of Massachusetts, Amherst, Mass.
37. For Pelletier's occupation, see 1920 census, series T625, roll 697, p. 189.
38. Perry, *Salem Fire Relief*, 14–15.
39. SEN, 3 July 1914, 6.
40. Ibid., 1 July 1914, 10.
41. Ibid., 2 July 1914, 12.
42. Perry, *Salem Fire Relief*, 14.
43. CdS, 24 July 1914, 1.

44. Ibid., 9 July 1914, 6.

45. John Porter Sumner, "Reminiscences of the Salem Fire of June 25, 1914," E S956 1970, PEM.

46. SEN, 2 July 1914, 11.

47. Ibid., 1 July 1914, 1.

48. Ibid., 7.

49. Ibid., 29 June 1914, 11.

50. Ibid., 3.

51. Ibid., 26 June 1914, 2.

52. Ibid., 15.

53. Ibid., 27 June 1914, 1.

54. Ibid., 1 July 1914, 8; 2 July 1914, 13.

55. Ibid., 11; CdS, 9 July 1914, 6.

56. "Number of Families Rehabilitated in Outlying Towns," undated list, third unlabeled folder, Miscellaneous collection of circulars, etc., 1914, box I, Salem Fire Collection, E S1 F6 1914$_2$, PEM.

57. CdS, 24 July 1914, 8.

58. Ibid., 31 July 1914, 8.

59. Ibid., 24 July 1914, 8; ibid., 31 July 1914, 8; SEN, 30 June 1914, 1.

60. For lists of major Franco-American centers measured in various ways, see Gérard J. Brault, "État présent des études sur les centres franco-americains de la Nouvelle-Angleterre," in *Vie Française: Situation de la recherche sur la Franco Américanie*, ed. Claire Quintal and André Vachon (Quebec: Le Conseil de la Vie Française en Amerique, 1980), 11–12.

61. CdS, 24 July 1914, 1, 8.

62. Ibid., 31 July 1914, 1; ibid., 7 August 1914, 8.

63. Report of T. V. Powderly to [William Wilson], 28 July 1914, reprinted in United States Department of Labor, *Reports of the Department of Labor, 1914* (Washington, D.C.: GPO, 1915), 297.

64. Vincent J. Falzone, *Terence V. Powderly: Middle-Class Reformer* (Washington, D.C.: University Press of America, 1978), 173, 176–78.

65. Henri T. LeDoux "Rapport Officiel du Président Général"; Élie Véniza, "Rapport Officiel du Secrétaire Général"; and Report and Audit of USJBA by Sinclair E. Allison to Charles C. Gray, insurance commissioner for Rhode Island, 16 August 1915; all in the proceedings of the Huitième Congrès, Worcester, Mass., le 14, 15–Septembre, USBJAS.

66. AN, 1 July 1914, 8; AN, 29 July 1914, 7.

67. Ibid., 26 June 1914, 1, 5.

68. *La Presse,* 26 June through 8 July 1914.

69. *Le Progrès du Golfe*, June through August 1914. On the geographic origins of Salem's French Canadians, see Yves Roby, *The Franco-Americans of New England: Dreams and Realities*, trans. Mary Ricard ([Sillery, Quebec]: Septentrion, 2004), 11; and Gerard J. Brault, *The French-Canadian Heritage in New England* (Hanover, N.H.: University Press of New England), 56; *La Revue de Salem,* December 2006, 44–45.

70. CdS, 17 July 1914, 8.

71. Robert J. Gagnon, "Family History of the Gagnon Family," 9 February 1989, CS 90.G34, Institut Français.

72. Interview of Paul Bossé by Elizabeth Duclos-Orsello with Elizabeth Blood, 8 July 2011, Franco-American Oral History Collection, Salem State University Archives, Salem, Mass.

73. Rapport de l'élection des officiers, 2 December 1914; Rapport de l'élection des officiers, 6 December 1916, folder "Laurier No. 72, Salem, Massachusetts, Election Results Only"; both in USJBA.

74. SEN, 11 May 1914; United States Census Bureau, *Fourteenth Census of the United States* (Washington, D.C.: GPO, 1921) 1:65.

75. CdS, 24 July 1914, 1; CdS, 31 July 1914, 1, 5; CdS, 1 October 1914, 1; CdS, 17 December 1914, 1.

76. Ibid., 11 September 1914, 8.

77. "Programme-Souvenir a l'occasion des Noces d'Argent du Conseil Laurier No. 72, Union St-Jean-Baptiste d'Amérique," 23 June 1929, in folder "[Laurier No. 72, Salem, Massachusetts,] Old Correspondence Only," USJBA. The souvenir program says only that "un bon nombre" of members transferred and "un certaine nombre" left the organization.

78. *SEN*, 29 June 1914, 5.

79. Ibid., 30 June 1914, 10.

80. Ibid., 1 July 1914, 7, 11.

81. Ibid., 2 July 1914, 1.

82. Ibid., 12.

83. Ibid., 3 July 1914, 4.

84. Ibid., 9; Account marked "Written by Dr. Walter G. Phippen, 31 Chestnut St. Salem in 1964," and [Phippen?] to William H. D. Barr, 21 July 1914; both in "Letters, Reports and Records Relating to the Salem Fire," E S1 F6 1914_{10}, PEM.

85. Minutes of Health Committee meeting, 28 June 1914, 5:00 PM, "Letters, Reports and Records Relating to the Salem Fire," E S1 F6 1914_{10}, PEM.

86. Minutes of Health Committee meeting, 29 June 1914, 5:00 PM, "Letters, Reports and Records Relating to the Salem Fire," E S1 F6 1914_{10}, PEM.

87. John F. Moors to Mr. [August] Cunningham, 3 September 1914, "File 835.08—MASSACHUSETTS, Salem Fire—Reports, 6/25/1914," ARC.

88. Herman Bernstein, ed., *American Jewish Year Book 5675*, vol. 16 (Philadelphia: Jewish Publication Society of America, 1914), 143.

89. Graves to Cole, 1 July 1914, MNGMA.

90. Minutes of Health Committee meeting, 28 June 1914, 5:00 PM, "Letters, Reports and Records Relating to the Salem Fire," E S1 F6 1914_{10}, PEM.

91. Graves to Cole, 1 July 1914, MNGMA.

92. Minutes of Health Committee meeting, 1 July 1914, 5:00 PM, "Letters, Reports and Records Relating to the Salem Fire," E S1 F6 1914_{10}, PEM; SEN, 2 July 1914, 13.

93. Minutes of Health Committee meeting, 2 July 1914, "Letters, Reports and Records Relating to the Salem Fire," E S1 F6 1914_{10}, PEM; SEN, 2 July 1914, 12.

94. SEN, 2 July 1914, 12. The specific numbers are a bit dodgy. Graves estimated fifty to sixty individuals living above Rome's Furniture Store, and eighty people eating there. The health committee twice reported that fifty-three individuals were moved from Lafayette Street to Bertram Field. The newspaper said it was eighty, plus forty more from Beverly.

95. SEN, 2 July 1914, 12.

96. SEN, 1 July 1914, 10; SEN, 3 July 1914, 4; CdS, 9 July 1914, 6. For the size of Forest

River Park, see City of Salem, Massachusetts, *Manual of the City Government 2001* (Salem, Mass., 2001), 77.

97. Robert Ellis Smith, *Ben Franklin's Web Site: Privacy and Curiosity from Plymouth Rock to the Internet* (Providence, R.I.: Privacy Journal, 2000), 55–56. Thanks to Deirdre Mulligan and Chris Hoofnagle for their help on the history of postal privacy.

98. SEN, 30 June 1914, 8; SEN, 2 July 1914, 12.

99. Smith, *Ben Franklin's Web Site*, 49–57.

100. SEN, 3 July 1914, 6.

101. Perry, *Salem Fire Relief*, 16.

102. SEN, 3 July 1914, 6.

103. Ibid., 6.

104. Ibid., 30 June 1914, 10.

105. CdS, 9 July 1914, 6.

106. SEN, 2 July 1914, 12.

107. CdS, 31 July 1914, 8.

108. Ibid., 17 July 1914, 8.

109. SEN, 29 June 1914, 12, 11.

110. Ibid., 30 June 1914, 10.

111. Ibid., 9.

112. Ibid., 10, 11.

113. Ibid., 1 July 1914, 10.

114. Ibid., 2 July 1914, 1, 8, 11, 9.

115. Ibid., 8.

116. Ibid., 1 July 1914, 11; ibid., 2 July 1914, 8.

117. Ibid., 29 June 1914, 5.

118. Perry, *Salem Fire Relief*, 16.

119. SEN, 1 July 1914, 11; SEN, 2 July 1914, 8.

120. Perry, *Salem Fire Relief*, 15.

121. Ibid., 69–70 (quotes on 70).

122. SEN, 2 July 14, 8; SEN, 3 July 1914, 2.

123. White armband reading "Officiale Italiano / de questo campo / Campe Police / Camp Police" in bag of ephemera, Miscellaneous collection of circulars, etc., 1914, box I, Salem Fire Collection, E S1 F6 1914_2, PEM.

124. Minutes of Health Committee meeting, 9 July 1914, 5:00 PM, "Letters, Reports and Records Relating to the Salem Fire," E S1 F6 1914_{10}, PEM. The Polish and Greek papers do not survive, and it seems possible that the health committee was referring to newspapers from neighboring cities, not Salem.

125. John P. Sullivan to William Cardinal O'Connell, 26 November 1919, folder 68:10, Immaculate Conception, PC.

126. John F. Moors to Mr. [August] Cunningham, 3 September 1914, "File 835.08—MASSACHUSETTS, Salem Fire—Reports, 6/25/1914," ARC.

127. SEN, 3 July 1914, 6.

128. Thomas J. Archdeacon, *Becoming American: An Ethnic History* (New York: Free Press, 1983), 138–40.

129. SEN, 8 July 1914, 2.

130. SEO, 25 July 1914, 1.
131. CdS, 24 July 1914, 8.
132. Ibid., 17 July 1914, 5.
133. Ibid., 24 July 1914, 8.
134. Ibid., 31 July 1914, 8.
135. Powderly to [Wilson] in United States Department of Labor, *Reports of the Department of Labor, 1914*, 297.
136. Perry, *Salem Fire Relief*, 18.

Chapter 4. "The Relief Would Have Had to Pay Someone"

1. Personal narrative of Frank A. Gillis, item 150, AMF, and narrative of Lt. Rod Macdonald, item 194, AMF; MC, 8 December 1917, 4; F. A. Gillis to R. P. Bell, 26 January 18, item 196.5, HRCC; "Chronology," folder 23.18, "Halifax Disaster Record Office—Research Notes," AMP.

2. Narrative of Lt. Rod Macdonald, item 194, AMF.

3. [Frank McKelvey Bell], "Notes of Medical Relief Committee of Halifax Disaster," item 10, AMF; memorandum, [Benson] to Secretary, Militia Council, 15 December 1917, file 86-2-1, "Explosion—Reports," box 4548, MD6 (also item 250, AMF).

4. Interview of Gertrude Young, née Hook, 2 July 1985, interview 29, KOH.

5. Personal narrative of Frank A. Gillis, item 150, AMF; [Personal narrative of] Charles J. Burchell, AMP.

6. Lisa Levenstein, *A Movement without Marches: African American Women and the Politics of Poverty in Postwar Philadelphia* (Chapel Hill: University of North Carolina Press, 2009), 24.

7. Janney Byron Deacon, *Disasters and the American Red Cross in Disaster Relief* (New York: Russell Sage Foundation, 1918), 137.

8. MC, 19 December 1917, 5.

9. There are about 14,000 of these files housed in the 184 boxes of HRCP. About two-thirds of the files, mostly at the end of the series, are from people seeking restitution for minor property damage and contain almost no information. I read the files in every sixth box of the first 119; I also read every file mentioned in the randomly selected files. This made for a total of 739 sampled files. Because the collection was rehoused after it was catalogued, the physical boxes are not congruent with the box numbers listed in the finding aid. Therefore, a particular box described in the finding aid might be stretched over two physical boxes, or it might be combined with a neighboring box, or both. When this was the case, I read through the entire first physical box that contained the described box number. My sampling technique was adapted from Konrad H. Jarausch and Kenneth A. Hardy, *Quantitative Methods for Historians* (Chapel Hill: University of North Carolina Press, 1991), 68–74.

10. Janet F. Kitz, *Shattered City: The Halifax Explosion and the Road to Recovery* (Halifax, Nova Scotia: Nimbus, 1989), 125–35; see also Garry Shutlak, "A Vision of Regeneration: Reconstruction after the Explosion, 1917–21," in *Ground Zero: A Reassessment of the 1917 Explosion in Halifax Harbour*, ed. Alan Ruffman and Colin D. Howell (Halifax, Nova Scotia: Nimbus/The Gorsebrook Research Institute for Atlantic Canada Studies at Saint Mary's

University, 1994), 421; Renée Élise de Gannes, "Better Suited to Deal with Women and Children: Pioneer Policewomen in Halifax, Nova Scotia" (MA thesis, Dalhousie University, 1999), 28.

11. For Cutten's outspoken views on several subjects, see "Education: Colgate's Cutten," *Time*, 2 February 1942.

12. Michelle Hébert Boyd, *Enriched by Catastrophe: Social Work and Social Conflict after the Halifax Explosion* (Halifax, Nova Scotia: Fernwood, 2007). Specific examples, including the cities from which individual women arrived, are listed on a staff list, c. 25 January 1918, item 162.1, HRCC.

13. HH, 10 January 1918, 3.

14. J. Howard T. Falk, "History of Rehabilitation Work since January 9, 1918," 26 February 1918, item 162.5, HRCC.

15. Susan Traverso, *Welfare Politics in Boston, 1910–1940* (Amherst: University of Massachusetts Press, 2003), 65.

16. Suzanne Morton, *Wisdom, Justice and Charity: Canadian Social Work through the Life of Jane B. Wisdom, 1884–1975* (Toronto: University of Toronto Press, 2014), 143–44, 119–21; on the Toynbee family, see Alon Kadish, *Apostle Arnold: The Life and Death of Arnold Toynbee, 1852–1883* (Durham, N.C.: Duke University Press, 1986), 1–23, 45.

17. File 1283, box 48, HRCP.

18. File 2298, box 66, HRCP.

19. In many turn-of-the-century American cities, a greater proportion of working-class families owned houses than middle-class families, and renting out a part of them for extra cash was common. Olivier Zunz, *The Changing Face of Inequality: Urbanization, Industrial Development, and Immigrants in Detroit, 1880–1920* (Chicago: University of Chicago Press, 1982), 152–58.

20. Jeanne Boydston, *Home and Work: Housework, Wages, and the Ideology of Labor in the Early Republic* (New York: Oxford University Press, 1990); Bettina Bradbury, *Working Families: Age, Gender, and Daily Survival in Industrializing Montreal* (Toronto: McClelland and Stewart, 1993); Jocelyn Olcott, "Introduction: Researching and Rethinking the Labors of Love," *Hispanic American Historical Review* 91, no. 1 (February 2011): 1–27; Eileen Boris and S. J. Kleinberg, "Mothers and Other Workers: (Re)Conceiving Labor, Maternalism, and the State," *Journal of Women's History* 15, no. 3 (Autumn 2003): 90–117.

21. Interview with Frank Brinton, 10 July 1985, interview 68, KOH.

22. File 3606, box 90, HRCP.

23. File 705, box 36, HRCP.

24. File 721, box 36, HRCP.

25. File 705, box 36, HRCP.

26. File 721, box 36, HRCP.

27. File 4392, box 107, HRCP.

28. File 2291, box 66, HRCP.

29. File 1886, box 53, HRCP.

30. File 1911, box 60, HRCP.

31. On this notice, see entry for 3 July 1918, file 3305, box 83, HRCP; and for 3 July 1918, box 2206, box 65, HRCP.

32. File 3049, box 78, HRCP; file 2206, box 65, HRCP; file 3305, box 83, HRCP.

33. File 3592, box 89, HRCP.
34. File 298, box 24, HRCP.
35. Memorandum from Judge Wallace, item 158.146a, HRCC.
36. David Frank, "The Miners' Financier: Women in the Cape Breton Coal Towns, 1917," *Atlantis* 8, no. 2 (Spring 1983): 137.
37. Catherine Cavanaugh, "The Limitations of Pioneering Partnership: The Alberta Campaign for Homestead Dower, 1909–25," *Canadian Historical Review* 74, no. 2 (1993): 198–225.
38. Linda Gordon, *Pitied but Not Entitled: Single Mothers and the History of Welfare 1890–1930* (New York: Free Press, 1994), 56–58, 52; Molly Ladd-Taylor, *Mother-Work: Women, Child Welfare, and the State, 1890–1930* (Urbana: University of Illinois Press, 1994). Veronica Strong-Boag, "'Wages for Housework': Mothers' Allowances and the Beginnings of Social Security in Canada," *Journal of Canadian Studies* 14, no. 1 (Spring 1979): 24–34
39. The Christians' case in the next two paragraphs is drawn from file 4083, box 101, HRCP, unless otherwise cited.
40. John Griffith Armstrong, *The Halifax Explosion and the Royal Canadian Navy: Inquiry and Intrigue* (Vancouver: University of British Columbia Press, 2002), 37–38, 42–43.
41. The Wassons' story in the next two paragraphs comes from file 304, box 24, HRCP. For Bessie's birthplace, see her entry in the "Halifax Explosion Remembrance Book," available at http://www.gov.ns.ca/nsarm/virtual/remembrance/list.asp?ID=1840, accessed 18 February 2010.
42. See Walter Nugent, *Crossings: The Great Transatlantic Migrations, 1870–1914* (Bloomington: Indiana University Press, 1992), chap. 14.
43. J. Stredder to T. MacIlreith, n.d., item 94.1e, HRCC.
44. Malcolm MacLeod, "Searching the Wreckage for Signs of Region: Newfoundland and the Halifax Harbour Explosion," in Ruffman and Howell, *Ground Zero*, 209–10; MC, 17 December 1917, 2.
45. Charles S. Damrell to [P. F. Martin], 8 December 1917, item 113.2d; Thomas M. Wilson to R. P. Bell, 21 December 1917, item 113.2i; both in HRCC.
46. MacLeod, "Searching the Wreckage," 213. J. Castell Hopkins, "Halifax Disaster and the War," in *Canadian Annual Review of Public Affairs 1917* (Toronto: Canadian Annual Review, 1918), 468, reports a $200,000 contribution from Ontario.
47. MacLeod, "Searching the Wreckage," 208. The population comparison is from 1921.
48. Interview of Bertha Sullivan, née Ryan, 5 July 1985, interview 43, KOH.
49. L. D. McCann, "'Living a Double Life': Town and Country in the Industrialization of the Maritimes," in *Geographical Perspectives on the Maritime Provinces*, ed. Douglas Day (Halifax, Nova Scotia: Saint Mary's University, 1988): 93–113.
50. File 294, box 24, HRCP.
51. Telegram, R. L. Borden to George Perley, 11 December 1917; telegram, N. W. Rowell to Borden, 10 December 1917; clipping from *Ottawa Citizen*, 13 December 1917; all in file OC 445, reel C-4352, RBP; MC, 13 December 1917, 6; Kitz, *Shattered City*, 83–84. For a celebratory and popular accounting of the imperial response to the explosion, see Blair Beed, *1917 Halifax Explosion and American Response* (Halifax, Nova Scotia: Dtours Visitors and Convention Service, 1998), 75–80. See also telegrams promising contributions in the folder labeled "1917 Halifax Explosion," vol. 19, Lieutenant Governor's Correspondence,

RG 1, NSA. For a recitation of large public and private donations from across Canada, North America, and the empire, see Hopkins, "Halifax Disaster and the War," 468–69.

52. West Indian Club of Queen's University to the Officer Administering the Government of the Colony of Barbados, 15 April 1918, enclosure in T. E. Fell to Walter H. Long, 4 June 1918, CO 28/293, the National Archives, Kew. My thanks to Paula Hastings for this citation. See Paula Hastings, "Territorial Spoils, Transnational Black Resistance, and Canada's Evolving Autonomy during the First World War," *Histoire sociale/Social History* 47, no. 94 (June 2014): 443–70.

53. George Hinkley Lyman, *The Story of the Massachusetts Committee on Public Safety* (Boston: Wright and Potter as state printers, 1919), 189–212. See also J. J. Phelan to Warren S. Stone, 21 March 1918, reel 11, M-HRC. On the public-health commission, see William J. Buxton, "Private Wealth and Public Health: Rockefeller Philanthropy and the Massachusetts-Halifax Relief Committee/Health Commission," in Ruffman and Howell, *Ground Zero*, 183–93; Leslie Baker, "'A Visitation of Providence': Public Health and Eugenic Reform in the Wake of the Halifax Disaster," *Canadian Bulletin of Medical History/Bulletin canadien d'histoire de la médecine* 31, no. 1 (2014): 99–122.

54. Quoted in Lyman, *Committee on Public Safety*, 191.

55. Lyman, *Committee on Public Safety*, 211.

56. Massachusetts Bureau of Statistics, *Decennial Census, 1915* (Boston: Wright and Potter as state printers, 1918), 290.

57. Alan A. Brookes, "Out-Migration from the Maritime Provinces, 1860–1900: Some Preliminary Considerations," *Acadiensis* 5, no. 2 (Spring 1976): 26–55; Patricia A. Thornton, "The Problem of Out-Migration from Atlantic Canada, 1871–1921: A New Look," *Acadiensis* 15, no. 1 (Autumn 1985): 3–34; Yves Otis and Bruno Ramirez, "Nouvelles perspectives sur le mouvement d'émigration des Maritimes vers les États-Unis, 1906–1930," *Acadiensis* 28, no. 1 (Autumn 1998): 27–46; Betsy Beattie, *Obligation and Opportunity: Single Maritime Women in Boston, 1870–1930* (Montreal and Kingston: McGill–Queen's University Press, 2000); Bruno Ramirez, *Crossing the 49th Parallel: Migration from Canada to the United States, 1900–1930* (Ithaca, N.Y.: Cornell University Press, 2001).

58. Wilson to R. P. Bell, 21 December 1917, item 113.2i, HRCC.

59. TS memorandum of telephone call from Mr. Reeves, 13 December 1917, reel 16, M-HRC.

60. Undated staff list (before 25 January 1918), item 162.1, HRCC.

61. Lyman, *Committee on Public Safety*, 204–5. On the varying quality of furniture, see G. F. Pearson to H. B. Endicott, 29 July 1918, and Pearson to J. A. Malone, 29 July 1918; both on reel 12, M-HRC.

62. W. T. Murphy to J. J. Phelan, n.d. but received 20 May 1918; Phelan to Murphy 20 May 1918, reel 12, M-HRC. For similar cases, see Phelan to G. F. Pearson, 3 January 1918 and 24 January 1918; and Phelan to Mrs. H. Stone, 5 December 1918; all on reel 12, M-HRC.

63. Pearson to Ratshesky, 6 September 1918, reel 12, M-HRC.

64. The archives in question here are that of the Boston committee, the records of which appear to have been kept rather haphazardly.

65. MacKenzie's story in this and the next paragraph is drawn from file 3819, box 96, HRCP. On Egan, see de Gannes, "Better Suited to Deal with Women and Children," 40–44.

66. The 1911 census found 830 "negroes" in Halifax. See Canada, *Fifth Census of Canada, 1911* (Ottawa: C. H. Parmelee as King's Printer, 1913), 2:194–97. Unless otherwise cited, Henry's story in the next three paragraphs is drawn from file 3337, box 84, HRCP.

67. Robin W. Winks, *The Blacks in Canada* (New Haven, Conn.: Yale University Press, 1971), 306–13; R. Bruce Shephard, *Deemed Unsuitable* (Toronto: Umbrella, 1997), chap. 6; Sarah-Jane Mathieu, *North of the Color Line: Migration and Black Resistance in Canada, 1870–1955* (Chapel Hill: University of North Carolina Press, 2010), 6–7.

68. File 1305, box 49, HRCP. See also file 950, file 41, HRCP.

69. G. F. Pearson to J. J. Phelan, 23 September 1918, reel 12, M-HRC.

70. Mariggi's story in the following four paragraph is drawn from file 1268, box 47, HRCP. There was no consistent spelling of her name.

71. On the relief camp, see Judith Fingard, Janet Guildford, and David Sutherland, *Halifax: The First 250 Years* (Halifax, Nova Scotia: Formac, 1999), 144.

Chapter 5. "A Desirable Measure of Responsibility"

1. On the morgue, see Arthur S. Barnstead, "Interim Report of Mortuary Committee, Halifax Relief, in Connection with the Explosion Disaster, December 6th, 1917," 4 February 1918, item 191, reel 15,123, HEC; statement from Police Officer Leo Tooke, item 258, AMF; Thomas H. Raddall, *In My Time: A Memoir* (Toronto: McClelland and Stewart, 1976), 38–41; Laura M. Mac Donald, *Curse of the Narrows* (New York: Walker, 2005), 160–64; Joseph Scanlon, "Dealing with Mass Death after a Community Catastrophe: Handling Bodies after the 1917 Halifax Explosion," *Disaster Prevention and Management* 7, no. 4 (1998): 288–304.

2. "Morgue at Chebucto Street," item 203, AMF; Janet F. Kitz, *Shattered City: The Halifax Explosion and the Road to Recovery* (Halifax, Nova Scotia: Nimbus, 1989), 108. On identifying the dead and the objects collected from the unknown bodies, see Janet F. Kitz, "The Explosion Mortuary Artifacts: A Look at the Victims," in *Ground Zero: A Reassessment of the 1917 Explosion in Halifax Harbour*, ed. Alan Ruffman and Colin D. Howell (Halifax, Nova Scotia: Nimbus/The Gorsebrook Research Institute for Atlantic Canada Studies at Saint Mary's University, 1994): 9–34.

3. Description of burial service for unidentified dead, J. H. Mitchell, 17 December 1918, item 282, AMF; Kitz, *Shattered City*, 108–9.

4. MC, 21 December 1917, 8.

5. John C. Weaver, "Reconstruction of the Richmond District in Halifax: A Canadian Episode in Public Housing and Town Planning, 1918–1921," *Plan Canada*, March 1976, 36–38; Michelle Hébert Boyd, *Enriched by Catastrophe: Social Work and Social Conflict after the Halifax Explosion* (Halifax, Nova Scotia: Fernwood, 2005).

6. The classic work on Canadian laborism is Craig Heron, "Labourism and the Canadian Working Class," *Labour/Le Travail* 13 (Spring 1984), 45–76 (quote on 72). On laborism in Halifax, see Suzanne Morton, "The Halifax Relief Commission and Labour Relations during the Reconstruction of Halifax, 1917–1919," *Acadiensis* 18, no. 2 (Spring 1989): 73–93 (quote on 75); Suzanne Morton, "Labourism and Economic Action: The Halifax Shipyards Strike of 1920," *Labour/Le Travail* 22 (Fall 1988), 67–98.

7. *Canadian Railroad Employees' Monthly* 4, no. 1 (March 1918): 578.

8. Personal testimony of Rev. Hugh Upham, item 234, AMF.

9. Personal narrative of Mr. L. A. Myers [Miles], item 205; story brought by Mary P. Freeman, item 276f; both in AMF.

10. Quote from a letter by W. B. Fielding, 15 December 1917; MC, 17 December 1917, 3. See also J. H. Tuck, "Nova Scotia and the Conscription Election of 1917" (MA thesis, Dalhousie University, 1968), 62; Judith Fingard, Janet Guildford, and David Sutherland, *Halifax: The First 250 Years* (Halifax, Nova Scotia: Formac, 1999), 128, 174–75; and Ernest R. Forbes, "Battles in Another War: Edith Archibald and the Halifax Feminist Movement," in *Challenging the Regional Stereotype: Essays on the 20th Century Maritimes* (Fredericton, New Brunswick: Acadiensis Press 1989), 71.

11. [Questionnaire and answers to questionnaire requested by W. C. Milner], 12 July 1920, item 7, HEC.

12. MC, 8 December 1917, 3.

13. ADR, 21 December 1917, 2; EM, 11 December 1917, 1.

14. Personal narrative of Miss Jessie Parker, item 214; W. A. N. Conaghan to MacMechan, 29 January 1918, item 1a; both AMF.

15. R. T. MacIlreith to D. MacGillivray, 14 December 1917, item 158.33f, HRCC.

16. [D. MacGillivray] to Worrell, 15 December 1917, item 158.23, HRCC. The Dartmouth Relief Committee, also acting on a request from clergy for a formal role in relief, recognized them as visitors to investigate applicants and advise the committee. See Minutes of the [Dartmouth] Registration & Relief Committee Meeting, 24 January 1918, Item 158.232h, HRCC.

17. MacGillivray to William Foley, 21 December 1917, item 158.34, HRCC. Foley was Catholic, but MacGillivray told him he was making the same request of Protestant clergy.

18. EM, 15 December 1917, 6.

19. MS list of "Contributions toward Relief Fund," EJM.

20. Rector's Address reprinted in Parish Notes, pasted in with minutes of annual meeting, 4 February 1918, Minute book 4 February 1918 to 15 December 1921, item 1, vol. 318, St.GAP.

21. H. Y. Payzant to "Bro. Reid," 16 December 1917, item 161.6, HRCC.

22. File 736, box 37, HRCP.

23. File 477, box 30, HRCP.

24. File 1587, box 53, HRCP.

25. File 4721, box 113, HRCP.

26. File 3818, box 96, HRCP.

27. File 1268, box 47, HRCP (Mariggi); file 960, box 41, HRCP (Swan).

28. The Carson story in the following five paragraphs is drawn from file 2647, box 71 (Carson); file 2285, box 66 (Gaudet); and file 403, box 28 (Major); all HRCP.

29. Rector's Address reprinted in Parish Notes, pasted in with minutes of annual meeting, 4 February 1918, Minute book 4 February 1918 to 15 December 1921, item 1, vol. 318, St.GAP.

30. File 963, box 41, HRCP.

31. MC, 8 December 1917, 3.

32. *St. Paul's Church Parish Magazine* 34, nos. 1–2 (January–February 1918): 1–3, in St. Paul's Church Parish Magazines, accession 2001-5-311, St.PP.

33. MC, 13 December 1917, 7.
34. Ibid., 14 May 1918, 4; personal narrative of Mrs. Annie Anderson, item 114, AMF.
35. ADR, 7 January 1918, 2.
36. See display ads from Oxford Street Methodist, J. Wesley Smith (Methodist), West End Baptist, North Baptist, Salvation Army, St. Matthew's, Cathedral of All Saints (Anglican), Robie Street Methodist, and Grafton Street Methodist that ran in the EM, 15 December 1917, 4; and an advertisement from St. Joseph's Catholic that ran in the MC, 12 December 1917, 3.
37. Personal narrative of Mr. L. A. Myers [Miles], item 205, AMF.
38. MC, 28 December 1917, 7.
39. Rector's Address reprinted in Parish Notes, pasted in with minutes of annual meeting, 4 February 1918, Minute book 4 February 1918 to 15 December 1921, item 1, vol 318, St.GAP.
40. *St. Paul's Church Parish Magazine* 33, no. 12 (December 1917): 6–7, in St. Paul's Church Parish Magazines, accession 2001-5-311, St.PP.
41. *St. Paul's Church Parish Magazine* 34, nos. 1–2 (January–February 1918): 1–3, in St Paul's Church Parish Magazines, accession 2001-5-311, St.PP.
42. ADR, 5 January 1918, 3; see also Scrapbook, October 25, 1917–June 1919, vol. 71, CCH.
43. Resolution from the Maritime Hierarchy, EJM.
44. MC, 29 December 1918, 1.
45. R. W. Ross, "What Church Union Will Mean for the City of Halifax," *New Outlook*, 10 June 1925, 37.
46. Report of the Committee of the Privy Council, 9 March 1918, item 23, AMF; "Churches," 31 December 1918, item A.6.1, Appraisal Board Records, series A, Halifax Relief Commission fonds, MG 36, NSA; "Damages Caused by Explosion," EJM.
47. The Devastated District of Dartmouth, personal observation of J. H. Mitchell, item 173, AMF; Edith Rowlings, *The Story of Emmanuel Church, Dartmouth, N.S. 1871–1987* (Dartmouth, Nova Scotia: Emmanuel Church, 1987), 32–36, NSA vertical file v.374 #6.
48. EM, 16 June 1920, 11–12; "Churches," 31 December 1918, item A.6.1, Appraisal Board Records, series A, Halifax Relief Commission fonds, MG 36, NSA; "The Council Reports, 1948–1949," TS perhaps by Jan Goeb, item 6, vol. 565, Jewish Historical Society fonds, MG 20, NSA; Beth Israel Synagogue, "About Us," available at http://www.jewish halifax.com/about/, accessed 8 June 2014.
49. Rev. C. J. Crowdis, MA, "A Common Sorrow and a Common Concern: Being a brief statement of the Halifax Disaster and the affairs, past and present of Kaye-Grove Church," pamphlet printed 1920, NSA vertical file v.113 #29.
50. John Webster Grant, *The Canadian Experience of Church Union* (Richmond, Va.: John Knox Press, 1967), 28.
51. The most important history of church union remains the encyclopedic Claris Edwin Silcox, *Church Union in Canada: Its Causes and Consequences* (New York: Institute of Social and Religious Research, 1933). See also W. E. Mann, "The Canadian Church Union, 1925" in *Institutionalism and Church Unity: A Symposium*, ed. Nils Ehrenstrom and Walter S. Muelder (New York: Association Press, 1963): 171–93; N. Keith Clifford, *The Resistance to Church Union in Canada, 1904–1939* (Vancouver: University of British Columbia Press,

1985), 14–25. On church union in the Maritimes, see James Cameron, "The Garden Distressed: Church Union and Dissent on Prince Edward Island, 1925," *Acadiensis* 21, no. 2 (Spring 1992): 108–31; and Twila F. Buttimer, "Great Expectations: The Maritime Methodist Church and Church Union, 1925" (MA thesis, University of New Brunswick, 1980).

52. For an entertaining and insightful description of how this worked at one church, see John Kenneth Galbraith, *The Scotch* (Cambridge, Mass.: Riverside Press, 1964), 100–103.

53. Grant, *Canadian Experience of Church Union*, 45–49 (quote on 46); Silcox, *Church Union in Canada*, 219–30.

54. Crowdis, "Common Sorrow and a Common Concern," 1920 pamphlet, NSA. The numbers of dead are from a questionnaire and answers to questionnaire requested by W. C. Milner, 12 July 1920, item 7, HEC. These numbers are questionable; Crowdis gave a combined total death toll of 239.

55. ADR, 9 January 1918, 3.

56. Crowdis, "Common Sorrow and a Common Concern," 1920 pamphlet, NSA.

57. HH, 10 January 1918, 3; ADR, 9 January 1918, 3.

58. Session meeting of 10 January 1918, 8:00 PM, Session Book, 11 January 1893–21 September 1920, vol. 60, St. Matthew's Church fonds, MG 4, NSA; Journal entry for 16 January 1918, item 77s, AMF.

59. Robert E. Inglis, Beverly Were, W. G. Grant, and Albert E. Kingsbury, "United Memorial Church: 1918–1975" (self published, March 1975), NSA vertical file v.237 #16.

60. Journal entry for 21 January 1918, item 77v, AMF.

61. Inglis et al., "United Memorial Church: 1918–1975," NSA; Crowdis, "Common Sorrow and a Common Concern," 1920 pamphlet, NSA.

62. Crowdis, "Common Sorrow and a Common Concern," 1920 pamphlet, NSA.

63. Inglis et al., "United Memorial Church: 1918–1975," NSA; D. W. Johnson, *History of Methodism in Eastern British America* (Sackville, New Brunswick: Tribune, [1925?]), 25–26, 96.

64. Ross, "What Church Union Will Mean," 37.

65. Crowdis, "Common Sorrow and a Common Concern," 1920 pamphlet, NSA.

66. See, for instance, Crowdis, "Common Sorrow and a Common Concern," 1920 pamphlet, NSA.

67. Kitz, *Shattered City*, 194–95, 210–12. On the development of Fort Needham into a memorial park, see Paul Alfred Erickson, "Halifax's Other Hill: Fort Needham from Earliest Times," *Occasional Papers in Anthropology* (Halifax, Nova Scotia: Department of Anthropology, Saint Mary's University, 1984), 34–39.

68. Bryan D. Palmer, *Working-Class Experience: Rethinking the History of Canadian Labour, 1800–1991*, 2nd ed. (Toronto: McClelland and Stewart, 1992), 207.

69. On Joy, see Morton, "Labourism and Economic Action," 72. On the crisis in workman's compensation, see meetings of 28 December 1917 and 14 March 1918, series 2, Workers' Compensation board of Nova Scotia fonds, reel 14,890, NSA.

70. See "Secretary of the Meeting" [Fred F. Mathers?] to each of the invitees, 22 December 1917, items 79–87, HEC.

71. [Mathers] to A. K. Maclean, 10 January 1918, item 162, HEC.

72. J. Howard T. Falk, "History of Rehabilitation Work Since January 9, 1918," 26

February 1918, item 162.5, HRCC; Jos. A. Garnett to J. H. Winfield, 14 January 1918, item 158.75, HRCC.

73. Warren S. Stone to F. A. Hallet, 16 March 1918, reel 12; J. J. Phelan to Warren S. Stone, 21 March 1918, reel 11; JFO'H [J. Frank O'Hare] to G. Fred Pearson, 15 June 1918, reel 12; all in M-HRC.

74. G. A. Stone to G. B. Cutten, 10 July 1918, item 161.91, HRCC. See also file 475, box 30, HRCP.

75. See file 483, box 30; file 478, box 30; file 967, box 41; and file 2392, box 67; all HRCP.

76. File 3035, box 77, HRCP.

77. Remarks by John T. Joy, *Seventh Annual Convention of the Atlantic Coast District of the International Longshoremen's Association* (Boston: I. H. Feinberg and Son, 1919), 11. Thanks to David Frank for this reference.

78. File 1268, box 47, HRCP.

79. File 2647, box 71, HRCP.

80. MC, 21 December 1917, 8.

81. *Canadian Railroad Employees' Monthly*, March 1918, 601. No copy of the magazine's volume 3 appears to have survived, and so there is no way of knowing what, if anything, the local union reported in the December 1917, January 1918, or February 1918 issues.

82. Ian McKay and Suzanne Morton, "The Maritimes: Expanding the Circle of Resistance," in *The Workers' Revolt in Canada*, ed. Craig Heron, 43–86; Sarah-Jane Mathieu, *North of the Color Line: Migration and Black Resistance in Canada, 1870–1955* (Chapel Hill: University of North Carolina Press, 2010), 121.

83. Maurine Weiner Greenwald, *Women, War, and Work: The Impact of World War I on Women Workers in the United States* (Westport, Conn.: Greenwood, 1980), 139–84.

84. Peter D. Lambly, "Working Conditions and Industrial Relations on Canada's Street Railways, 1900–1920" (MA thesis, Dalhousie University, 1983), 134–36.

85. Ibid., 137–41.

86. Greenwald, *Women, War, and Work*, 176–80.

87. Sarah-Jane (Saje) Mathieu, "North of the Colour Line: Sleeping Car Porters and the Battle against Jim Crow on Canadian Rails, 1880–1920," *Labour/Le Travail* 47 (Spring 2001), 38–40 (quote on 40).

88. Mathieu, *North of the Color Line*, 97.

89. *Canadian Railroad Employees' Monthly*, April 1918, 119.

90. William E. Greening and Murdoch MacKay Maclean, *It Was Never Easy 1908–1958: A History of the Canadian Brotherhood of Railway, Transport and General Workers* (Ottawa: Mutual Press, 1961), 58–60; Mathieu, *North of the Color Line*, 133–34.

91. Morton, "Halifax Relief Commission," 81–83 (Eisnor quote on 83). Low began signing his telegrams as reconstruction manager by December 10, a mere four days after the explosion. See his telegrams in item 19900192, 58A 1 2.11, Col. Robert S. Low papers, in the Air Vice Marshall Clifford Mackey McEwan papers, George Metcalf Archival Collection, Canadian War Museum, Ottawa, Ontario.

92. Morton, "Halifax Relief Commission," 78, 84–88; Ian McKay, *The Craft Transformed: An Essay on the Carpenters of Halifax, 1885–1985* (Halifax, Nova Scotia: Holdfast Press, 1985), 64–68.

93. D. J. Bercuson, "Organized Labour and the Imperial Munitions Board," *Relations industrielles/Industrial Relations* 28, no. 3 (July 1973): 602–16. On the postwar "labor revolt," see Gregory S. Kealey, "1919: The Canadian Labour Revolt," *Labour/Le Travail* 13 (Spring 1984): 11–44; and Craig Heron, ed., *The Workers' Revolt in Canada, 1917–1925* (Toronto: University of Toronto Press, 1998).

94. Morton, "Halifax Relief Commission," 88–91 (quote [from the *Citizen,* 22 January 1920] on 90); McKay, *Craft Transformed*, 69–71; Morton, "Labourism and Economic Action," 75, 78–79, 81.

95. Morton, "Labourism and Economic Action," 83, 89–90, 85–87.

96. Ibid., 90–93; Fingard, Guildford, and Sutherland, *Halifax*, 139–40. The Halifax Shipyards had been owned since 1918 by Roy M. Wolvin, and they would form the nucleus of Wolvin's British Empire Steel Corporation, or Besco. Later in the decade, Besco gained notoriety for poor financial practices and vicious antiunionism in the steel factories and coalfields of Cape Breton. See David Frank, "The Cape Breton Coal Industry and the Rise and Fall of the British Empire Steel Corporation," *Acadiensis* 7, no. 1 (Autumn 1977): 3–34.

97. Morton, "Labourism and Economic Action," 95–96; Ernest R. Forbes, *The Maritime Rights Movement, 1919–1927: A Study in Canadian Regionalism* (Montreal and Kingston: McGill–Queen's University Press, 1979); McKay and Morton, "The Maritimes," 77–79.

Chapter 6. "The Sufferings of This Time Are Not Worthy to Be Compared with the Glory That Is to Come"

1. *Pilot,* 4 July 1914, 2; SEN, 29 June 1914, 3.
2. Rom. 8:18. The translation is the Rheims-Douay Version, which was used by Anglophone Catholics until the mid-twentieth century; Binette, of course, would have read the passage in French. June 28, 1914, was the fourth Sunday after Pentecost, and following the Roman missal, Catholics around the world read Rom. 8:18–23 and Luke 5:1–11.
3. Luke 5:5–7, quote from verse 6, Rheims-Douay translation.
4. *Pilot,* 4 July 1914, 2.
5. Horace Miner, *St. Denis: A French-Canadian Parish* (Chicago: University of Chicago Press, 1939), 105.
6. SEN, 27 June 1914, 1. The SEO used precisely the same phrase, 27 June 1914, 4.
7. SEN, 29 June 1914, 3. His text was Joel 2:25, quoted here in the King James translation.
8. SEN, 6 July 1914, 12. Langdale's text was Heb. 11:10.
9. *Pilot,* 4 July 1914, 2.
10. SEN, 1 July 1914, 5; SEN, 6 July 1914, 8; SEN, 10 July 1914, 8; SEN, 11 July 1914, 8; *Pilot,* 18 July 1914, 1. The churches I designate as "Irish" were technically territorial parishes without national designations. In practice, this meant they were Irish. The archdiocese included most of eastern Massachusetts, stretching from the border with New Hampshire south to Cape Cod and as far west as the Worcester County line. For its Catholic population—estimated at 850,000 in 1907—see *Catholic Encyclopedia,* s.v. "Boston" (New York: Encyclopedia Press, 1913–1914), available at http://www.newadvent.org/cathen/02703a.htm, accessed 2 April 2010.

11. Montayne Perry, *The Salem Fire Relief* (Salem, Mass.: Milo A. Newhall, 1915), 35.
12. Ibid., 50.
13. John P. Sullivan to William Cardinal O'Connell, 4 February 1915, folder 68:9, PC.
14. SEN, 1 July 1914, 10.
15. Perry, *Salem Fire Relief*, 29–30.
16. Ibid., 37.
17. Ibid., 38–44 (quote on 43).
18. SEN, 1 July 1914, 10.
19. 1910 census, series T624, roll 587, p. 71.
20. SEN, 26 June 1914, 2.
21. SEN, 30 June 1914, 5; J. L. Simon to William Cardinal O'Connell, 26 November 1919, folder 68:23, PC.
22. *Courrier de Lawrence*, 30 June 1914, 1.
23. SEN, 30 June 1914, 5. For an illustrated history of the Polish community, see Felicia L. Wilczenski and Emily A. Murphy, *The Polish Community of Salem* (Charleston S.C.: Arcadian, 2012).
24. SEN, 2 July 1913, 13; 1920 census, series T625, roll 737, p. 204. Though the census listed Brogi as a telephone clerk in 1920, by the next year the city directory listed him as a publisher. He also served the Mazzini Association—which was headquartered at his print shop—as its welfare director. See *Boston Register and Business Directory* (Boston: Sampson and Murdock, 1921), 47, 322, 724.
25. SEN, 8 July 1914, 8; SEN, 15 July 1914, 5; *Naumkeag Directory*, 257, 424.
26. SEN, 3 July 1914, 6.
27. Minutes of Executive Committee meeting, 31 August 1914, American Jewish Committee minute book, Blaustein Library and Historical Archive, American Jewish Committee, New York. Thanks to the Blaustein's director, Michele Anish, for sending the minutes.
28. SEN, 8 July 1914, 9.
29. Gerard J. Brault, *The French-Canadian Heritage in New England* (Hanover, N.H.: University Press of New England, 1986), 9, 7, 66–67, 70–73;Yves Roby, *The Franco-Americans of New England: Dreams and Realities*, trans. Mary Ricard ([Sillery, Quebec]: Septentrion, 2004), 167–88; Richard S. Sorrell, "The Sentinelle Affair (1924–1929): Religion and Militant Survivance in Woonsocket, Rhode Island," *Rhode Island History* 36, no. 3 (August 1977): 67–80. Historians have begun to turn away from this narrative, noting *survivance* as an elite concern. See Mark Paul Richard, *Loyal but French: The Negotiation of Identity by French-Canadian Descendants in the United States* (East Lansing: Michigan State University Press, 2008), 29–30.
30. The quote is from Gary Gerstle, *Working-Class Americanism: The Politics of Labor in a Textile City, 1914–1960*, rev. ed. (1989; repr. Princeton, N.J.: Princeton University Press, 2002), 40. See also Philip T. Silvia Jr., "Neighbors from the North: French-Canadian Immigrants vs. Trade Unionism in Fall River, Massachusetts," in *Steeples and Smokestacks: A Collection of Essays on the Franco-American Experience in New England*, ed. Claire Quintal (Worcester, Mass.: Institut Français, Assumption College, 1996); Pierre Anctil, "Aspects of Class Ideology in a New England Ethnic Minority: The Franco-Americans of Woonsocket, Rhode Island (1865–1929)" (PhD diss., New School for Social Research, 1980). It is, of

course, not unusual for historians to portray Catholics as priest-ridden, precisely because they have believed claims of clerical authority without noticing that they were contested. See Leslie Woodcock Tentler, "On the Margins: The State of American Catholic History," *American Quarterly* 45, no. 1 (March 1993): 104–27. On the history of lay leadership within the American church and the bishops' contested rise to power, see Jay P. Dolan, *The American Catholic Experience: A History from Colonial Times to the Present* (1985; repr. Notre Dame, Ind.: University of Notre Dame Press, 1992), esp. chap. 6.

31. Richard, *Loyal but French*, 60, 116, 117–18, 188.

32. Julie Byrne, *O God of Players: The Story of the Immaculata Mighty Macs* (New York: Columbia University Press, 2003), 10–11; James P. McCartin and Joseph A. McCartin, "Working-Class Catholicism: A Call for New Investigations, Dialogue, and Reappraisal," *Labor: Studies in Working-Class History of the Americas* 4, no. 1 (Spring 2007): 99–110.

33. Evelyn Savidge Sterne, *Ballots and Bibles: Ethnic Politics and the Catholic Church in Providence* (Ithaca, N.Y.: Cornell University Press, 2003).

34. Jennifer Guglielmo, "Italian Women's Proletarian Feminism in the New York City Garment Trades, 1890s–1940s," in *Women, Gender, and Transnational Lives: Italian Workers of the World*, ed. Donna R. Gabaccia and Franca Iacovetta (Toronto: University of Toronto Press, 2002): 247–98.

35. *Guide Franco-Américain des Etats de Nouvelle-Angleterre, 1916* (Fall River, Mass.: A. A. Belanger, 1916), 99.

36. France Ariel, *Canadiens et Américains chez eux: Journals, lettres, impressions d'une artiste française* (Montreal: Granger Frères, 1920), 272.

37. SEN, 8 July 1914, 9.

38. *Guide Franco-Américain*, 98, 99; *La Paroisse Saint-Joseph: 1873–1948: Soixante-quinzième anniversaire* ([Salem, Mass.?]: Association Laurier, 1948), 8, 12; William Byrne et al., *History of the Catholic Church in the New England States* (Boston: Hurd and Everts, 1899), 271, 342–43, 316–17.

39. SEN, 27 June 1914, 5.

40. Ibid., 8 July 1914, 6.

41. CdS, 17 June 1914, 5.

42. SEN, 30 June 1914, 10; SEN, 1 July 1914, 10. For a normal Sunday schedule, see SEO, 20 June 1914, 4, which advertised Low Mass at 7:00, 8:00, and 9:15, plus High Mass at 10:30.

43. SEN, 8 July 1914, 6; SEN, 26 June 1914, 14.

44. Perry, *Salem Fire Relief*, 17.

45. Edward Dunbar Johnson, "The Salem Fire," after June 1915, E S1 F6 1914$_{11}$, PEM.

46. SEN, 1 July 1914, 11.

47. AN, 6 July 1914, 1.

48. SEN, 1 July 1914, 5.

49. Ibid., 29 June 1914, 5.

50. *La Revue de Salem*, August 2003, 24–25.

51. On Baptist missionary activity in Quebec, see Jean-Louis Lalonde, *Des loups dans la bergerie: Les protestants de langue française au Québec, 1534–2000* ([Montreal]: Fides, 2002), 205–12. On Francophone Protestants in Quebec generally, see also Marie-Claude Rocher, "Double traïtrise ou double appartenance? Le patrimoine des protestants fran-

cophones au Québec," *Ethnologies* 25, no. 2 (2003): 215–33; Denis Remon, ed., *L'identité des protestants francophones au Québec: 1834–1997* (Montreal: ACFAS, 1998); David-Thiery Ruddel, *Le protestantisme français au Québec, 1840–1919: "Images" et témoignages*, Mercury Series History 36 (Ottawa: Musée national de l'homme, 1983); Roberto Perin, "French-Speaking Canada from 1840," in *A Concise History of Christianity in Canada*, ed. Terrence Murphy (Toronto: Oxford University Press, 1996), 191–96.

52. SEN, 30 June 1914, 2.

53. Report of the American Baptist Foreign Mission Society, in *Annual of the Northern Baptist Convention, 1922* (N.p.: American Baptist Publication Society, 1922), 531; *Annual Report of the Essex Institute for the Year Ending May 5, 1919* (Salem, Mass.: Printed for the Essex Institute by Newcomb and Gauss, 1919), 13; Robert G. Torbet, *Venture of Faith: The Story of the American Baptist Foreign Mission Society and the Woman's American Baptist Foreign Mission Society, 1814–1954* (Philadelphia: Judson Press, 1955), 581.

54. Richard, *Loyal but French*, 94–95; Michael Guignard, "Maine's Corporation Sole Controversy," *Maine Historical Society Newsletter* 12 (Winter 1973): 111–30; Claire Quintal, "Les institutions franco-américaines: Pertes et progrès," in *Le Québec et les francophones de la Nouvelle-Angleterre*, ed. Dean Louder (Sainte-Foy, Quebec: Presses de l'Université Laval, 1991), 67; George Neil Emery and John Charles Herbert Emery, *A Young Man's Benefit: The Independent Order of Odd Fellows and Sickness Insurance in the United States and Canada, 1860–1929* (Montreal and Kingston: McGill–Queen's University Press, 1998), 28.

55. André Boucher, "Le rôle joué par les marguilliers," in *Le laïc dans l'Église canadienne-française de 1830 à nos jours,* by Pierre Hurtubise et al. (Montreal: Fides, 1972), 163–73; Brault, *French-Canadian Heritage*, 9; Dolan, *American Catholic Experience*, 178–79; Miner, *St. Denis*, 54–56.

56. David J. Blow, "The Establishment and Erosion of French-Canadian Culture in Winooski, Vermont, 1867–1900," *Vermont History* 43, no. 1 (January 1975): 63–65.

57. Dolan, *American Catholic Experience*, 179–80, 189–92; Guignard, "Maine's Corporation Sole Controversy"; James B. O'Hara, "The Modern Corporation Sole," *Dickinson Law Review* 93 (Fall 1988): 23–39.

58. Douglas J. Slawson, *Ambition and Arrogance: Cardinal William O'Connell of Boston and the American Catholic Church* (San Diego, Calif.: Cobalt Productions, 2007), 61–62.

59. On French Canadian devotionalism, see Perin, "French-Speaking Canada," 197–203; Susan Mann, *The Dream of Nation: A Social and Intellectual History of Quebec*, 2nd ed. (1982; repr. Montreal and Kingston: McGill–Queen's University Press, 2002), 122. For an example of Cardinal William O'Connell standing in the way of French Canadian religious practice, see James M. O'Toole, *Militant and Triumphant: William Henry O'Connell and the Catholic Church in Boston, 1859–1944* (Notre Dame, Ind.: University of Notre Dame Press, 1992), 154.

60. Gerstle, *Working-Class Americanism*, 40.

61. O'Toole, *Militant and Triumphant*, 171.

62. Richard, *Loyal but French*, 94–95; Roby, *Franco-Americans*, 176–80, 187; Slawson, *Ambition and Arrogance*, 35. Guignard reports that O'Connell left Portland "before French dislike for him had crystallized" but notes that "several of his policies had drawn sharp criticism from Franco-Americans"; Guignard, "Maine's Corporation Sole Controversy," 113.

63. Guignard, "Maine's Corporation Sole Controversy," 115.
64. Roby, *Franco-Americans*, 171–73.
65. Quoted in Slawson, *Ambition and Arrogance*, 7.
66. CdS, 24 July 1914, 1; CdS, 31 July 1914, 5; CdS, 11 September 1914, 8; CdS, 1 October 1914, 1; CdS, 17 December 1914, 1.
67. F. J. Maley to O'Connell, 18 January 1915, folder 38:14, PC.
68. CdS, 31 July 1914, 1.
69. J. L. Boureault, "Cinquantième Anniversaire de la Paroisse de Sainte-Anne de Salem, Mass.," 11 November 1951, Institut Français, Assumption College, Worcester, Mass.; Richard J. Haberlin to Joseph A. Peltier, 15 April 1916, folder 68:14, PC.
70. R. J. Sullivan to Rainville, 23 March 1915, folder 68:1; Maley to O'Connell, 20 March 1915, folder 38:14; both PC.
71. Rainville to O'Connell, 24 March 1915, folder 68:1, PC.
72. Sullivan to Rainville, 23 [26] March 1915, folder 68:1, PC.
73. Sullivan to Rainville, 9 June 1915, folder 68:1, PC.
74. Rainville to O'Connell, 10 June 1915, folder 68:1, PC.
75. [Lorraine St. Pierre], *Saint Joseph Parish: 1873–1973* (Salem, Mass.: Printed by Compass Press, 1973), 10; Miss Marguerite-Marie Deschamps to O'Connell, 22 June 1915, folder 68:1, PC; *Paroisse Saint-Joseph*, 12–13. *Cortons*, usually *cretons*, are crackers covered with a pork spread.
76. Rainville to O'Connell, 14 April 1916; Haberlin to Rainville, 17 April 1916; both in folder 68:1, PC.
77. [St. Pierre], *Saint Joseph Parish*, 10; *Paroisse Saint-Joseph*, 13–14; Philip E. Jalbert, "The French-Canadian in South Salem, Massachusetts," seminar paper, Salem State College, 1975, reprinted in *Immigration in American Life: Graduate Seminar Research Papers on North Shore Selected Topics*, ed. Adele L. Younis, vol. 5 (Salem, Mass: Salem State College, 1977), 15.
78. R. J. Sullivan to Maley, 31 May 1915; Daniel F. Horgan to O'Connell, 5 December 1915; Sullivan to Horgan, 4 December 1915; Horgan to Haberlin, 18 October 1918; all in folder 38:14, PC.
79. John A. Degan to O'Connell, 20 October 1916, folder 28:1, PC.
80. *Paroisse Saint-Joseph*, 12.
81. Affidavit by Joseph Lemary, 29 March 1916; Alberg Goguerd [?] to O'Connell, 10 July 1916; all in folder 28:1, PC; Rainville to Haberlin, 8 February 1917; Haberlin to Rainville, 24 February 1917; Haberlin to Rainville, 26 March 1917; Rainville to O'Connell, 24 May 1917; all in folder 68:1, PC. On the church's bad condition, see, e.g., L. C. Bedard to Haberlin, 30 August 1917; Bedard to Haberlin, 4 March 1918; both in folder 28:1, PC.
82. Haberlin to Rainville, 11 June 1917, folder 68:1, PC.
83. The school survived but moved across town to become "St. Joseph School at St. James Parish." By the time the parish closed, many of its parishioners were Spanish-speaking, reflecting the changed demographics of the Point. Nonetheless, the memory of its Francophone heritage remained. See these articles by Kathy McCabe: "Closing Parish Offers Up Its Light: Church in Salem Hosts Final Mass after 131 Years," *Boston Globe*, 15 August 2004, North section, 1; "Catholic Elementary School Survives," *Boston Globe*, 15 August 2004, North section, 1; "'Our New Home,'" *Boston Globe*, 29 May 2005,

North section, 1; and "Past to Shape Church's Future," *Boston Globe*, 26 June 2005, North section 6.

84. Michael Paulson and Kathy McCabe, "A Rebirth for Beverly Church: Worshipers Start Parish as Gift in Dominican Republic," *Boston Globe*, 29 August 2004, A1.

85. *Paroisse Saint-Joseph*, 13. Translated in Elaine Call, "The Church in Salem, Massachusetts: A Study of the Spiritual and Physical Significance of Immigrant Churches in America," seminar paper, Salem State College, 1975, reprinted in *Immigration in American Life: Graduate Seminar Research Papers on North Shore Selected Topics*, ed. Adele L. Younis, vol. 1 (Salem, Mass.: Salem State College, 1977), 45.

86. *Paroisse Saint-Joseph*, 14.

87. Roby, *Franco-Americans*, 264; Sorrell, "Sentinelle Affair."

88. Slawson, *Ambition and Arrogance*, 5–7; Perin, "French-Speaking Canada," 257; Mann, *Dream of Nation*, 115–30. Ultramontanists believed in the supreme episcopal, religious, and temporal authority of the pope. Their national iterations were theologically and politically linked: in France and Quebec they opposed *laïcité* and urged a greater role for the Church in secular life; in Italy they demanded the return of the Papal States; and in the United States they stood against the liberal "American heresy."

89. Dolan, *American Catholic Experience*, 110, 114–16.

90. Brault, *French-Canadian Heritage*, 8–11.

91. Binette claimed that fifty-two of fifty-five Franco-American priests in the Boston archdiocese sided with the *Sentinellistes*, and an opponent agreed that Massachusetts priests were "almost unanimous." Roby, *Franco-Americans*, 264. On deindustrialization, see Shaun S. Nicholls, "Crisis Capital: The Making and Unmaking of Industrial Massachusetts, 1873–present" (PhD diss., Harvard University, in progress).

92. Pauline Chase-Harrell, Carol Ely, and Stanley Moss, "Administrative History of the Salem Maritime National Historic Site," report prepared for the National Park Service North Atlantic Regional Office, 9 February 1993, 7–19; James M. Lindgren, *Preserving Historic New England: Preservation, Progressivism, and the Remaking of Memory* (New York: Oxford University Press, 1995), 168–70.

93. Richard Carter Nyman, *Union-Management Coöperation in the "Stretch Out": Labor Extension at the Pequot Mills* (New Haven, Conn.: Yale University Press, 1934), 210; Aviva Chomsky, *Linked Labor Histories: New England, Colombia, and the Making of a Global Working Class* (Durham, N.C.: Duke University Press, 2008), 85.

94. *Naumkeag Directory*, 475; Massachusetts Bureau of Statistics, *Directory of Labor Organizations in Massachusetts, 1914*, no. 98, vol. 2, 9 March 1914, 47–48; SEN, 29 June 1914, 3.

95. The SEN estimated 3,000 workers, 26 June 1914, 5; T. V. Powderly estimated 1,500 in his report to [William Wilson], 28 July 1914, reprinted in United States Department of Labor, *Reports of the Department of Labor, 1914* (Washington, D.C.: GPO, 1915), 296.

96. SEN, 6 July 1914, 1.

97. Chomsky, *Linked Labor Histories*, 37, 54.

98. Entries for 21 June 1918 and 18 September 1918, NDR. For an account of the strike, see Chomsky, *Linked Labor Histories*, 49–51.

99. Entries for 21 June 1918 and 18 September 1918, NDR (quote from the latter entry); Nyman, *Union-Management Coöperation*, 3, 6. On wartime inflation—consumer

prices doubled nationally from 1914 to 1920—see Frank Stricker, "The Wages of Inflation: Worker's Earnings in the World War One Era," *Mid-America* 63, no. 2 (April 1981): 93–105.

100. "Joint Research in the Pequot Mills. Confidential memorandum prepared by Frances Goodell, former industrial engineer for the Pequot Mills," [August? 1933], binder labeled "Naumkeag Steam Cotton Co. (Pequot Mills.) Working Data. Stretch Out and General," box 9, SNP; William F. Hartford, *Where Is Our Responsibility? Unions and Economic Change in the New England Textile Industry, 1870–1960* (Amherst: University of Massachusetts Press, 1996), 40–42, 51, 54; Cletus E. Daniel, *Culture of Misfortune: An Interpretive History of Textile Unionism in the United States* (Ithaca, N.Y.: ILR Press, 2001), 13–28.

101. Entry for 25 November 1918, NDR.

102. Entry for 19 December 1918, NDR.

103. Chomsky, *Linked Labor Histories*, 51; Nyman, *Union-Management Coöperation*, 1 (n. 2), 6.

104. O. S. Beyer, "Proposed Program of Cooperation for the United Textile Workers of America and the Naumkeag Steam Cotton Company," 25 February 1930 in binder marked on inside "Joint Research Studies Loaned by Mr. J. P. O'Connell," box 8, SNP. The loom fixers may have joined the UTW in 1925; see pamphlet by Thomas F. McMahon, "United Textile Workers of America," Workers' Education Organization Series no. 2 (New York: Workers' Education Bureau Press, 1926), 39.

105. Chomsky, *Linked Labor Histories*, 52–54.

106. R. D. Smith[?], "Report of Visit to the Naumkeag Mill," 20 May 1930, binder labeled "Naumkeag Steam Cotton Co. (Pequot Mills.) Working Data. Stretch Out and General," box 9, SNP; Nyman, *Union-Management Coöperation*, 6–8; Chomsky, *Linked Labor Histories*, 54. The Irish seem to have mostly worked in the Danvers Bleachery.

107. Warren Cooke, "Report of a Visit to the Naumkeag Mill, Salem, Mass., October 3d to 6th, 1930"; "Joint Research in the Pequot Mills. Confidential memorandum prepared by Frances Goodell, former industrial engineer for the Pequot Mills," [August? 1933]; R. D. Smith [?], "Report of Visit to the Naumkeag Mill," 20 May 1930; all in binder labeled "Naumkeag Steam Cotton Co. (Pequot Mills.) Working Data. Stretch Out and General" box 9, SNP; Jean Christie, "Morris Llewellyn Cooke: Progressive Engineer" (PhD diss., Columbia University, 1963), 13, 52–54; Kenneth E. Trombley, *The Life and Times of a Happy Liberal: A Biography of Morris Llewellyn Cooke* (New York: Harper, 1954), 101; Chomsky, *Linked Labor Histories*, 55.

108. "Memoranda of Trade Union Meeting and Conferences with Trade Union Officials," [May 1933], binder labeled "Naumkeag Steam Cotton Co. (Pequot Mills). Working Data. Data Relating to Strike of 1933," box 6; "Summary of Interview with C. L. Pattee, in Charge of Research, Pequot Mill," 29 May 1933; "Memorandum of Interview with Wilfred Levesque, Vice. Pres. of Local No. 33 U.T.W. and Nominal Leader of the Pro-strike Group," 25 May 1933, in binder labeled "Naumkeag Steam Cotton Co. (Pequot Mills). Working Data. Data Relating to Strike of 1933," box 8; "Interview with Francis Goodell, formerly head of Pequot Mill Research Staff," 1 August 1933, in binder labeled "Naumkeag Steam Cotton Co. (Pequot Mills). Working Data. Stretch Out and General," box 9; all in SNP.

109. Trombley, *Life and Times of a Happy Liberal*, 101.

110. Chomsky, *Linked Labor Histories*, 56–57.

111. On the 1933 strike, see Chomsky, *Linked Labor Histories*, 58–74; Nyman, *Union-Management Coöperation*, 119–81.

112. "Memorandum of Interview with Wilfred Levesque, Business Agent, Local No. 1, Independent Sheeting Workers of America, concerning Pequot Mills Strike of 1935 and Events Immediately Following," 27 November 1935, notebook labeled "Interview Memoranda, Newspaper Clippings, etc," box 7, SNP.

113. "Situation at Mill and Interviews with Mill Officials: Agent J. Foster Smith, Treasurer Ernest N. Hood," 10 May 1933; "Memoranda of Trade Union Meeting and Conferences with Trade Union Officials"; both in binder labeled "Naumkeag Steam Cotton Co (Pequot Mills). Working Data. Data Relating to Strike of 1933," box 8, SNP; "Memorandum of Interview with Fred. M. Knight, Labor Relations Adjuster, Concerning Strike of 1935,' 27 November 1935, notebook labeled "Interview Memoranda, Newspaper Clippings, etc," box 6, SNP.

114. "Memorandum of Interview with Fred. M. Knight, Labor Relations Adjuster, Concerning Strike of 1935," 27 November 1935; "Memorandum of Interview with Wilfred Levesque, Business Agent, Local No. 1, Independent Sheeting Workers of America, concerning Pequot Mills Strike of 1935 and Events Immediately Following," 27 November 1935; "Interview with James Burke, Secretary, Independent Sheeting Workers' Union, Local No. 1, 60 Washington Street, Salem, Mass.," 4 November 1935; all in notebook labeled "Interview Memoranda, Newspaper Clippings, etc," box 6, SNP; R. C. Nyman, "Union-Management Cooperation in the Stretch-Out Postscript," 11 December 1935, box 8, SNP.

Conclusion

1. *Chicago Tribune*, 10 June 1919, 18; *Atlanta Constitution*, 24 June 1919, 18.

2. Basil King, *The City of Comrades* (New York: Harper and Brothers, 1919), 191. So far as I have been able to find, no print of the film remains, and my description of the film comes from the novel, contemporary reviews, and the synopsis in the *AFI Catalog*, s.v. "The City of Comrades," available at http://www.afi.com/members/catalog/DetailView.aspx?s=1&Movie=17442, accessed 31 December 2013.

3. *Sun* (Baltimore), 21 September 1919, B7.

4. Randall Stewart, "William Benjamin Basil King, 1859–1928," in *Dictionary of American Biography*, vol. 1928 (New York: American Council of Learned Societies, 1928–1936), reproduced in the Biography Resource Center (Farmington Hills, Mich.: Gale, 2010), available at http://galenet.galegroup.com/servlet/BioRC, accessed 19 June 2010.

5. Samuel Henry Prince, *Catastrophe and Social Change: Based upon a Sociological Study of the Halifax Disaster*, Studies in History, Economics, and Public Law no. 212 (New York: Columbia University, 1920), 57. On Prince, see Russell R. Dynes and E. L. Quarantelli, "The Place of the Explosion in the History of Disaster Research: The Work of Samuel H. Prince," in *Ground Zero: A Reassessment of the 1917 Explosion in Halifax Harbour*, ed. Alan Ruffman and Colin D. Howell (Halifax, Nova Scotia: Nimbus/The Gorsebrook Research Institute for Atlantic Canada Studies at Saint Mary's University, 1994), 55–67.

6. On Prince and Kropotkin, see Rebecca Solnit, *A Paradise Built in Hell: The Extraordinary Communities That Arise in Disasters* (New York: Viking, 2009), 73–97.

7. Personal narrative of Miss Florence J. Murray, item 192, AMF. On Murray's life as a medical missionary, see Ruth Compton Brouwer, *Modern Women Modernizing Men: The Changing Missions of Three Professional Women in Asia and Africa, 1902–69* (Vancouver: University of British Columbia Press, 2002).

8. King, *City of Comrades*, 7.

9. Colin Ward, *Anarchy in Action*, 4th ed. (1973; repr. London: Freedom Press, 1996), 18, 31–39; Aristide R. Zolberg, "Moments of Madness," *Politics and Society* 2 (1972): 183–207.

10. J. Grove Smith, *Fire Waste in Canada* (Ottawa: Commission of Conservation, 1918), 99, cited in B[radley] E. S. Rudachyk, "The Most Tyrannous of Masters: Fire in Halifax, Nova Scotia, 1830–1850" (MA thesis, Dalhousie University, 1984), 1.

11. Rudachyk, "Most Tyrannous of Masters," 30; John C. Weaver and Peter de Lottinville, "The Conflagration and the City: Disaster and Progress in British North America during the Nineteenth Century," *Histoire sociale/Social History* 13 (November 1980): 417–49.

12. After 1914, only one urban fire even came close to Salem's in the number of buildings burned or the estimated dollar amount of damages. "Conflagrations in America since 1900," part 2, *Quarterly of the National Fire Protection Association* 44 (April 1951): 23–33.

13. Bertram E. Ames, "Report No. 150 on Conflagration, Salem, Mass., June 25, 1914," report to Underwriters' Bureau of New England, E S1 F6 1914_{15}, PEM.

14. Sven Lindqvist, *A History of Bombing*, trans. Linda Haverty Rugg (New York: New Press, 2001); Megan Kate Nelson, *Ruin Nation: Destruction and the American Civil War* (Athens: University of Georgia Press, 2012); Kenneth Hewitt, "'When the Great Planes Came and Made Ashes of Our City . . .': Towards an Oral Geography of the Disasters of War," *Antipode* 26, no. 1 (January 1994): 4–5.

15. Robert J. Nicholls et al., "Coastal Systems and Low-Lying Areas," in *Climate Change 2007: Impacts, Adaptation and Vulnerability. Contribution of Working Group II to the Fourth Assessment Report of the Intergovernmental Panel on Climate Change*, ed. M. L. Parry et al. (Cambridge: Cambridge University Press, 2007), 333; Ted Steinberg, *Acts of God: The Unnatural History of Natural Disasters in America*, 2nd ed. (New York: Oxford University Press, 2006).

16. Kevin E. Trenberth et al., "Surface and Atmospheric Climate Change," in *Climate Change 2007: The Physical Science Basis. Contribution of Working Group I to the Fourth Assessment Report of the Intergovernmental Panel on Climate Change*, ed. S. Solomon et al. (Cambridge: Cambridge University Press, 2007), 299–316; IPCC, "Summary for Policymakers," in *Managing the Risks of Extreme Events and Disasters to Advance Climate Change Adaptation*, Special Report of Working Groups I and II of the Intergovernmental Panel on Climate Change (Cambridge: Cambridge University Press, 2012), 1–19; Robert Mendelsohn et al., "The Impact of Climate Change on Global Tropical Cyclone Damage," *Nature Climate Change* 2 (March 2012): 205–9.

17. Nicholls et al. "Coastal Systems," 337–38.

18. Roger Bilham, "The Seismic Future of Cities," *Bulletin of Earthquake Engineering* 7, no. 4 (1 November 2009): 839–87; Roger Bilham, "Urban Earthquake Fatalities: A Safer World, or Worse to Come?," *Seismological Research Letters* 75, no. 6 (November 2004): 706–12; James Jackson, "Fatal Attraction: Living with Earthquakes, the Growth of Villages into Megacities, and Earthquake Vulnerability in the Modern World," *Philosophical*

Transactions of the Royal Society A: Mathematical, Physical and Engineering Sciences 364, no. 1845 (2006): 1911–25; Linda Rowan, "Seismology: Urban Hazards," *Science* 307, no. 5706 (7 January 2005): 18e; Friedemann Wenzel, Fouad Bendimerad, and Ravi Sinha, "Megacities—Megarisks," *Natural Hazards* 42, no. 3 (2007): 481–91.

19. R. R. Frerichs, P. S. Kim, R. Barrais, and R. Piarroux, "Nepalese Origin of Cholera Epidemic in Haiti," *Clinical Microbiology and Infection* 18, no. 6 (June 2012): E158–E163; Jonathan M. Katz, *The Big Truck That Went By: How the World Came to Save Haiti and Left Behind a Disaster* (New York: Palgrave Macmillan, 2013), 217–44.

20. Naomi Klein, *The Shock Doctrine: The Rise of Disaster Capitalism* (New York: Picador, 2007); Evgeny Morozov, *To Save Everything, Click Here: The Folly of Technological Solutionism* (New York: Public Affairs, 2014); Steinberg, *Acts of God*, 149–72.

21. Jacob A. C. Remes, "Solidarity, Citizenship, and the Opportunities of Disasters," in *Labor Rising: The Past and Future of Working People in America*, ed. Daniel Katz and Richard Greenwald (New York: New Press, 2012), 152–53; Solnit, *Paradise Built in Hell*.

22. scott crow, *Black Flags and Windmills: Hope, Anarchy, and the Common Ground Collective* (Oakland, Calif.: PM Press, 2011); Sue Hilderbrand, scott crow, and Lisa Fithian, "Common Ground Relief," in *What Lies Beneath: Katrina, Race, and the State of the Nation*, ed. South End Press Collective (Cambridge, Mass.: South End, 2006), 80–98. For a more critical view, see also Rachel E. Luft, "Looking for Common Ground: Relief Work in Post-Katrina New Orleans as an American Parable of Race and Gender Violence," *NWSA Journal* 20, no. 3 (Fall 2008): 5–31. On other new forms of civic engagement, see David G. Ortiz and Stephen F. Ostertag, "Katrina Bloggers and the Development of Collective Civic Action: The Web as a Mobilizing Structure," *Sociological Perspectives* 57, no. 1 (2014): 52–78.

23. Beverly Bell, *Fault Lines: Views across Haiti's Divide* (Ithaca, N.Y.: Cornell University Press, 2014).

24. Benjamin Shepard, "From Flooded Neighborhoods to Sustainable Urbanism: A New York Diary," *Socialism and Democracy* 27, no. 3 (2013): 42–64.

BIBLIOGRAPHY

Primary Sources

ARCHIVAL COLLECTIONS CONSULTED

Amherst, Mass.
   W. E. B. DuBois Library, University of Massachusetts
      United Brotherhood of Carpenters and Joiners of America, Massachusetts State Council records, 1892–1980
Boston, Mass.
   American Jewish Historical Society
      Abraham C. Ratshesky Collection
   Special Collections Division, State Library of Massachusetts
      Massachusetts-Halifax Relief Committee correspondence and papers
   Baker Business Library, Harvard University
      Naumkeag Steam Cotton Company records
Braintree, Mass.
   Archives of the Roman Catholic Archdiocese of Boston
      Parish correspondence files, 1907–1940
      Parish Histories Collection
Brunswick, Maine
   George J. Mitchell Department of Special Collections and Archives, Bowdoin College
      Alumni records
College Park, Md.
   National Archives II
      American Red Cross, Central Files
      Halifax consular records
Halifax, Nova Scotia
   Anglican Diocese of Nova Scotia and Prince Edward Island Archives
   Archives and Special Collections, Killam Library, Dalhousie University
      Halifax Relief Commission fonds

        Archibald MacMechan papers
        Technical University of Nova Scotia papers
    Catholic Archdiocese of Halifax Archives
        Edward J. McCarthy fonds
    Dartmouth Heritage Museum
        Court and police records
        Explosion of 1917 Collection
        Oral History Collections
    Halifax Regional Municipality Archives
        Police records
    Maritime Museum of the Atlantic
        Halifax Explosion Photographic Collection
        Halifax Explosion Research Files
    Nova Scotia Archives
        Canadian Red Cross Society, Nova Scotia Division, fonds
        Commercial Club of Halifax fonds
        Helen Creighton fonds
        Jean Davis Collection
        Michael D. Coolen diary
        Agnes Dennis fonds
        Halifax Explosion Collection
        Halifax Explosion Memorial Bells Committee Collection
        Halifax Ladies' College fonds
        Halifax Relief Commission fonds
        Halifax School for the Blind fonds
        R. V. Harris Collection
        A. C. Hawkins fonds
        (Halifax) Jewish Historical Society fonds
        Janet Kitz fonds
        Lieutenant Governor's correspondence
        Local Council of Women of Halifax fonds
        Archibald MacMechan fonds
        Map Collection
        Nova Scotia Society for the Prevention of Cruelty fonds
        Rotary Club (Halifax) records
        St. George's Anglican Parish fonds
        St. John's Ambulance Brigade (Nova Scotia Council) fonds
        St. Matthew's Church fonds
        St. Matthias Anglican Parish fonds
        Society for the Prevention of Cruelty fonds
        W. E. Tibbs private papers
        Victorian Order of Nurses, Halifax branch, fonds
        Workers' Compensation Board of Nova Scotia fonds
    St. Paul's Parish Archives
Lexington, Mass.
    Van Gorden–William Library, National Heritage Museum

Madison, Wisc.
　Wisconsin Historical Society
　　Textile Workers' Union of America Papers
Manchester, N.H.
　Franco-American Centre
Mason, Ohio
　Christian Lantz papers, privately held by Maurice Umble
New Haven, Conn.
　Knights of Columbus Supreme Council Archives
　Manuscripts and Archives, Sterling Memorial Library, Yale University
　　Smith-Nyman papers
　　Yale University Institute for Human Relations papers
New York, N.Y.
　Blaustein Library and Historical Archive, American Jewish Committee
　　Executive Committee minute books
　Frick Art Reference Library, Frick Collection
　　Henry Clay Frick papers
　Brooklyn Historical Society
　　Greenpoint YMCA Congress Alumni scrapbooks
Ottawa, Ontario
　Library and Archives Canada
　　Account Related to the Halifax Explosion
　　Sir Robert Borden Collection
　　Canadian Brotherhood of Railway, Transport and General Workers fonds
　　Canadian Expeditionary Force personnel files
　　Department of Indian Affairs and Northern Development fonds
　　Department of Marine fonds
　　Department of National Defence fonds
　　Department of Transport fonds
　　Victor Christian William Cavendish, Ninth Duke of Devonshire fonds
　　Fred Longland fonds
　　Flora McDonnell fonds
　　Edward Atcherley Eckersall Nixon fonds
　　N. L. Power fonds
　George Metcalf Archival Collection, Canadian War Museum
　　Sgt. Samuel Mansfield papers
　　Air Vice Marshall Clifford Mackey McEwan papers
　　Tom Kenneth Triggs papers
Salem, Mass.
　Clerk's Office, Essex County District Court
　Phillips Library, Peabody Essex Museum
　　Lydia Ropes Dow Findlay typescript
　　John F. Hurley papers
　　Knights of Pythias, Starr King Lodge, No. 81, [Essex, Mass.] Records, 1911–1914
　　St. Joseph's Church Miscellany
　　Salem Club records

Salem Fires scrapbooks
Salem, Mass. Fire Collection E S1 F6 1914$_{1-16}$
Salem Rebuilding Commission records
Records of the Société St.-Jean-Baptiste
Starr King Lodge, A F and A M Masons records
John Porter Sumner typescript
Winslow Lewis Commandery, Knights Templar records
Salem City Hall
Archives and Special Collections, Salem State University Library
  Franco-American oral histories
  Nelson Dionne Collection
Salem YMCA
  Scrapbook Collection
Washington, D.C.
  National Archives I
    Records of the Office of the Quartermaster General
    Naval Attache Reports
    (Formerly) Confidential Correspondence, Office of Naval Intelligence, 1913–1924
    Naval Records Collection of the Office of Naval Records and Library
    Naval Vessel Log Books
    General Records of the Navy, Secretary of the Navy General correspondence
  Library of Congress Manuscripts Division
    US Work Projects Administration, Federal Writers' Project, Folklore Project
  Archives Center, National Museum of American History, Smithsonian Institution
    United Shoe Machinery Corporation records
Woonsocket, R.I.
  Union St.-Jean-Baptiste d'Amérique Corporate Archives
Worcester, Mass.
  Emmanuel D'Alzon Library, Assumption College
    Institut Français
  Archives and Special Collections, College of the Holy Cross
    David I. Walsh papers
  Massachusetts Military Museum and Archives
    Salem Fire correspondence

NEWSPAPERS AND MAGAZINES

Acadian Daily Recorder (Halifax, Nova Scotia)
Atlanta Constitution
L'Avenir National (Manchester, N.H.)
Boston Globe
Canadian Forward (Toronto)
Canadian Labor Leader (Sydney, Nova Scotia)
Canadian Railroad Employees' Monthly (Halifax, Nova Scotia)
Le Canado Américain (Manchester, N.H.)
Chicago Tribune

The Citizen (Halifax, Nova Scotia)
Courrier de Lawrence
Courrier de Salem
The Columbiad (New Haven, Conn.)
Daily Echo (Halifax, Nova Scotia)
Dalhousie Gazette (Halifax, Nova Scotia)
Evening Mail (Halifax, Nova Scotia)
Halifax Herald
Morning Chronicle (Halifax, Nova Scotia)
La Presse (Montreal)
Le Progrès du Golfe (Rimouski, Quebec)
New York Times
The Pilot (Boston)
La Revue de Salem
Salem Evening News
Salem Register
Saturday Evening Observer (Salem, Mass.)
Sun (Baltimore)
The Survey (New York)

REPORTS, PAMPHLETS, MEMOIRS, AND OTHER PUBLISHED PRIMARY SOURCES

Abstract of the Proceedings of the Most Worshipful Grand Lodge of Free and Accepted Masons of the Commonwealth of Massachusetts for the Year 1915. Cambridge, Mass.: Caustic-Claflin, 1916.

Ames, Bertram Eugene. "An Investigation into the Causes, Development and Results of the Conflagration of Salem, Massachusetts, June 25 and June 26, 1914, with Conclusions and Recommendations for the Prevention of a Repetition of a Similar Catastrophe." Civil engineering thesis, University of Maine, 1915.

Annual Report of the Essex Institute for the Year Ending May 5, 1919. Salem, Mass.: Printed for the Essex Institute by Newcomb and Gauss, 1919.

Ariel, France. *Canadiens et Américains chez eux: Journals, lettres, impressions d'une artiste française*. Montreal: Granger Frères, 1920.

"Baptisms." *Baptist Home Mission Monthly* 37 no. 6 (June 1905): 248.

Bernstein, Herman, ed. *American Jewish Year Book 5675*. Vol. 16. Philadelphia: Jewish Publication Society of America, 1914.

Bird, Ethel. "Review of *Disasters and the American Red Cross in Disaster Relief*, by J. Byron Deacon." *American Journal of Sociology* 24 (July 1918): 112–13.

Borden, Robert Laird. *Robert Laird Borden: His Memoirs*. 2 vols. Edited by Henry Borden. Toronto: MacMillan, 1938.

Boston Register and Business Directory. Boston: Sampson and Murdock, 1921.

Bruce, Herbert Putnam. "An Investigation of the Public Water Supplies, Building Construction, and Automatic Sprinkler Protection as Affected by the Conflagration of Salem, Massachusetts, June 25 and 26, 1914." Civil engineering thesis, University of Maine, 1915.

Byrne, William, William A. Leahy, J. J. McCoy, Jas. H. O'Donnell, A. Dowling, John E.

Finen, Edmund J. A. Young, and John S. Michaud. *History of the Catholic Church in the New England States*. Boston: Hurd and Everts, 1899.

Canada. *Fifth Census of Canada, 1911*. Vol. 2. Ottawa: C. H. Parmelee as King's Printer, 1913.

Canada. *Sixth Census of Canada, 1921*. Vol. 1. *Population*. Ottawa: F. A. Acland as King's Printer, 1924.

Catholic Encyclopedia, s.v. "Boston." New York: Encyclopedia Press, 1913–1914. Available at http://www.newadvant.org/cathen/02703a.htm.

[Chambers, Bertram M.]. "Halifax Explosion." *Naval Review* 7, no. 1 (1920): 445–57.

Chancellor, William E. *Reading and Language Lessons for Evening Schools*. New York: American Book Company, 1912.

Creighton, Helen. *Helen Creighton: A Life in Folklore*. Toronto: McGraw-Hill Ryerson, 1975.

Deacon, Janney Byron. *Disasters and the American Red Cross in Disaster Relief*. New York: Russell Sage Foundation, 1918.

Deacon, J. Byron. "The Future of the Red Cross Home Service." *Proceedings of the National Conference of Social Work* 46 (1920): 365–71.

"Education: Colgate's Cutten." *Time*, 2 February 1942. Available at http://205.188.238.109/time/magazine/article/0,9171,849752,00.html.

Egan, Maurice Francis, and John B. Kennedy. *Knights of Columbus in Peace and War*. Vol. 1. New Haven, Conn.: Knights of Columbus, 1920.

Fenwick, Thomas. "Salem: From Puritan to Progressive." *New England Magazine*, June 1914, 161–77.

Galbraith, John Kenneth. *The Scotch*. Cambridge, Mass.: Riverside Press, 1964.

Gardner, Augustus Peabody. *Some Letters of Augustus Peabody Gardner*. Edited by Constance Lodge Gardner. Boston: Houghton Mifflin, 1920.

Gardner, Constance Lodge. *Augustus Peabody Gardner, Major, United States National Guard, 1865–1918*. Cambridge: Privately printed at the Riverside press, 1919.

Guide Franco-Américain des Etats de Nouvelle-Angleterre, 1916. Fall River, Mass.: A. A. Belanger, 1916.

Gunn, Selskar M., and Samuel M. Schmidt. "An Investigation of Housing Conditions in Salem, Mass." Salem, Mass.: Housing Committee of the Associated Charities and the Committee on Nuisances of the Civic League, [December 1911].

Heustis, George F. "The Night of the Salem Fire." *Tenney Service* 1, no. 9 (September 1914): 3–7.

Hopkins, J. Castell. "Halifax Disaster and the War." In *Canadian Annual Review of Public Affairs 1917*, 468–69. Toronto: Canadian Annual Review, 1918.

Hopkins, Robert M. Letter to the editor. *Christian Century*, 10 January 1918, 20.

"John Farwell Moors, '83, Elected Fellow." *Harvard Alumni Bulletin* 20, no. 16 (17 January 1918): 280–81.

Jones, Arthur Bennett. *The Salem Fire*. Boston: Gorham Press, 1914.

King, Basil. *The City of Comrades*. New York: Harper and Brothers, 1919.

Lyman, George Hinkley. *The Story of the Massachusetts Committee on Public Safety*. Boston: Wright and Potter as state printers, 1919.

MacLennan, Hugh. *Barometer Rising*. 1941. Repr. Toronto: McClelland and Stewart, 1989.

Massachusetts. *Annual Report of the Adjutant General of the Commonwealth of Massachusetts for the Year Ending December 31, 1914*. Public Document 7. Boston: Wright and Potter as state printers, 1915.

———. *Annual Report of the Adjutant General of the Commonwealth of Massachusetts for the Year Ending December 31, 1917*. Public Document 7. Boston: Wright and Potter as state printers, 1918.

Massachusetts Bureau of Statistics. *Decennial Census, 1915*. Boston: Wright and Potter as state printers, 1918.

———. *Directory of Labor Organizations in Massachusetts, 1914*. No. 98. 9 March 1914

———. *Directory of Labor Organizations in Massachusetts, 1919*. No. 127. May 1919.

McAlpine's Halifax City Directory. Vol 49. Halifax, Nova Scotia: Royal Print and Litho, 1917.

McMahon, Thomas F. "United Textile Workers of America." Workers' Education Organization Series no. 2. New York: Workers' Education Bureau Press, 1926.

Moors, John F. "Review of *Disasters and the American Red Cross in Disaster Relief*, by J. Byron Deacon." *The Survey* 39, no. 27 (26 January 1918): 472.

"Moors, John Farwell." In *National Cyclopaedia of American Biography*, 41:382–83. Clifton, N.J.: James T. White, 1891– .

Naumkeag Directory for Salem, Beverly, Danvers, Marblehead, Peabody, Hamilton, Manchester, Middleton, Topsfield and Wenham. Vol. 21. Salem, Mass: Henry M. Meek, 1913.

North Shore Blue Book and Social Register. Boston: A. E. Foss, 1917.

Nova Scotia Archives and Records Management and the Halifax Foundation. "Halifax Explosion Remembrance Book," December 3, 2002. Available at http://www.gov.ns.ca/nsarm/virtual/remembrance/.

Nyman, Richard Carter. *Union-Management Coöperation in the "Stretch-Out": Labor Extension at the Pequot Mills*. New Haven, Conn.: Yale University Press, 1934.

Perry, Montayne. *The Salem Fire Relief*. Salem, Mass.: Milo A. Newhall, 1915.

Prince, Samuel Henry. *Catastrophe and Social Change: Based upon a Sociological Study of the Halifax Disaster*. Studies in History, Economics, and Public Law no. 212. New York: Columbia University, 1920.

Proceedings of the Grand Commandery of Knights Templar and Appendant Orders of Massachusetts and Rhode Island for the Year Ending Oct. 29, 1914. Boston: Griffith-Stillings Press, 1914.

Proceedings of the M. W. Grand Lodge of Free and Accepted Masons of California. October A. L. 5906 (1906).

Proceedings of the Massachusetts Council of Deliberation at the Session Held at Boston, June 18, 1915. Boston: The Council, 1915.

Proceedings of the Most Worshipful Grand Lodge of Ancient Free and Accepted Masons of the Commonwealth of Massachusetts for the Year 1914. Boston: Poole, 1915.

Proceedings of the Most Worshipful Grand Lodge of Ancient, Free and Accepted Masons of Nova Scotia and the Fifty-Third Annual Communication. Halifax, Nova Scotia: Weeks, 1918.

Prominent People of the Maritime Provinces. Montreal: Canadian Publicity, 1922.

Putnam, Eben. *Report of the Commission on Massachusetts' Part in the World War: History*. Vol. 1. Boston: Commonwealth of Massachusetts, 1931.

Raddall, Thomas H. *In My Time: A Memoir*. Toronto: McClelland and Stewart, 1976.

"Ratshesky, Abraham C." In *National Cyclopaedia of American Biography*, 33:307–8. Clifton, N.J.: James T. White, 1891- .

Report of the American Baptist Foreign Mission Society. In *Annual of the Northern Baptist Convention, 1922*. N.p.: American Baptist Publication Society, 1922.

"Review of *Disasters and the American Red Cross in Disaster Relief*, by J. Byron Deacon." *The Bookman* 46 (February 1918): 735.

———. *Journal of the Association of Collegiate Alumnae* 4 (April 1918): 551–52.

———. *National Municipal Review* 7 (March 1918): 194.

———. *New Republic*, 16 March 1918, 214.

———. *South Atlantic Quarterly* 17 (April 1918), 182.

Ripley, William Z. "Race Factors in Immigration." *Atlantic Monthly* 93, no. 558 (March 1904): 299–308.

"The Recall in Salem, Mass." *National Municipal Review*, April 1915, 304.

Ross, R. W. "What Church Union Will Mean for the City of Halifax." *New Outlook*, 10 June 1925, 37.

Seventh Annual Convention of the Atlantic Coast District of the International Longshoremen's Association. Boston: I. H. Feinberg and Son, 1919.

United Masonic Relief. Washington, D.C.: Masonic Service Association of the United States, 1931.

United States Census. Digitized and searched on *Heritage Quest Online*. Available at http://www.heritagequestonline.com.

United States Census Bureau. *Fourteenth Census of the United States*. Washington, D.C.: GPO, 1921.

United States Congress. *Congressional Record*. 63rd Cong., 2nd sess., 1914. Vol. 51, pt. 12.

United States Department of Labor. *Reports of the Department of Labor, 1914*. Washington, D.C.: GPO, 1915.

Secondary Sources

BOOKS, ARTICLES, AND CHAPTERS

"A. R. Mosher: A Biographical Sketch." *Canadian Unionist*, September 1950, 210–12.

Abrams, Richard M. *Conservatism in a Progressive Era: Massachusetts Politics, 1900–1912*. Cambridge, Mass.: Harvard University Press, 1964.

"After the Flood." *This American Life*. Episode 296. 9–11 September 2005. Available at http://www.thisamericanlife.org/radio-archives/episode/296/After-the-Flood.

Aldrich, Daniel P. *Building Resilience: Social Capital in Post-Disaster Recovery*. Chicago: University of Chicago Press, 2012.

American Film Institute. "The City of Comrades." *AFI Catalog*, n.d. Available at http://www.afi.com/members/catalog/DetailView.aspx?s=1&Movie=17442.

Anctil, Pierre. *A Franco-American Bibliography: New England*. [Bedford, N.H.: National Materials Development Center], 1979.

Anderson, Benedict. *Imagined Communities: Reflections on the Origin and Spread of Nationalism*. Rev. ed. London: Verso, 1991.

Archdeacon, Thomas J. *Becoming American: An Ethnic History*. New York: Free Press, 1983.

Armstrong, Frederick H. "The Second Great Fire of Toronto, 19–20 April, 1904." *Ontario History* 70 (March 1978): 3–38.

Armstrong, John G. "Letters from Halifax: Reliving the Halifax Explosion through the Eyes of My Grandfather, a Sailor in the Royal Canadian Navy." *Northern Mariner/Le marin du nord* 8, no. 4 (October 1998): 55–74.

Armstrong, John Griffith. *The Halifax Explosion and the Royal Canadian Navy: Inquiry and Intrigue*. Vancouver: University of British Columbia Press, 2002.

Auf der Heide, Erik. "Common Misconceptions about Disasters: Panic, the 'Disaster Syndrome,' and Looting." In *The First 72 Hours: A Community Approach to Disaster Preparedness*, edited by Margaret O'Leary, 340–80. New York: iUniverse, 2004.

Baker, Janet E. *Archibald MacMechan: Canadian Man of Letters*. Lockeport, Nova Scotia: Roseway, 2000.

Baker, Leslie. "'A Visitation of Providence': Public Health and Eugenic Reform in the Wake of the Halifax Disaster." *Canadian Bulletin of Medical History/Bulletin canadien d'histoire de la médecine* 31, no. 1 (2014): 99–122.

Barkan, Elliott Robert. "French Canadians." In *Harvard Encyclopedia of American Ethnic Groups*, edited by Stephan Thernstrom, 388–401. Cambridge, Mass.: Belknap Press, 1980.

Barratt, A. J. "Atlantic Crossings." *Ships Monthly*, February 2002, 22–25.

Barry, John M. *Rising Tide: The Great Mississippi Flood of 1927 and How It Changed America*. New York: Simon and Schuster, 1997.

Barth, Gunther Paul. *City People: The Rise of Modern City Culture in Nineteenth-Century America*. New York: Oxford University Press, 1980.

Beal, John P. "It's Déjà Vu All Over Again: Lay Trusteeism Rides Again." *Jurist* 68 (2008): 497–568.

Beall, Irl V. "The Halifax Explosion and the Cutter *Morrill*." *Inland Seas* 23, no. 3 (1967): 179.

Beattie, Betsy. *Obligation and Opportunity: Single Maritime Women in Boston, 1870–1930*. Montreal and Kingston: McGill–Queen's University Press, 2000.

Beed, Blair. *1917 Halifax Explosion and American Response*. Halifax, Nova Scotia: Dtours Visitors and Convention Service, 1998.

Beito, David T. *From Mutual Aid to the Welfare State: Fraternal Societies and Social Services, 1890–1967*. Chapel Hill: University of North Carolina Press, 2000.

Bell, Beverly, *Fault Lines: Views across Haiti's Divide*. Ithaca, N.Y.: Cornell University Press, 2014.

Bercuson, D. J. "Organized Labour and the Imperial Munitions Board." *Relations industrielles/Industrial Relations* 28, no. 3 (July 1973): 602–16.

Bergman, Jonathan C. "A New Deal for Disaster: The 'Hurricane of 1938' and Federal Disaster Relief Operations, Suffolk County, New York." *Long Island Historical Journal* 20 nos. 1–2 (Fall 2007/Spring 2008): 15–39.

Beth Israel Synagogue. "About Us." Available at http://www.jewishhalifax.com/about/.

Bhidé, Amar H. "Building the Professional Firm: McKinsey and Co.: 1939–1968." *International Journal of Entrepreneurship Education* 1, no. 2 (2002): 247–76.

Biel, Steven, ed. *American Disasters*. New York: New York University Press, 2001.

Bilham, Roger. "Death Toll from Earthquakes." *Geotimes* 47, no. 7 (July 1998): 4.

———. "The Seismic Future of Cities." *Bulletin of Earthquake Engineering* 7, no. 4 (1 November 2009): 839–87.

———. "Urban Earthquake Fatalities: A Safer World, or Worse to Come?" *Seismological Research Letters* 75, no. 6 (November 2004): 706–12.

Bird, Michael J. *The Town That Died: The True Story of the Greatest Man-Made Explosion before Hiroshima.* 1967. Repr. Halifax, Nova Scotia: Nimbus, 1992.

Bixel, Patricia Bellis, and Elizabeth Hayes Turner. *Galveston and the 1900 Storm: Catastrophe and Catalyst.* Austin: University of Texas Press, 2000.

Blewett, Mary H. *Men, Women, and Work: Class, Gender, and Protest in the New England Shoe Industry, 1780–1910.* Urbana: University of Illinois Press, 1988.

Bliss, Michael. *A Canadian Millionaire: The Life and Business Times of Sir Joseph Flavelle, Bart., 1858–1939.* Toronto: Macmillan of Canada, 1978.

Blodgett, Geoffrey. "Yankee Leadership in a Divided City, 1860–1910." In *Boston, 1700–1980: The Evolution of Urban Politics,* edited by Ronald P. Formisano and Constance K. Burns, 87–110. Westport, Conn.: Greenwood Press, 1984.

Bloomfield, Elizabeth. "Boards of Trade and Canadian Urban Development." *Urban History Review* 12 (October 1983): 77–99.

Blow, David J. "The Establishment and Erosion of French-Canadian Culture in Winooski, Vermont, 1867–1900." *Vermont History* 43, no. 1 (January 1975): 59–74.

Boris, Eileen, and S. J. Kleinberg. "Mothers and Other Workers: (Re)Conceiving Labor, Maternalism, and the State." *Journal of Women's History* 15, no. 3 (Autumn 2003): 90–117.

Bothwell, Robert. *The Penguin History of Canada.* Toronto: Penguin Canada, 2006.

Boucher, André. "Le rôle joué par les marguilliers." In *Le laïc dans l'Église canadienne-française de 1830 à nos jours,* by Pierre Hurtubise et al., 163–73. Montreal: Fides, 1972.

Bourdieu, Pierre. "The Forms of Capital." In *Handbook of Theory and Research for the Sociology of Education,* edited by J. G. Richardson, 241–58. New York: Greenwood, 1985.

Bourdreau, Michael. *City of Order: Crime and Society in Halifax, 1918–35.* Vancouver: University of British Columbia Press, 2012.

Boyd, Michelle Hébert. *Enriched by Catastrophe: Social Work and Social Conflict after the Halifax Explosion.* Halifax, Nova Scotia: Fernwood, 2007.

Boydston, Jeanne. *Home and Work: Housework, Wages, and the Ideology of Labor in the Early Republic.* New York: Oxford University Press, 1990.

Bradbury, Bettina. *Working Families: Age, Gender, and Daily Survival in Industrializing Montreal.* Toronto: McClelland and Stewart, 1993.

Bradshaw, Larry, and Lorrie Beth Slonsky. "The Real Heroes and Sheroes of New Orleans." *Socialist Worker,* 9 September 2005, 4–5.

Brandes, Stuart D. *American Welfare Capitalism, 1880–1940.* Chicago: University of Chicago Press, 1976.

Brault, Gérard J. "État présent des études sur les centres franco-américains de la Nouvelle-Angleterre." In *Vie Française: Situation de la recherche sur la Franco-Américanie,* edited by Claire Quintal and André Vachon, 9–25. Quebec: Le Conseil de la Vie Française en Amerique, 1980.

Brault, Gerard J. *The French-Canadian Heritage in New England.* Hanover, N.H.: University Press of New England, 1986.

Bridges, Amy. *Morning Glories: Municipal Reform in the Southwest.* Princeton, N.J.: Princeton University Press, 1999.

Brinkley, Douglas. *The Great Deluge: Hurricane Katrina, New Orleans, and the Mississippi Gulf Coast.* New York: William Morrow, 2006.

Brison, Jeffrey D. *Rockefeller, Carnegie, and Canada: American Philanthropy and the Arts and Letters in Canada.* Montreal and Kingston: McGill-Queen's University Press, 2005.

Brookes, Alan A. "Out-Migration from the Maritime Provinces, 1860–1900: Some Preliminary Considerations." *Acadiensis* 5, no. 2 (Spring 1976): 26–55.

Brouwer, Ruth Compton. "'Home Lessons, Foreign Tests': The Background and First Missionary Term of Florence Murray, Maritime Doctor in Korea." *Journal of the CHA*, n.s., 6 (1996): 103–28.

———. *Modern Women Modernizing Men: The Changing Missions of Three Professional Women in Asia and Africa, 1902–69.* Vancouver: University of British Columbia Press, 2002.

Bukowczyk, John J., Nora Faires, David R. Smith, and Randy William Widdis. *Permeable Border: The Great Lakes Basin as Transnational Region, 1650–1990.* Pittsburgh University of Pittsburgh Press, 2005.

Bumsted, J. M. *The Peoples of Canada: A Post-Confederation History.* 2nd ed. Don Mills, Ontario: Oxford University Press, 2004.

Burns, Constance K. "The Irony of Progressive Reform: Boston, 1898–1910." In *Boston, 1700–1980: The Evolution of Urban Politics*, edited by Ronald P. Formisano and Constance K. Burns, 133–64. Westport, Conn.: Greenwood Press, 1984.

Burns, Robin B. "The Montreal Irish and the Great War." [Canadian Catholic Historical Association] *Historical Studies* 52 (1985): 67–81.

Burton, Antoinette. "Who Needs the Nation? Interrogating 'British' History." *Journal of Historical Sociology* 10 (September 1997): 227–48.

Buxton, William J. "Private Wealth and Public Health: Rockefeller Philanthropy and the Massachusetts-Halifax Relief Committee/Health Commission." In *Ground Zero: A Reassessment of the 1917 Explosion in Halifax Harbour*, edited by Alan Ruffman and Colin D. Howell, 183–93. Halifax, Nova Scotia: Nimbus/The Gorsebrook Research Institute for Atlantic Canada Studies at Saint Mary's University, 1994.

Byrne, Julie. *O God of Players: The Story of the Immaculata Mighty Macs.* New York: Columbia University Press, 2003.

Calliste, Agnes. "The Struggle for Employment Equity by Blacks on American and Canadian Railroads." *Journal of Black Studies* 25, no. 3 (January 1995): 297–317.

Cameron, James. "The Garden Distressed: Church Union and Dissent on Prince Edward Island, 1925." *Acadiensis* 21, no. 2 (Spring 1992): 108–31.

Campbell, Douglas F. "A Group, a Network and the Winning of Church Union in Canada: A Case Study in Leadership." *Canadian Review of Sociology and Anthropology* 25, no. 1 (February 1988): 41–66.

———. "Engaging Third Parties: Canadian Church Unionists and Their Opponents in the Secular Forum." *Journal of Church and State* 33, no. 1 (December 1, 1991): 75–94.

Campbell, W. Joseph. *Getting It Wrong: Ten of the Greatest Misreported Stories in American Journalism.* Berkeley: University of California Press, 2010.

"Canadian Klondike Club Is Hub of French Community in Salem." *Lynn Sunday Post*, 28 May 1967, 13.

Cannon, S. "On parle Français à Salem, Massachusetts." *Etudes canadiennes* 22 (1987): 103–22.

Capozzola, Christopher. *Uncle Sam Wants You: World War I and the Making of the Modern American Citizen*. New York: Oxford University Press, 2008.

Carnes, Mark C. *Secret Ritual and Manhood in Victorian America*. New Haven, Conn.: Yale University Press, 1989.

Cavallo, Dom. "Social Reform and the Movement to Organize Children's Play during the Progressive Era." *History of Childhood Quarterly* 3 (1976): 508–33.

Cavanaugh, Catherine. "The Limitations of the Pioneering Partnership: The Alberta Campaign for Homestead Dower, 1909–25." *Canadian Historical Review* 74, no. 2 (1993): 198–225.

Certeau, Michel de. *The Practice of Everyday Life*. Berkeley: University of California Press, 1984.

Chambers, Clarke A. *Paul U. Kellogg and the Survey: Voices for Social Welfare and Social Justice*. Minneapolis: University of Minnesota Press, 1971.

Chambers, John Whiteclay. *The Tyranny of Change*. New York: St. Martin's Press, 1980.

Chapman, Harry. *In the Wake of the Alderney: Dartmouth, Nova Scotia, 1750–2000*. Halifax, Nova Scotia: Nimbus, 2001.

Chauncey, George. *Gay New York: Gender, Urban Culture, and the Making of the Gay Male World, 1890–1940*. New York: Basic Books, 1994.

Chomsky, Aviva. *Linked Labor Histories: New England, Colombia, and the Making of a Global Working Class*. Durham, N.C.: Duke University Press, 2008.

———. "Salem as Global City, 1850–2004." In *Salem: Place, Myth, and Memory*, edited by Dane Anthony Morrison and Nancy Lusignan Schultz, 219–47. Boston: Northeastern University Press, 2004.

Clancey, Gregory. *Earthquake Nation: The Cultural Politics of Japanese Seismicity, 1868–1930*. Berkeley: University of California Press, 2006.

———. "Toward a Spatial History of Emergency: Notes from Singapore." In *Beyond Description: Singapore Space Historicity*, edited by Ryan Bishop, John Phillips, and Wei-Wei Yoo, 30–59. London: Routledge, 2004.

Clarke, Brian. "English-Speaking Canada from 1854." In *A Concise History of Christianity in Canada*, edited by Terrance Murphy, 261–359. Toronto: Oxford University Press, 1996.

Clawson, Mary Ann. *Constructing Brotherhood: Class, Gender, and Fraternalism*. Princeton, N.J.: Princeton University Press, 1989.

Clifford, N. Keith. *The Resistance to Church Union in Canada, 1904–1939*. Vancouver: University of British Columbia Press, 1985.

———. "The Interpreters of the United Church of Canada." *Church History* 46, no. 2 (June 1977): 203–14.

Clow, Barbara. *Negotiating Disease: Power and Cancer Care, 1900–1950*. Montreal and Kingston: McGill–Queen's University Press, 2001.

Cohen, Lizabeth. *Making a New Deal: Industrial Workers in Chicago, 1919–1939*. Cambridge: Cambridge University Press, 1991.

"Conflagrations in America since 1900." Part 2. *Quarterly of the National Fire Protection Association* 44 (April 1951): 23–33.

Conflagrations in America since 1914: A Record of the Principal Conflagrations in the United

States and Canada since the Burning of Salem. Boston: National Fire Protection Association, 1942.

Cooper, Jerry M. *The Rise of the National Guard*. Lincoln: University of Nebraska Press, 1997.

Corry, J. A. "The Growth of Government Activities in Canada, 1914–1921." *Canadian Historical Association Annual Report* (1940): 63–73.

Cronon, William. *Nature's Metropolis: Chicago and the Great West*. New York: Norton, 1991

crow, scott. *Black Flags and Windmills: Hope, Anarchy, and the Common Ground Collective* Oakland, Calif.: PM Press, 2011.

Cumbler, John T. *Working-Class Community in Industrial America: Work, Leisure, and Struggle in Two Industrial Cities, 1880–1930*. Westport, Conn.: Greenwood Press, 1979.

D'Agostino, Peter. *Rome in America: Transnational Catholic Ideology from the Risorgimento to Fascism*. Chapel Hill: University of North Carolina Press, 2004.

Dane, Perry. "The Corporation Sole and the Encounter of Law and Church." In *Sacred Companies: Organizational Aspects of Religion and Religious Aspects of Organizations*, edited by N. J. Demerth III, Peter Dobkin Hall, Terry Schmitt, and Rhys H. Williams, 50–61. New York: Oxford University Press, 1998.

Daniel, Cletus E. *Culture of Misfortune: An Interpretive History of Textile Unionism in the United States*. Ithaca, N.Y.: ILR Press, 2001.

Dauber, Michele Landis. "The Sympathetic State." *Law and History Review* 23 (2005): 387–442.

———. *The Sympathetic State: Disaster Relief and the Origins of the American Welfare State*. Chicago: University of Chicago Press, 2013.

Davies, Andrea Rees. *Saving San Francisco: Relief and Recovery after the 1906 Disaster*. Philadelphia: Temple University Press, 2012.

Davies, Gareth. "The Emergence of a National Politics of Disaster, 1865–1900." *Journal of Policy History* 26, no. 3 (2014): 305–26.

Davis, Mike. *Ecology of Fear: Los Angeles and the Imagination of Disaster*. New York: Vintage, 1999.

Derthick, Martha. *The National Guard in Politics*. Cambridge, Mass.: Harvard University Press, 1965.

Deutsch, Sarah. *Women and the City: Gender, Space, and Power in Boston, 1870–1940*. New York: Oxford University Press, 2000.

DeVault, Marjorie L. *Feeding the Family: The Social Organization of Caring as Gendered Work*. Chicago: University of Chicago Press, 1991.

Dolan, Jay P. *The American Catholic Experience: A History from Colonial Times to the Present*. 1985. Repr. Notre Dame, Ind.: University of Notre Dame Press, 1992.

Donnell, Robert. "Geographical Analysis of Urban Structural Fire Problems: The Case of the Salem, Massachusetts, Conflagration of June 25, 1914." *Proceedings of the New England–St. Lawrence Valley Geographical Society* 27 (1997): 69–89.

———. "Locational Response to Catastrophe." *Essex Institute Historical Collections* 113, no. 2 (April 1977): 105–16.

Doty, C. Stewart. *The First Franco-Americans: New England Life Histories from the Federal Writers' Project, 1938–1939*. Orono: University of Maine at Orono Press, 1985.

"Douglas Brinkley: Chronicling 'The Great Deluge.'" *Fresh Air*, 16 September 2008. Available at http://www.npr.org/templates/story/story.php?storyId=4851087.

Driskell, Jay. *Schooling Jim Crow: The Fight for Atlanta's Booker T. Washington High School and the Roots of Black Protest Politics*. Charlottesville: University of Virginia Press, 2014.

Dynes, Russell R., and E. L. Quarantelli. "The Place of the Explosion in the History of Disaster Research: The Work of Samuel H. Prince." In *Ground Zero: A Reassessment of the 1917 Explosion in Halifax Harbour*, edited by Alan Ruffman and Colin D. Howell, 55–67. Halifax, Nova Scotia: Nimbus/The Gorsebrook Research Institute for Atlantic Canada Studies at Saint Mary's University, 1994.

Earle, Michael. *Workers and the State in Twentieth Century Nova Scotia*. Fredericton, New Brunswick: Acadiensis Press, 1989.

Ebert, Susan. "Community and Philanthropy." In *The Jews of Boston*, edited by Jonathan D. Sarna, Ellen Smith, and Scott-Martin Kosofsky, 221–47. 2nd ed. New Haven, Conn.: Yale University Press, 2005.

Emery, George Neil, and John Charles Herbert Emery. *A Young Man's Benefit: The Independent Order of Odd Fellows and Sickness Insurance in the United States and Canada, 1860–1929*. Montreal and Kingston: McGill-Queen's University Press, 1998.

Emmons, David M. *The Butte Irish: Class and Ethnicity in an American Mining Town, 1875–1925*. Urbana: University of Illinois Press, 1990.

English, John. *The Decline of Politics: The Conservatives and the Party System 1901–20*. 2nd ed. Toronto: University of Toronto Press, 1993.

Erickson, Paul A. *Halifax's North End: An Anthropologist Looks at the City*. Hantsport, Nova Scotia: Lancelot Press, 1986.

Erikson, Kai T. *A New Species of Trouble: Explorations in Disaster, Trauma, and Community*. New York: Norton, 1994.

———. *Everything in Its Path: Destruction of Community in the Buffalo Creek Flood*. New York: Simon and Schuster, 1976.

Falzone, Vincent J. *Terence V. Powderly: Middle-Class Reformer*. Washington, D.C.: University Press of America, 1978.

Fear, John. "'The Lumber Piles Must Go': Ottawa's Lumber Interests and the Great Fire of 1900." *Urban History Review* 8, no. 1 (June 1979): 38–65.

Feigenbaum, Anna, Fabian Frenzel, and Patrick McCurdy. *Protest Camps*. London: Zed, 2013.

Ferdinand, Theodore N. "Politics, the Police, and Arresting Policies in Salem, Massachusetts, since the Civil War." In *Deviant Behavior and Social Process*, edited by William A. Rushing, 2nd ed., 144–57. Chicago: Rand McNally College Publishing, 1975.

Fergusson, C[harles] Bruce. "Jewish Communities in Nova Scotia." *Nova Scotia Journal of Education* 5th ser. 11, no. 1 (October 1961): 45–48.

Ferland, Jacques. "Canadiens, Acadiens, and Canada: Knowledge and Ethnicity in Labour History." *Labour/Le Travail* 50 (Fall 2002): 101–15.

Finke, Roger, and Rodney Stark. *The Churching of America, 1776–2005: Winners and Losers in Our Religious Economy*. New Brunswick, N.J.: Rutgers University Press, 2005.

Fingard, Judith. "From Sea to Rail: Black Transportation Workers and Their Families in Halifax, c. 1870–1916." *Acadiensis* 24, no. 2 (Spring 1995): 49–64.

———. "Race and Respectability in Victorian Halifax." *Journal of Imperial and Commonwealth History* 20 (1992): 169–95.

———. *The Dark Side of Life in Victorian Halifax*. Porters Lake, Nova Scotia: Pottersfield Press, 1989.

Fingard, Judith, Janet Guildford, and David Sutherland. *Halifax: The First 250 Years*. Halifax, Nova Scotia: Formac, 1999.

Flink, James J. *The Automobile Age*. Cambridge, Mass.: MIT Press, 1988.

———. *The Car Culture*. Cambridge, Mass.: MIT Press, 1975.

Fogelson, Robert M. *America's Armories: Architecture, Society, and Public Order*. Cambridge, Mass.: Harvard University Press, 1989.

Foner, Philip S. *The Industrial Workers of the World, 1905–1917*. Vol. 4 of *History of the Labor Movement in the United States*. New York: International, 1965.

Forbes, Ernest R. "Battles in Another War: Edith Archibald and the Halifax Feminist Movement." In *Challenging the Regional Stereotype: Essays on the 20th Century Maritimes*, 67–89. Fredericton, New Brunswick: Acadiensis Press, 1989.

———. *The Maritime Rights Movement, 1919–1927: A Study in Canadian Regionalism*. Montreal and Kingston: McGill–Queen's University Press, 1979.

Fox, John. *Insuring the Future: The Holyoke Mutual Insurance Company in Salem, 1843–1993*. Acton, Mass.: Tapestry Press, 1993.

Frank, David. *J. B. McLachlan: A Biography*. Toronto: J. Lorimer, 1999.

———. "The Cape Breton Coal Industry and the Rise and Fall of the British Empire Steel Corporation." *Acadiensis* 7, no. 1 (Autumn 1977): 3–34.

———. "The Miners' Financier: Women in the Cape Breton Coal Towns, 1917." *Atlantis* 8, no. 2 (Spring 1983): 137–43.

Franz, Kathleen. *Tinkering: Consumers Reinvent the Early Automobile*. Philadelphia: University of Pennsylvania Press, 2005.

Frerichs, R. R., P. S. Keim, R. Barrais, and R. Piarroux. "Nepalese Origin of Cholera Epidemic in Haiti." *Clinical Microbiology and Infection* 18, no. 6 (June 2012): E158–E163.

Gabaccia, Donna R. "Is Everywhere Nowhere? Nomads, Nations, and the Immigrant Paradigm of United States History." *Journal of American History* 86 (December 1999): 1115–34.

Gauvreau, Michael, and Ollivier Hubert. "Introduction: Beyond Church History: Recent Developments in the History of Religion in Canada." In *The Churches and Social Order in Nineteenth- and Twentieth-Century Canada*, edited by Michael Gauvreau and Ollivier Hubert, 3–45. Montreal and Kingston: McGill–Queen's University Press, 2006.

Gerstle, Gary. *Working-Class Americanism: The Politics of Labor in a Textile City, 1914–1960*. Rev. ed. 1989. Repr. Princeton, N.J.: Princeton University Press, 2002.

Gilbert, Emily, and Corey Ponder. "Between Tragedy and Farce: 9/11 Compensation and the Value of Life and Death." *Antipode* 46, no. 2 (March 2014): 404–25.

Gilfoyle, Timothy J. "The Moral Origins of Political Surveillance: The Preventive Society in New York City, 1867–1918." *American Quarterly* 38, no. 4 (1986): 637–52.

Glasner, Joyce. *The Halifax Explosion: Surviving the Blast That Shook a Nation*. Canmore, Alberta: Altitude, 2003.

Glazer, Nona Y. "Servants to Capital: Unpaid Domestic Labor and Paid Work." *Review of Radical Political Economics* 16, no. 1 (1984): 61–87.

Glenn, David. "Lost (and Found) in the Flood." *Chronicle of Higher Education*, 7 October 2005, A14.

Glenn, John M., Lilian Brandt, and F. Emerson Andrews. *Russell Sage Foundation 1907–1946*. Vol. 1. New York: Russell Sage Foundation, 1947.

Glenn, Susan A. *Daughters of the Shtetl: Life and Labor in the Immigrant Generation*. Ithaca, N.Y.: Cornell University Press, 1990.

Goodhue, Albert, Jr., ed. "The Salem Fire." *Essex Institute Historical Collections* 100 (1964): 183–94.

Gordon, Linda. *Pitied but Not Entitled: Single Mothers and the History of Welfare, 1890–1930*. New York: Free Press, 1994.

Graeber, David. *Fragments of an Anarchist Anthropology*. Chicago: Prickly Paradigm Press, 2004.

Graham, Cooper C., et al. *D. W. Griffith and the Biograph Company*. Metuchen, N.J.: Scarecrow Press, 1985.

Grant, Jill L., Leifka Vissers, and James Haney. "Early Town Planning Legislation in Nova Scotia: The Roles of Local Reformers and International Experts." *Urban History Review* 40, no. 2 (Spring 2012): 3–14.

Grant, John Webster. *The Church in the Canadian Era*. Rev. ed. Vancouver: Regent College Publishing, 1998.

———. *The Canadian Experience of Church Union*. Ecumenical Studies in History 8. Richmond, Va.: John Knox Press, 1967.

Greene, Julie. *The Canal Builders: Making America's Empire at the Panama Canal*. New York: Penguin Press, 2009.

Greening, William E., and Murdoch MacKay Maclean. *It Was Never Easy 1908–1958: A History of the Canadian Brotherhood of Railway, Transport and General Workers*. Ottawa: Mutual Press, 1961.

Greenwald, Maurine Weiner. *Women, War, and Work: The Impact of World War I on Women Workers in the United States*. Westport, Conn.: Greenwood, 1980.

Griffith, Ernest S. *A History of American City Government: The Progressive Years and Their Aftermath, 1900–1920*. New York: Praeger, 1974.

Guglielmo, Jennifer. "Italian Women's Proletarian Feminism in the New York City Garment Trades, 1890s–1940s." In *Women, Gender, and Transnational Lives: Italian Workers of the World*, edited by Donna R. Gabaccia and Franca Iacovetta, 247–98. Toronto: University of Toronto Press, 2002.

———. *Living the Revolution: Italian Women's Resistance and Radicalism in New York City, 1880–1945*. Chapel Hill: University of North Carolina Press, 2010.

Guignard, Michael. "Maine's Corporation Sole Controversy." *Maine Historical Society Newsletter* 12 (Winter 1973): 111–30.

Haas, Edward F. "'Don't Believe Any False Rumors . . .': Mayor Victor H. Schiro, Hurricane Betsy, and Urban Myths." *Louisiana History* 45, no. 4 (2004): 463–68.

Hale, Grace Elizabeth. *Making Whiteness: The Culture of Segregation in the South, 1890–1940*. New York: Vintage Books, 1999.

Hall, David D., ed. *Lived Religion in America: Toward a History of Practice*. Princeton, N.J.: Princeton University Press, 1997.

Ham, Edward Billings. "French National Societies in New England." *New England Quarterly* 12, no. 2 (June 1939): 315–32.

Hanington, J. Brian. *Every Popish Person: The Story of Roman Catholicism in Nova Scotia and the Church of Halifax, 1604–1984*. [Halifax, Nova Scotia]: Archdiocese of Halifax, 1984.

Hansen, Marcus Lee, and John Bartlet Brebner. *The Mingling of the Canadian and American People*. New Haven, Conn.: Yale University Press, 1940.

Hareven, Tamara K. "An Ambiguous Alliance: Some Aspects of American Influences on Canadian Social Welfare." *Histoire sociale/Social History* 2 (April 1969): 82–98.

———. *Family Time and Industrial Time: The Relationship between the Family and Work in a New England Industrial Community*. Cambridge: Cambridge University Press, 1982.

———. "Family Time and Industrial Time: Family and Work in a Planned Corporation Town, 1900–1924." *Journal of Urban History* 1 (1975): 365–89.

———. "The Laborers of Manchester, New Hampshire, 1900–1940: The Role of Family and Ethnicity in Adjustment to Industrial Life." *Labor History* (1975): 249–65.

Hareven, Tamara, and Randolph Langenbach. *Amoskeag: Life and Work in an American Factory-City*. New York: Pantheon, 1978.

Harney, Robert F. "Franco-Americans and Ethnic Studies: Notes on a Mill Town." In *The Quebec and Acadian Diaspora in North America*, edited by Raymond Breton and Pierre Savard, 77–88. Toronto: Multicultural History Society of Ontario, 1982.

Hartford, William F. *Where Is Our Responsibility? Unions and Economic Change in the New England Textile Industry, 1870–1960*. Amherst: University of Massachusetts Press, 1996.

[Harvey, O. L.]. *The Anvil and the Plow: A History of the United States Department of Labor*. Washington, D.C.: GPO, [1963].

Hastings, Paula. "Territorial Spoils, Transnational Black Resistance, and Canada's Evolving Autonomy during the First World War." *Histoire sociale/Social History* 47, no. 94 (June 2014): 443–70.

Hendrickson, Dyke. *Quiet Presence: Dramatic, First-Person Accounts: The True Stories of Franco-Americans in New England*. Portland, Maine: G. Gannet, 1980.

Heron, Craig. "Labourism and the Canadian Working Class." *Labour/Le Travail* 13 (Spring 1984): 45–76.

———, ed. *The Workers' Revolt in Canada, 1917–1925*. Toronto: University of Toronto Press, 1998.

Heron, Craig, and Myer Siemiatycki. "The Great War, the State, and Working-Class Canada." In *The Workers' Revolt in Canada, 1917–1925*, edited by Craig Heron, 11–42. Toronto: University of Toronto Press, 1998.

Hersey, John. *Hiroshima*. New York: Bantam Books, 1948.

Hewitt, Kenneth. "The Idea of Calamity in a Technocratic Age." In *Interpretations of Calamity from the Viewpoint of Human Ecology*, edited by Kenneth Hewitt, 3–32. Boston: Allen and Unwin, 1983.

———. "'When the Great Planes Came and Made Ashes of Our City . . .': Towards an Oral Geography of the Disasters of War." *Antipode* 26, no. 1 (January 1994): 1–34.

Higginbotham, Evelyn Brooks. *Righteous Discontent: The Women's Movement in the Black Baptist Church, 1880–1920*. Cambridge, Mass.: Harvard University Press, 1994.

Hilderbrand, Sue, scott crow, and Lisa Fithian. "Common Ground Relief." In *What Lies Beneath: Katrina, Race, and the State of the Nation*, edited by South End Press Collective, 80–98. Cambridge, Mass.: South End, 2006.

Hill, Frances. "Salem as Witch City." In *Salem: Place, Myth, and Memory*, edited by Dane Anthony Morrison and Nancy Lusignan Schultz, 283–96. Boston: Northeastern University Press, 2004.

Hogan, Brian F. "The Guelph Novitiate Raid: Conscription, Censorship, and Bigotry during the Great War." *Canadian Catholic Historical Association Study Sessions* 45 (1978): 57–80.

Hogarty, Richard A. *Massachusetts Politics and Public Policy: Studies in Power and Leadership*. Amherst: University of Massachusetts Press, 2002.

Hornsby, Stephen J., Victor A. Konrad, and James J. Herlan, eds. *The Northeastern Borderlands: Four Centuries of Interaction*. Orono: Canadian-American Center at the University of Maine/Fredericton, New Brunswick: Acadiensis Press, 1989.

Huthmacher, J. Joseph. *Massachusetts People and Politics 1919–1933*. Cambridge, Mass.: Belknap Press of Harvard University Press, 1959.

IPCC. "Summary for Policymakers." In *Managing the Risks of Extreme Events and Disasters to Advance Climate Change Adaptation*. Special Report of Working Groups I and II of the Intergovernmental Panel on Climate Change, 1–19. Cambridge: Cambridge University Press, 2012.

Irwin, Julia F. *Making the World Safe: The American Red Cross and a Nation's Humanitarian Awakening*. Oxford: Oxford University Press, 2013.

Jackson, Hugh J. M. *The Voyage of the* Komagata Maru: *The Sikh Challenge to Canada's Colour Bar*. Expanded ed. Vancouver: University of British Columbia Press, 2014.

Jackson, James. "Fatal Attraction: Living with Earthquakes, the Growth of Villages into Megacities, and Earthquake Vulnerability in the Modern World." *Philosophical Transactions of the Royal Society A: Mathematical, Physical and Engineering Sciences* 364, no. 1845 (2006): 1911–25.

Jacobs, Jane. *The Death and Life of Great American Cities*. 1961. Repr. New York: Vintage, 1992.

Jarausch, Konrad H., and Kenneth A. Hardy. *Quantitative Methods for Historians*. Chapel Hill: University of North Carolina Press, 1991.

Jefferson, Cord. "Ask the Expert: The Looting Lie." *Campus Progress*, 15 January 2010. Available at http://www.campusprogress.org/asktheexpert/4982/the-looting-lie.

Johnson, D. W. *History of Methodism in Eastern British America*. Sackville, New Brunswick: Tribune, [1925?].

Johnston, Robert D. "Re-Democratizing the Progressive Era: The Politics of Progressive Era Political Historiography." *Journal of the Gilded Age and Progressive Era* 1, no. 1 (2002): 68–92.

———. *The Radical Middle Class: Populist Democracy and the Question of Capitalism in Progressive Era Portland, Oregon*. Princeton, N.J.: Princeton University Press, 2003.

Jones, Esyllt. "Politicizing the Laboring Body: Working Families, Death, and Burial in Winnipeg's Influenza Epidemic, 1918–1919." *Labor: Studies in Working-Class History of the Americas* 3 no. 3 (Fall 2006): 57–75.

Jones, Marian Moser. *The American Red Cross from Clara Barton to the New Deal*. Baltimore: Johns Hopkins University Press, 2013.

Kadish, Alon. *Apostle Arnold: The Life and Death of Arnold Toynbee, 1852–1883*. Durham, N.C.: Duke University Press, 1986.

Kampas, Barbara Pero. *The Great Salem Fire of 1914*. Charleston, S.C.: History Press, 2008.

Katz, Jonathan M. *The Big Truck That Went By: How the World Came to Save Haiti and Left Behind a Disaster*. New York: Palgrave Macmillan, 2013.

Katz, Michael B. *In the Shadow of the Poorhouse: A Social History of Welfare in America*. 10th anniv. ed. New York: Basic Books, 1996.

Kealey, Gregory S. "1919: The Canadian Labour Revolt." *Labour/Le Travail* 13 (Spring 1984): 11–44.

Kelley, Robin D. G. *Race Rebels: Culture, Politics, and the Black Working Class*. New York: Free Press, 1996.

Kerr, Donald A. "Another Calamity: The Litigation." In *Ground Zero: A Reassessment of the 1917 Explosion in Halifax Harbour*, edited by Alan Ruffman and Colin D. Howell, 365–75. Halifax, Nova Scotia: Nimbus/The Gorsebrook Research Institute for Atlantic Canada Studies at Saint Mary's University, 1994.

Kessler-Harris, Alice. *Out to Work: A History of Wage-Earning Women in the United States*. 20th ed. New York: Oxford University Press, 2003.

———. "The Wages of Patriarchy: Some Thoughts about the Continuing Relevance of Class and Gender." *Labor: Studies in Working-Class History of the Americas* 3 no. 3 (Fall 2006): 7–21.

Keyssar, Alexander. *Out of Work: The First Century of Unemployment in Massachusetts*. Cambridge: Cambridge University Press, 1986.

King, William M. "Guardian of the Public Safety: Garrett A. Morgan and the Lake Erie Crib Disaster." *Journal of Negro History* 70 (Winter/Spring 1985): 1–13.

Kitz, Janet F. "The Explosion Mortuary Artifacts: A Look at the Victims." In *Ground Zero: A Reassessment of the 1917 Explosion in Halifax Harbour*, edited by Alan Ruffman and Colin D. Howell, 9–34. Halifax, Nova Scotia: Nimbus/The Gorsebrook Research Institute for Atlantic Canada Studies at Saint Mary's University, 1994.

———. "The Halifax Explosion, December 6, 1917." *Canadian Oral History Association Journal* 12 (1992): 6–11.

———. Kitz, Janet F. *Shattered City: The Halifax Explosion and the Road to Recovery*. Halifax, Nova Scotia: Nimbus, 1989.

———. "The Inquiry into the Halifax Explosion of December 6, 1917: The Legal Aspects." *Journal of the Royal Nova Scotia Historical Society* 5 (2002): 64–78.

Klein, Naomi. *The Shock Doctrine: The Rise of Disaster Capitalism*. New York: Picador, 2007.

Kline, Bernard G. "Post Cards of the 1917 Explosion." In *Ground Zero: A Reassessment of the 1917 Explosion in Halifax Harbour*, edited by Alan Ruffman and Colin D. Howell, 139–61. Halifax, Nova Scotia: Nimbus/The Gorsebrook Research Institute for Atlantic Canada Studies at Saint Mary's University, 1994.

Klinenberg, Eric. *Heat Wave: A Social Autopsy of Disaster in Chicago*. Chicago: University of Chicago Press, 2002.

Klug, Thomas A. "The Immigration and Naturalization Service (INS) and the Making of a Border-Crossing Culture on the U.S.-Canada Border, 1891–1941." *American Review of Canadian Studies* 40, no. 3 (2010): 395–415.

Knowles, Scott Gabriel. *The Disaster Experts: Mastering Risk in Modern America*. Philadelphia: University of Pennsylvania Press, 2011.

Konvitz, Josef. "Représentations urbaines et bombardements stratégiques, 1914–1945." *Annales: Économies, sociétés, civilisations* 44, no. 4 (August 1989): 823–47.

Ladd-Taylor, Molly. *Mother-Work: Women, Child Welfare, and the State, 1890–1930.* Urbana: University of Illinois Press, 1994.

Lalonde, Jean-Louis. *Des loups dans la bergerie: Les protestants de langue française au Québec, 1534–2000.* [Montreal]: Fides, 2002.

Langer, William L. "The Next Assignment." *American Historical Review* 63, no. 2 (January 1958): 283–304.

Lazar, M. M., and Sheva Medjuck. "In the Beginning: A Brief History of Jews in Atlantic Canada." *Canadian Jewish Historical Society Journal* 5, no. 2 (Fall 1981): 91–108.

LeBlanc, Robert G. "The Franco-American Response to the Conscription Crisis in Canada, 1916–1918." *American Review of Canadian Studies* 23, no. 3 (Autumn 1993): 343–72.

Ledesma, Irene. "Natural Disasters and Community Survival in Texas and Louisiana in the Gilded Age." *Gulf Coast Historical Review* 10 (Fall 1994): 72–84.

Levenstein, Lisa. *A Movement without Marches: African American Women and the Politics of Poverty in Postwar Philadelphia.* Chapel Hill: University of North Carolina Press, 2009.

Lindgren, James M. *Preserving Historic New England: Preservation, Progressivism, and the Remaking of Memory.* New York: Oxford University Press, 1995.

Lindqvist, Sven. *A History of Bombing.* Translated by Linda Haverty Rugg. New York: New Press, 2001.

Lombardi, John. *Labor's Voice in the Cabinet: A History of the Department of Labor from Its Origin to 1921.* New York: Columbia University Press, 1942.

Luft, Rachel E. "Looking for Common Ground: Relief Work in Post-Katrina New Orleans as an American Parable of Race and Gender Violence." *NWSA Journal* 20, no. 3 (Fall 2008): 5–31.

Lynd, Staughton. "Communal Rights." *Texas Law Review* 62, no. 8 (May 1984): 1417–41.

Mac Donald, Laura M. *Curse of the Narrows.* New York: Walker, 2005.

Macgillivray, Don. "Military Aid to the Civil Power: The Cape Breton Experience in the 1920s." In *Cape Breton Historical Essays*, edited by Brian Tennyson and Don Macgillivray, 95–109. Sydney, Nova Scotia: College of Cape Breton Press, 1980.

MacLeod, Malcolm. "Searching the Wreckage for Signs of Region: Newfoundland and the Halifax Harbour Explosion." In *Ground Zero: A Reassessment of the 1917 Explosion in Halifax Harbour*, edited by Alan Ruffman and Colin D. Howell, 207–18. Halifax, Nova Scotia: Nimbus/The Gorsebrook Research Institute for Atlantic Canada Studies at Saint Mary's University, 1994.

MacMechan, Archibald. *The Halifax Explosion: December 6, 1917.* Edited by Graham Metson. Toronto: McGraw-Hill Ryerson, 1978.

Maloney, Joan M. "John F. Hurley: Salem's First Hurrah." *Essex Institute Historical Collections* 128 (January 1992): 27–58.

Mann, Susan. *The Dream of Nation: A Social and Intellectual History of Quebec.* 2nd ed. 1982; Repr. Montreal and Kingston: McGill–Queen's University Press, 2002.

Mann, W. E. "The Canadian Church Union, 1925." In *Institutionalism and Church Unity: A Symposium*, edited by Nils Ehrenstrom and Walter S. Muelder, 171–93. New York: Association Press, 1963.

Marble, Allan E., and Verilea D. Ellis. *The House That Sexton Built: A Century of Outstanding Graduates*. Tantallon, Nova Scotia: Glen Margaret, 2007.
March, William. *Red Line: The Chronicle-Herald and the Mail-Star, 1875–1954*. Halifax, Nova Scotia: Chebucto Agencies, 1986.
Maritime Museum of the Atlantic. "Ships of the Halifax Explosion." Last updated 18 February 2009. Available at http://museum.gov.ns.ca/mma/AtoZ/expships.html.
Martin, Sandra. "Leonard Arthur Kitz, Lawyer and Politician 1916–2006." *Globe and Mail*, 4 February 2006, S9.
Martland, Samuel. "Reconstructing the City, Constructing the State: Government in Valparaiso after the Earthquake of 1906." *Hispanic American Historical Review* 87, no. 2 (May 2007): 221–54.
Mason, Julie, and Michael Hedges. "Washington Sends More Troops, Begs for Patience." *Houston Chronicle*, 2 September 2005.
Mathieu, Sarah-Jane. *North of the Color Line: Migration and Black Resistance in Canada, 1870–1955*. Chapel Hill: University of North Carolina Press, 2010.
Mathieu, Sarah-Jane (Saje). "North of the Colour Line: Sleeping Car Porters and the Battle against Jim Crow on Canadian Rails, 1880–1920." *Labour/Le Travail* 47 (Spring 2001): 9–41.
Maxwell, Kenneth. *Pombal: Paradox of the Enlightenment*. Cambridge: Cambridge University Press, 1995.
McCabe, Kathy. "Catholic Elementary School Survives." *Boston Globe*, 15 August 2004, North section, 1.
———. "Closing Parish Offers Up Its Light: Church in Salem Hosts Final Mass after 131 Years." *Boston Globe*, 15 August 2004, North section, 1.
———. "'Our New Home.'" *Boston Globe*, 29 May 2005, North section, 1.
———. "Past to Shape Church's Future." *Boston Globe*, 26 June 2005, North section, 6.
McCann, L. D. "'Living a Double Life': Town and Country in the Industrialization of the Maritimes." In *Geographical Perspectives on the Maritime Provinces*, edited by Douglas Day, 93–113. Halifax, Nova Scotia: Saint Mary's University, 1988.
McCarthy, Kathleen D. *Noblesse Oblige: Charity and Cultural Philanthropy in Chicago, 1849–1929*. Chicago: University of Chicago Press, 1982.
McCartin, James P., and Joseph A. McCartin. "Working-Class Catholicism: A Call for New Investigations, Dialogue, and Reappraisal." *Labor: Studies in Working-Class History of the Americans* 4, no. 1 (Spring 2007): 99–110.
McGowan, Mark. "The Maritimes Region and the Building of a Canadian Church: The Case of the Diocese of Antigonish after Confederation." *Canadian Catholic Historical Association Historical Studies* 70 (2004): 46–67.
———. *The Waning of the Green: Catholics, the Irish, and Identity in Toronto, 1887–1922*. Montreal and Kingston: McGill–Queen's University Press, 1999.
McKay, Ian. *The Craft Transformed: An Essay on the Carpenters of Halifax, 1885–1985*. Halifax, Nova Scotia: Holdfast Press, 1985.
McKay, Ian, and Robin Bates. *In the Province of History: The Making of the Public Past in Twentieth-Century Nova Scotia*. Montreal and Kingston: McGill–Queen's University Press, 2010.
McKay, Ian, and Suzanne Morton. "The Maritimes: Expanding the Circle of Resistance."

In *The Workers' Revolt in Canada, 1917–1925*, edited by Craig Heron, 43–86. Toronto: University of Toronto Press, 1998.

McManus, Sheila. *The Line Which Separates: Race, Gender, and the Making of the Alberta-Montana Borderlands*. Lincoln: University of Nebraska Press, 2005.

McWilliams, Carey. *Factories in the Field: The Story of Migratory Farm Labor in California*. Boston: Little, Brown, 1939.

Medjuck, Sheva. *Jews of Atlantic Canada*. St. John's, Newfoundland: Breakwater, 1986.

Mendelsohn, Robert, Kerry Emanuel, Shun Chonabayashi, and Laura Bakkesen. "The Impact of Climate Change on Global Tropical Cyclone Damage." *Nature Climate Change* 2 (March 2012): 205–9.

Miller, David C. *Introduction to Collective Behavior and Collective Action*. 2nd ed. Prospect Heights, Ill.: Waveland, 2000.

Miner, Horace. *St. Denis: A French-Canadian Parish*. Chicago: University of Chicago Press, 1939.

Montgomery, David. *The Fall of the House of Labor: The Workplace, the State, and American Labor Activism, 1865–1925*. Cambridge: Cambridge University Press, 1987.

Morozov, Evgeny. *To Save Everything, Click Here: The Folly of Technological Solutionism*. New York: Public Affairs, 2014.

Morrison, Dane Anthony, and Nancy Lusignan Schultz, eds. *Salem: Place, Myth, and Memory*. Boston: Northeastern University Press, 2004.

Morton, Desmond. *A Short History of Canada*. 2nd rev. ed. Toronto: McClelland and Stewart, 1994.

Morton, Suzanne. "The Halifax Relief Commission and Labour Relations during the Reconstruction of Halifax, 1917–1919." *Acadiensis* 18, no. 2 (Spring 1989): 73–93.

———. *Ideal Surroundings: Domestic Life in a Working-Class Suburb in the 1920s*. Toronto: University of Toronto Press, 1995.

———. "Labourism and Economic Action: The Halifax Shipyards Strike of 1920." *Labour/Le Travail* 22 (Fall 1988): 67–98.

———. "'Never Handmaidens': The Victorian Order of Nurses and the Massachusetts-Halifax Health Commission." In *Ground Zero: A Reassessment of the 1917 Explosion in Halifax Harbour*, edited by Alan Ruffman and Colin D. Howell, 195–205. Halifax, Nova Scotia: Nimbus/The Gorsebrook Research Institute for Atlantic Canada Studies at Saint Mary's University, 1994.

———. *Wisdom, Justice and Charity: Canadian Social Welfare through the Life of Jane B. Wisdom, 1884–1975*. Toronto: University of Toronto Press, 2014.

Morton, William Lewis. *The Progressive Party in Canada*. Toronto: University of Toronto Press, 1950.

Mott, Morris. "Tobias C. Norris." In *Manitoba Premiers of the 19th and 20th Centuries*, edited by Barry Ferguson and Robert Vardhaugh, 139–63. Regina, Saskatchewan: Canadian Plains Research Center Press, 2010.

Muise, Del. "'The Great Transformation': Changing the Urban Face of Nova Scotia, 1871–1921." *Nova Scotia Historical Review* 11 (1991): 1–42.

Murphy, Dan. "Cutten's Ugly Legacy: Colgate Should Rename Residence Hall." *Colgate Maroon-News*, 24 February 2006. Available at http://www.maroon-news.com/2.5266/cutten-s-ugly-legacy-1.804949.

Murphy, Terrence, ed. *A Concise History of Christianity in Canada*. Toronto: Oxford University Press, 1996.

Murphy, Terrence, and Gerald J. Stortz, eds. *Creed and Culture: The Place of English-Speaking Catholics in Canadian Society, 1750–1930*. Montreal and Kingston: McGill-Queen's University Press, 1993.

Murray, T. J. "Medical Aspects of the Disaster: The Missing Report of Dr. David Fraser Harris." In *Ground Zero: A Reassessment of the 1917 Explosion in Halifax Harbour*, edited by Alan Ruffman and Colin D. Howell, 229–44. Halifax, Nova Scotia: Nimbus/The Gorsebrook Research Institute for Atlantic Canada Studies at Saint Mary's University, 1994.

Nelson, Megan Kate. *Ruin Nation: Destruction and the American Civil War*. Athens: University of Georgia Press, 2012.

New Weld, Elizabeth. "St. Joseph's Church Has Changed with the Point." *Boston Globe*, 13 November 1994, North weekly section, 16.

Newton, John. "Federal Legislation for Disaster Mitigation: A Comparative Assessment between Canada and the United States." *Natural Hazards* 16, no. 2 (November 1, 1997): 219–41.

Ngai, Mae M. *Impossible Subjects: Illegal Aliens and the Making of Modern America*. Princeton, N.J.: Princeton University Press, 2004.

Nicholls, Robert J., et al. "Coastal Systems and Low-Lying Areas." In *Climate Change 2007: Impacts, Adaptation and Vulnerability. Contribution of Working Group II to the Fourth Assessment Report of the Intergovernmental Panel on Climate Change*, edited by M. L. Perry et al., 315–56. Cambridge: Cambridge University Press, 2007.

Nolan, Stephen. *Leaving Newfoundland: A History of Out-Migration*. St. John's, Newfoundland: Flanker Press, 2007.

Nugent, Walter. *Crossings: The Great Transatlantic Migrations, 1870–1914*. Bloomington: Indiana University Press, 1992.

Nye, David. *When the Lights Went Out: A History of Blackouts in America*. Cambridge, Mass.: MIT Press, 2010.

O'Connor, Alice. *Social Science for What? Philanthropy and the Social Question in a World Turned Rightside Up*. New York: Russell Sage Foundation, 2007.

O'Connor, Thomas H. *The Boston Irish: A Political History*. Boston: Northeastern University Press, 1995.

O'Hara, James B. "The Modern Corporation Sole." *Dickinson Law Review* 93 (Fall 1988): 23–39.

Olcott, Jocelyn. "Introduction: Researching and Rethinking the Labors of Love." *Hispanic American Historical Review* 91, no. 1 (February 2011): 1–27.

Orihara, Minami, and Gregory Clancey. "The Nature of Emergency: The Great Kanto Earthquake and the Crisis of Reason in Late Imperial Japan." *Science in Context* 25, no. 1 (2012): 103–26.

Orsi, Robert A. *Between Heaven and Earth: The Religious Worlds People Make and the Scholars Who Study Them*. Princeton, N.J.: Princeton University Press, 2005.

———. *The Madonna of 115th Street: Faith and Community in Italian Harlem, 1880–1950*. New Haven, Conn.: Yale University Press, 1985.

Ortiz, David G., and Stephen F. Ostertag. "Katrina Bloggers and the Development of

Collective Civic Action: The Web as a Mobilizing Structure." *Sociological Perspectives* 57, no. 1 (2014): 52–78.

Otis, Yves, and Bruno Ramirez. "Nouvelles perspectives sur le mouvement d'émigration des Maritimes vers les États-Unis, 1906–1930." *Acadiensis* 28, no. 1 (Autumn 1998): 27–46.

O'Toole, James M. *Militant and Triumphant: William Henry O'Connell and the Catholic Church in Boston, 1859–1944*. Notre Dame, Ind.: University of Notre Dame Press, 1992.

Pachai, Bridglal, and Henry Bishop. *Historic Black Nova Scotia*. Halifax, Nova Scotia: Nimbus, 2006.

Page, Max. *The City's End: Two Centuries of Fantasies, Fears, and Premonitions of New York's Destruction*. New Haven, Conn.: Yale University Press, 2008.

Palmer, Bryan D. *Working-Class Experience: Rethinking the History of Canadian Labour, 1800–1991*. 2nd ed. Toronto: McClelland and Stewart, 1992.

Parker, Geoffrey. "Crisis and Catastrophe: The Global Crisis of the Seventeenth Century Reconsidered." *American Historical Review* 113, no. 4 (2008): 1053–79.

Paulson, Michael, and Kathy McCabe. "A Rebirth for Beverly Church: Worshipers Start Parish as Gift in Dominican Republic." *Boston Globe*, 29 August 2004, A1.

Peiss, Kathy. *Cheap Amusements: Working Women and Leisure in Turn-of-the-Century New York*. Philadelphia: Temple University Press, 1986.

Péloquin-Faré, Louise. *L'identité culturelle: Les franco-américains de la Nouvelle-Angleterre*. Paris: Crédif and Didier, 1983.

Perin, Roberto. "French-Speaking Canada from 1840." In *A Concise History of Christianity in Canada*, edited by Terrance Murphy, 190–259. Toronto: Oxford University Press, 1996.

Peterson, Sara Jo. "Voting for Play: The Democratic Potential of Progressive Era Playgrounds." *Journal of the Gilded Age and Progressive Era* 3, no. 2 (2004): 147–75.

Peterson, Thomas C., Peter A. Stott, and Stephanie Herring. "Explaining Extreme Events of 2011 from a Climate Perspective." *Bulletin of the American Meteorological Society* 93 no. 7 (July 2012): 1041–67.

Petrin, Ronald A. *French Canadians in Massachusetts Politics, 1885–1915: Ethnicity and Political Pragmatism*. Philadelphia: Balch Institute Press, 1990.

Petryna, Adriana. *Life Exposed: Biological Citizenship after Chernobyl*. Princeton, N.J.: Princeton University Press, 2002.

Phillips, Harlan B. *Felix Frankfurter Reminisces*. New York: Reynal, 1960.

Phillips, Ronnie J. "Coping with Financial Catastrophe: The San Francisco Clearinghouse during the Earthquake of 1906." *Research in Economic History* 21 (2003): 79–104.

Piccato, Pablo. *City of Suspects: Crime in Mexico City, 1900–1931*. Durham, N.C.: Duke University Press, 2001.

Pielke, R. A., Jr., et al. "Hurricanes and Global Warming." *Bulletin of the American Meteorological Society* 86 (2005): 1571–75.

Plender, Richard. *International Migration Law*. 2nd rev. ed. Dordrecht: Martinus Nijhoff, 1988.

Portelli, Alessandro. "The Death of Luigi Trastulli: Memory and the Event." In *The Death of Luigi Trastulli and Other Stories: Form and Meaning in Oral History*, 1–26. Albany: State University of New York Press, 1991.

Post, Robert C., and Nancy L. Rosenblum. "Introduction." In *Civil Society and Government*, edited by Nancy L. Rosenblum and Robert C. Post, 1–25. Princeton, N.J.: Princeton University Press, 2002.
Powell, Lawrence N. "What Does American History Tell Us about Katrina and Vice Versa?" *Journal of American History* 94, no. 3 (December 2007): 863–76.
Prosper, Kenny. "Mi'kmaw at the Halifax Explosion." *Mi'kmaq-Maliseet Nations News*. December 2002, 5.
Putnam, Robert. *Making Democracy Work: Civic Traditions in Modern Italy.* Princeton, N.J.: Princeton University Press, 1993.
Quintal, Claire. "Les institutions franco-américaines: Pertes et progrès." In *Le Québec et les francophones de la Nouvelle-Angleterre*, edited by Dean Louder, 61–84. Sainte-Foy, Quebec: Presses de l'Université Laval, 1991.
Quintal, Claire, and André Vachon, eds. *Situation de la recherche sur la Franco-Américanie.* Quebec: Le Conseil de la vie française en Amerique, 1980.
Ramirez, Bruno. *Crossing the 49th Parallel: Migration from Canada to the United States, 1900–1930.* Ithaca, N.Y.: Cornell University Press, 2001.
Remes, Jacob. "Movable Type: Toronto's Transnational Printers, 1867–1872." In *Workers across the Americas: The Transnational Turn in Labor History*, edited by Leon Fink, 384–408. New York: Oxford University Press, 2011.
Remes, Jacob A. C. "'Committed as Near Neighbors': The Halifax Explosion and Border-Crossing People and Ideas." *American Review of Canadian Studies* 45, no. 1 (Spring, 2015): 26–43.
———. "Solidarity, Citizenship, and the Opportunities of Disasters." In *Labor Rising: The Past and Future of Working People in America*, edited by Daniel Katz and Richard Greenwald, 143–53. New York: New Press, 2012.
Remon, Denis, ed. *L'identité des protestants francophones au Québec, 1834–1997.* Cahiers de l'ACFAS 94. Montreal: ACFAS, 1998.
Reverby, Susan. "A Caring Dilemma: Womanhood and Nursing in Historical Perspective." In *The Sociology of Heath and Illness: Critical Perspectives*, edited by Peter Conrad and Rochelle Kern, 184–95. 3rd ed. New York: St. Martin's, 1990.
Rice, Bradley R. "The Galveston Plan of City Government by Commission: The Birth of a Progressive Idea." *Southwestern Historical Quarterly* 78, no. 4 (April 1975): 365–408.
Richard, Mark Paul. *Loyal but French: The Negotiation of Identity by French-Canadian Descendants in the United States.* East Lansing: Michigan State University Press, 2008.
Robertson, Allen B. "After the Storm: The Church and Synagogue Response." In *Ground Zero: A Reassessment of the 1917 Explosion in Halifax Harbour*, edited by Alan Ruffman and Colin D. Howell, 219–27. Halifax, Nova Scotia: Nimbus/The Gorsebrook Research Institute for Atlantic Canada Studies at Saint Mary's University, 1994.
Roby, Yves. "Quebec in the United States: A Historiographical Survey." *Maine Historical Society Quarterly* 26 (Winter 1987): 126–59.
———. *The Franco-Americans of New England: Dreams and Realities.* Translated by Mary Ricard. [Sillery, Quebec]: Septentrion, 2004.
Rocher, Marie-Claude. "Double traïtrise ou double appartenance? Le patrimoine des protestants francophones au Québec." *Ethnologies* 25, no. 2 (2003): 215–33.
Rodgers, Daniel T. *Atlantic Crossings: Social Politics in a Progressive Age.* Cambridge, Mass.: Belknap Press, 1998.

———. "In Search of Progressivism." *Reviews in American History* 10, no. 4 (December 1982): 113–32.
Roper, H. "The Strange Political Career of A. C. Hawkins, Mayor of Halifax, 1918–1919." *Collections of the Royal Nova Scotia Historical Society* 41 (1982): 141–64.
Roper, Henry. "Archibald MacMechan and the Writing of 'The Halifax Disaster.'" In *Ground Zero: A Reassessment of the 1917 Explosion in Halifax Harbour*, edited by Alan Ruffman and Colin D. Howell, 85–99. Halifax, Nova Scotia: Nimbus/The Gorsebrook Research Institute for Atlantic Canada Studies at Saint Mary's University, 1994.
———. "The Halifax Board of Control: The Failure of Municipal Reform, 1906–1919." *Acadiensis* 14, no. 2 (1985): 46–95.
Rosen, Christine Meisner. *The Limits of Power: Great Fires and the Process of City Growth in America*. Cambridge: Cambridge University Press, 1986.
Rosenzweig, Roy. "Boston Masons, 1900–1935: The Lower Middle Class in a Divided Society." *Journal of Voluntary Action Research* 6, no. 3 (October 1977): 119–26.
———. *Eight Hours for What We Will: Workers and Leisure in an Industrial City, 1870–1920*. Cambridge: Cambridge University Press, 1983.
Rouillard, Jacques. *Histoire du syndicalisme au Québec: Des origines à nos jours*. Montreal: Boréal, 1989.
———. *Le syndicalisme québécois: Deux siècles d'histoire*. Montreal: Boréal, 2004.
Rowan, Linda. "Seismology: Urban Hazards." *Science* 307, no. 5706 (7 January 2005): 18e.
Rozario, Kevin. *The Culture of Calamity: Disaster and the Making of Modern America*. Chicago: University of Chicago Press, 2007.
Ruddel, David-Thiery. *Le protestantisme français au Québec, 1840–1919: "Images" et témoignages*. Mercury Series History 36. Ottawa: Musée national de l'homme, 1983.
Ruffman, Alan, David A. Greenberg, and Tad S. Murty. "The Tsunami from the Explosion in Halifax Harbour." In *Ground Zero: A Reassessment of the 1917 Explosion in Halifax Harbour*, edited by Alan Ruffman and Colin D. Howell, 327–44. Halifax, Nova Scotia: Nimbus/The Gorsebrook Research Institute for Atlantic Canada Studies at Saint Mary's University, 1994.
Ruffman, Alan, and Colin D. Howell. *Ground Zero: A Reassessment of the 1917 Explosion in Halifax Harbour*. Halifax, Nova Scotia: Nimbus/The Gorsebrook Research Institute for Atlantic Canada Studies at Saint Mary's University, 1994.
Ruffman, Alan, and David Simpson. "Realities, Myths, and Misconceptions of the Explosion." In *Ground Zero: A Reassessment of the 1917 Explosion in Halifax Harbour*, edited by Alan Ruffman and Colin D. Howell, 301–25. Halifax, Nova Scotia: Nimbus/The Gorsebrook Research Institute for Atlantic Canada Studies at Saint Mary's University, 1994.
Sarna, Jonathan D., Ellen Smith, and Scott-Martin Kosofsky, eds. *The Jews of Boston*. 2nd ed. New Haven, Conn.: Yale University Press, 2005.
Savage, Neil J. *Extraordinary Tenure: Massachusetts and the Making of the Nation: From President Adams to Speaker O'Neill*. Worcester, Mass.: Ambassador Books, 2004.
Saville, John. "The Ideology of Labourism." In *Knowledge and Belief in Politics: The Problem of Ideology*, edited by Robert Benewick, R. N. Berki, and Bhikhu C. Parekh, 213–26. New York: St. Martin's, 1973.

Sawislak, Karen. "September 11 and New York City: Patterns of Urban Disaster in the United States." *The Urban Lawyer* 34, no. 3 (Summer 2002): 599–607.

———. *Smoldering City: Chicagoans and the Great Fire, 1871–1874*. Chicago: Chicago University Press, 1995.

Scanlon, Joseph. "Dealing with Mass Death after a Community Catastrophe: Handling Bodies after the 1917 Halifax Explosion." *Disaster Prevention and Management* 7, no. 4 (1998): 288–304.

———. "Rescuers or Troublemakers? The Massachusetts Response to the 1917 Halifax Catastrophe." *Journal of the American Society of Professional Planners* (1999): 55–69.

———. "Rewriting a Living Legend: Researching the 1917 Halifax Explosion." *International Journal of Mass Emergencies and Disasters* 15, no. 1 (March 1997): 147–78.

———. "Source of Threat and Source of Assistance: The Maritime Aspects of the 1917 Halifax Explosion." *Northern Mariner/Le marin du nord* 10, no. 4 (2000): 39.

Scanlon, T. Joseph. "Disaster's Little Known Pioneer: Canada's Samuel Henry Prince." *International Journal of Mass Emergencies and Disasters* 6, no. 3 (November 1988): 213–32.

Scott, James C. *Domination and the Arts of Resistance: Hidden Transcripts*. New Haven, Conn.: Yale University Press, 1990.

———. *Seeing Like a State: How Certain Schemes to Improve the Human Condition Have Failed*. New Haven, Conn.: Yale University Press, 1998.

Secombe, Wally. "The Housewife and Her Labour under Capitalism." *New Left Review* 83 (February 1974): 3–24.

Semple, Neil. *The Lord's Dominion: The History of Canadian Methodism*. Montreal and Kingston: McGill–Queen's University Press, 1996.

Skocpol, Theda. *Protecting Soldiers and Mothers: The Political Origins of Social Policy in the United States*. Cambridge, Mass.: Belknap Press, 1992.

Shah, Nayan. *Contagious Divides: Epidemics and Race in San Francisco's Chinatown*. Berkeley: University of California Press, 2001.

Shepard, Benjamin. "From Flooded Neighborhoods to Sustainable Urbanism: A New York Diary." *Socialism and Democracy* 27, no. 3 (2013): 42–64.

Shephard, R. Bruce. *Deemed Unsuitable*. Toronto: Umbrella, 1997.

Shugerman, Jed Handelsman. "The Floodgates of Strict Liability: Bursting Reservoirs and the Adoption of *Fletcher v. Reynolds* in the Gilded Age." *Yale Law Journal* 110 (2000): 333–77.

Shutlak, Garry. "A Vision of Regeneration: Reconstruction after the Explosion, 1917–21." In *Ground Zero: A Reassessment of the 1917 Explosion in Halifax Harbour*, edited by Alan Ruffman and Colin D. Howell, 421–26. Halifax, Nova Scotia: Nimbus/The Gorsebrook Research Institute for Atlantic Canada Studies at Saint Mary's University, 1994.

Silcox, Claris Edwin. *Church Union in Canada: Its Causes and Consequences*. New York: Institute of Social and Religious Research, 1933.

Silvia, Philip T., Jr. "Neighbors from the North: French-Canadian Immigrants vs. Trade Unionism in Fall River, Massachusetts." In *Steeples and Smokestacks: A Collection of Essays on the Franco-American Experience in New England*, edited by Claire Quintal, 144–61. Worcester, Mass.: Institut Français, Assumption College, 1996.

Simpson, Michael. *Thomas Adams and the Modern Planning Movement: Britain, Canada, and the United States, 1900–1940*. London: Mansell, 1985.

Slawson, Douglas J. *Ambition and Arrogance: Cardinal William O'Connell of Boston and the American Catholic Church*. San Diego: Cobalt Productions, 2007.

Smith, Carl. *Urban Disorder and the Shape of Belief: The Great Chicago Fire, the Haymarket Bomb, and the Model Town of Pullman*. Chicago: University of Chicago Press, 1995.

Smith, David Edward. "Emergency Government in Canada." *Canadian Historical Review* 50, no. 4 (December 1969): 429–48.

Smith, Ellen. "'Israelites in Boston,' 1840–1880." In *The Jews of Boston*, edited by Jonathan D. Sarna, Ellen Smith, and Scott-Martin Kosofsky, 45–66. 2nd ed. New Haven, Conn.: Yale University Press, 2005.

Smith, Goldwin. *The Treaty of Washington, 1871: A Study in Imperial History*. Ithaca, N.Y.: Cornell University Press, 1941.

Smith, Robert Ellis. *Ben Franklin's Web Site: Privacy and Curiosity from Plymouth Rock to the Internet*. Providence, R.I.: Privacy Journal, 2000.

Solnit, Rebecca. *A Paradise Built in Hell: The Extraordinary Communities That Arise in Disasters*. New York: Viking, 2009.

Sorrell, Richard S. "The Sentinelle Affair (1924–1929): Religion and Militant Survivance in Woonsocket, Rhode Island." *Rhode Island History* 36, no. 3 (August 1977): 67–80.

Starr, Paul. *The Social Transformation of American Medicine*. New York: Basic Books, 1982.

Steedman, Carolyn. *Landscape for a Good Woman: A Story of Two Lives*. New Brunswick, N.J.: Rutgers University Press, 1987.

Steinberg, Ted. *Acts of God: The Unnatural History of Natural Disasters in America*. 2nd ed. Oxford: Oxford University Press, 2006.

Stentiford, Barry M. "The Meaning of a Name: The Rise of the National Guard and the End of a Town Militia." *Journal of Military History* 72, no. 3 (July 2008): 727–54.

Sterne, Evelyn Savidge. *Ballots and Bibles: Ethnic Politics and the Catholic Church in Providence*. Ithaca, N.Y.: Cornell University Press, 2003.

Stewart, Randall. "William Benjamin Basil King, 1859–1928." In *Dictionary of American Biography*. Vol. 1928. New York: American Council of Learned Societies, 1928–1936. Reproduced in the Biography Resource Center. Farmington Hills, Mich.: Gale, 2010. Available at http://galenet.galegroup.com/servlet/BioRC.

Stonebridge, Bob. "List of Halifax County Churches." *Halifax County Genealogy (GenWeb)*. 2004. Available at http://www.rootsweb.ancestry.com/~nshalifa/Churches.html.

Strange, Carolyn. *Toronto's Girl Problem: The Perils and Pleasures of the City, 1880–1930*. Toronto: University of Toronto Press, 1995.

Stricker, Frank. "The Wages of Inflation: Worker's Earnings in the World War One Era." *Mid-America* 63, no. 2 (April 1981): 93–105.

Stromquist, Shelton. *Reinventing "The People": The Progressive Movement, the Class Problem, and the Origins of Modern Liberalism*. Urbana: University of Illinois Press, 2006.

Strong-Boag, Veronica. "'Wages for Housework': Mothers' Allowances and the Beginnings of Social Security in Canada." *Journal of Canadian Studies* 14, no. 1 (Spring 1979): 24–34.

Struthers, James. *No Fault of Their Own: Unemployment and the Canadian Welfare State, 1914–1941*. Toronto: University of Toronto Press, 1983.

Sutherland, D. A. "The Personnel and Policies of the Halifax Board of Trade, 1890–1914." In *The Enterprising Canadians: Entrepreneurs and Economic Development in Eastern Canada, 1820–1914*, edited by Lewis R. Fischer and Eric W. Sager, 205–29. St. John's: Maritime History Group, Memorial University of Newfoundland, 1979.

Sutherland, David. "Halifax Harbour, December 6, 1917: Setting the Scene." In *Ground Zero: A Reassessment of the 1917 Explosion in Halifax Harbour*, edited by Alan Ruffman and Colin D. Howell, 3–8. Halifax, Nova Scotia: Nimbus/The Gorsebrook Research Institute for Atlantic Canada Studies at Saint Mary's University, 1994.

———. "Halifax, 1815–1914: 'Colony to Colony.'" *Urban History Review* 75, no. 1 (1975): 7–11.

"Tearful Rites Mark Closings of 3 Churches." *Boston Herald*, 30 August 2004, 5.

Teisch, Jessica B. *Engineering Nature: Water, Development, and the Global Spread of American Environmental Expertise*. Chapel Hill: University of North Carolina Press, 2011.

Tentler, Leslie Woodcock. "On the Margins: The State of American Catholic History." *American Quarterly* 45, no. 1 (March 1993): 104–27.

Thompson, E. P. "The Moral Economy of the English Crowd in the 18th Century." *Past and Present* 50 (1971): 76–136.

Thompson, John H. "American Muckrakers and Western Canadian Reformers." *Journal of Popular Culture* 4, no. 4 (Spring 1971): 1060–70.

Thompson, John Herd, and Stephen J. Randall. *Canada and the United States: Ambivalent Allies*. 4th ed. Athens: University of Georgia Press, 2008.

Thornton, Patricia A. "The Problem of Out-Migration from Atlantic Canada, 1871–1921 A New Look." *Acadiensis* 15, no. 1 (Autumn 1985): 3–34.

Toft, Jessica, and Laura S. Abrams. "Progressive Maternalists and the Citizenship Status of Low-Income Single Mothers." *Social Science Review* 78, no. 3 (September 2004): 447–65.

Torbet, Robert G. *Venture of Faith: The Story of the American Baptist Foreign Mission Society and the Woman's American Baptist Foreign Mission Society, 1814–1954*. Philadelphia: Judson Press, 1955.

Traverso, Susan. *Welfare Politics in Boston, 1910–1940*. Amherst: University of Massachusetts Press, 2003.

Treaster, Joseph B. "At Stadium, a Haven Quickly Becomes an Ordeal." *New York Times*, 1 September 2005, A18.

Trenberth, Kevin E., et al. "Surface and Atmospheric Climate Change." In *Climate Change 2007: The Physical Science Basis. Contribution of Working Group I to the Fourth Assessment Report of the Intergovernmental Panel on Climate Change*, edited by S. Solomon et al., 235–336. Cambridge: Cambridge University Press, 2007.

Trombley, Kenneth E. *The Life and Times of a Happy Liberal: A Biography of Morris Llewellyn Cooke*. New York: Harper, 1954.

Tweed, Thomas A. *Our Lady of the Exile: Diasporic Religion at a Cuban Catholic Shrine in Miami*. Oxford: Oxford University Press, 1997.

Ubiera-Minaya, Rosario, producer. *What's the Point: The Hope of a Growing Community*. DVD. Salem, Mass.: Peabody Essex Museum, 2004.

Vale, Lawrence, and Thomas Campanella, eds. *The Resilient City: How Modern Cities Recover from Disaster*. New York: Oxford University Press, 2005.

Van der Linden, Marcel. "Transnationalizing American Labor History." *Journal of American History* 86 (December 1999): 1078–92.
Venkatesh, Sudhir Alladi. *Off the Books: The Underground Economy of the Urban Poor*. Cambridge, Mass.: Harvard University Press, 2006.
Walkowitz, Daniel J. *Working with Class: Social Workers and the Politics of Middle-Class Identity*. Chapel Hill: University of North Carolina Press, 1999.
Wallace, Anthony F. C. *Tornado in Worcester: An Exploratory Study of Individual and Community Behavior in an Extreme Situation*. Washington, D.C.: National Research Council, National Academy of Sciences, 1956.
Wallace, Elisabeth. "The Origin of the Social Welfare State in Canada, 1867–1900." *Canadian Journal of Economics and Political Science* 16, no. 3 (August 1950): 383–93.
Ward, Colin. *Anarchy in Action*. 4th ed. 1973. Repr. London: Freedom Press, 1996.
Wayman, Dorothy G. *David I. Walsh, Citizen Patriot*. Milwaukee, Wisc.: Bruce, 1952.
Weaver, John C. "Reconstruction of the Richmond District in Halifax: A Canadian Episode in Public Housing and Town Planning, 1918–1921." *Plan Canada*, March 1976, 36–47.
Weaver, John C., and Peter de Lottinville. "The Conflagration and the City: Disaster and Progress in British North America during the Nineteenth Century." *Histoire sociale/Social History* 13 (1980): 417–49.
Weil, François. *Les franco-américains: 1860–1980*. Tours: Belin, 1989.
Wenzel, Friedemann, Fouad Bendimerad, and Ravi Sinha. "Megacities—Megarisks." *Natural Hazards* 42, no. 3 (2007): 481–91.
White, Jay. "Exploding Myths: The Halifax Harbour Explosion in Historical Context." In *Ground Zero: A Reassessment of the 1917 Explosion in Halifax Harbour*, edited by Alan Ruffman and Colin D. Howell, 251–74. Halifax, Nova Scotia: Nimbus/The Gorsebrook Research Institute for Atlantic Canada Studies at Saint Mary's University, 1994.
Widdis, Randy W. *With Scarcely a Ripple: Anglo-Canadian Migration into the United States and Western Canada, 1880–1920*. Montreal and Kingston: McGill–Queen's University Press, 1998.
Wiebe, Robert H. *The Search for Order, 1877–1920*. New York: Hill and Wang, 1967.
Wilczenski, Felicia L., and Emily A. Murphy. *The Polish Community of Salem*. Charleston S.C.: Arcadian, 2012.
Willrich, Michael. *City of Courts: Socializing Justice in Progressive Era Chicago*. Cambridge: Cambridge University Press, 2003.
Winks, Robin W. *The Blacks in Canada*. New Haven, Conn.: Yale University Press, 1971.
Wolfe, Alan. *The Future of Liberalism*. New York: Knopf, 2009.
Ydstie, John. "New Orleans Begins Probe into Katrina Deaths." *All Things Considered*, 28 September 2005.
Yorke, Lois K. "Best, Edna May Williston (Sexton)." In *Dictionary of Canadian Biography*. Vol. 15. Toronto/Quebec: University of Toronto/Université Laval, 2005.
Zolberg, Aristide R. "Moments of Madness." *Politics and Society* 2 (1972): 183–207.
Zunz, Olivier. *The Changing Face of Inequality: Urbanization, Industrial Development, and Immigrants in Detroit, 1880–1920*. Chicago: University of Chicago Press, 1982.

DISSERTATIONS, THESES, AND OTHER UNPUBLISHED
SECONDARY SOURCES

Anctil, Pierre. "Aspects of Class Ideology in a New England Ethnic Minority: The Franco-Americans of Woonsocket, Rhode Island, 1865–1929." PhD diss., New School for Social Research, 1980.

Boudreau, Michael S. "Crime and Society in a City of Order: Halifax, 1918–1935." PhD diss., Queen's University, 1996.

Buttimer, Twila F. "Great Expectations: The Maritime Methodist Church and Church Union, 1925." MA thesis, University of New Brunswick, 1980.

Call, Elaine. "The Church in Salem, Massachusetts: A Study of the Spiritual and Physical Significance of Immigrant Churches in America." Seminar paper, Salem State College, 1975. Reprinted in *Immigration in American Life: Graduate Seminar Research Papers on North Shore Selected Topics*, edited by Adele L. Younis. Vol. 1. Salem, Mass.: Salem State College, 1977.

Chase-Harrell, Pauline, Carol Ely, and Stanley Moss. "Administrative History of the Salem Maritime National Historic Site." Report prepared for the National Park Service North Atlantic Regional Office, 9 February 1993.

[Chomsky, Aviva]. "Stopping the Clock: A Time to Remember Salem's Pequot Mill Strike." Pamphlet for exhibition. [Salem, Mass.], 2004.

Christie, Jean. "Morris Llewellyn Cooke: Progressive Engineer." PhD diss., Columbia University, 1963.

City of Salem, Massachusetts. *Manual of the City Government 2001*. Salem, Mass., 2001.

Cuomo, Donna Fournier. "America's Immigrant Heritage—A Living, Pulsating Phenomenon: The French-Canadian Immigrant Experience in Lawrence." Seminar paper, Salem State College, 1974. Reprinted in *Immigration in American Life: Graduate Seminar Research Papers on North Shore Selected Topics*. Edited by Adele L. Younis. Vol. 10. Salem, Mass.: Salem State College, 1977.

de Gannes, Renée Élise. "Better Suited to Deal with Women and Children: Pioneer Policewomen in Halifax, Nova Scotia." MA thesis, Dalhousie University, 1999.

Donnell, Robert Phippen. "Locational Response to Catastrophe: The Dynamics of Locational Change in the Shoe and Leather Industry of Salem, Massachusetts, and after the Conflagration of June 25, 1914." MA thesis, Clark University, 1971.

Erickson, Paul Alfred. "Halifax's Other Hill: Fort Needham from Earliest Times." *Occasional Papers in Anthropology*. Halifax, Nova Scotia: Department of Anthropology, Saint Mary's University, 1984.

Erikson, Kai. "Reflections on Katrina." Talk at the William and Ida Friday Center for Continuing Education, University of North Carolina, Chapel Hill, 14 April 2008. Notes in author's possession.

Ferris, Russell L. "Habitants in Yankee Land: The French-Canadians in Southbridge, Massachusetts." Seminar paper, Salem State College, 1974. Reprinted in *Immigration in American Life: Graduate Seminar Research Papers on North Shore Selected Topics*. Edited by Adele L. Younis. Vol. 8. Salem, Mass.: Salem State College, 1977.

Jalbert, Philip E. "The French-Canadian in South Salem, Massachusetts." Seminar paper, Salem State College, 1975. Reprinted in *Immigration in American Life: Graduate*

Seminar Research Papers on North Shore Selected Topics. Edited by Adele L. Younis. Vol. 5. Salem, Mass.: Salem State College, 1977.

Kopf, Edward J. "The Intimate City: A Study in Urban Social Order: Chelsea, Massachusetts, 1906–1915." PhD diss., Brandeis University, 1974.

Kretzmann, Stephen Paul. "A House Built Upon the Sand: Race, Class, Gender, and the Galveston Hurricane of 1900." PhD diss., University of Wisconsin–Madison, 1995.

Lambly, Peter D. "Toward a 'Living Wage': The Minimum-Wage Campaign for Women in Nova Scotia, 1920–1935." BA honours essay, Dalhousie University, 1977.

———. "Working Conditions and Industrial Relations on Canada's Street Railways, 1900–1920." MA thesis, Dalhousie University, 1983.

Meyer, Michelle Annette. "Social Capital and Collective Efficacy for Disaster Resilience: Connecting Individuals with Communities and Vulnerability with Resilience in Hurricane-Prone Communities in Florida." PhD diss., Colorado State University, 2013.

[Murphy, Emily A.]. "Merchants, Clerks, Citizens, and Soldiers: The Second Corps of Cadets in Salem, Massachusetts." Pamphlet. Salem, Mass.: Salem Maritime National Historic Site, National Park Service, US Department of the Interior, [2005?].

Myers, Sharon. "I Can Manage My Own Business Affairs: Female Industrial Workers in Halifax at the Turn of the Twentieth Century." MA thesis, Saint Mary's University, 1989.

Nicholls, Shaun S. "Crisis Capital: The Making and Unmaking of Industrial Massachusetts, 1873–present." PhD diss., Harvard University, in progress.

Nicholson, Andrew. "Dreaming of the 'Perfect City': The Halifax Civic Improvement League 1905–1949." MA thesis, Saint Mary's University, 2000.

La Paroisse Saint-Joseph: 1873–1948: Soixante-quinzième anniversaire. [Salem, Mass.?]: Association Laurier, 1948.

Rudachyk, B[radley] E. S. "The Most Tyrannous of Masters: Fire in Halifax, Nova Scotia, 1830–1850." MA thesis, Dalhousie University, 1984.

Settle, Victor. "Halifax Shipyards 1918–1978: The Economic Impact." MA thesis, Saint Mary's University, 1994.

Smith, Michael. "Female Reformers in Nova Scotia: Architects of a New Womanhood." MA thesis, Saint Mary's University, 1986.

[St. Pierre, Lorraine]. *Saint Joseph Parish: 1873–1973.* Salem, Mass.: Printed by Compass Press, 1973.

Superstorm Research Lab. "A Tale of Two Sandys." White paper. December 2013.

Tuck, J. H. "Nova Scotia and the Conscription Election of 1917." MA thesis, Dalhousie University, 1968.

Umble, Maurice. "Individual Narrative of Christian Lantz." Unpublished manuscript in author's possession, n.d.

Waite, Catherine Ann. "The Longshoremen of Halifax, 1900–1930: Their Living and Working Conditions." MA thesis, Dalhousie University, 1977.

Vicero, Ralph Dominic. "Immigration of French Canadians to New England, 1840–1900: A Geographical Analysis." PhD diss., University of Wisconsin, 1968.

Wilk, Daniel Levinson. "Cliff Dwellers: Modern Service in New York City, 1800–1945." PhD diss., Duke University, 2005.

INDEX

Acadia University, 108
Archibald, Edith, 26, 41–44. *See also* Halifax explosion; Halifax Local Council of Women; relief managers
autonomy: limits on, 83, 192; of relief recipients, 3, 19, 82, 84; struggle for, 104, 107; of the working class, 20, 85, 128. *See also* labor; refugee camps; solidarity; survivors

Bertram Field. *See* refugee camps
Beverly (Massachusetts), 56; French Canadian church in, 181–82; as refuge from Salem fire, 78, 88, 90–93, 178
Bicknell, Ernest, 64, 68, 86, 169. *See also* Red Cross; relief managers
Binette, Donat, 98, 100, 165–67; as supporter of French Canadian autonomy, 181, 239n91; as translator and broker, 167–68, 174–75. *See also* churches; clergy; French Canadians; Rainville, Georges
Boardman, Mabel, 64–65. *See also* Red Cross
Borden, Robert (Prime Minister), 7, 26, 40, 43, 122
border (lands): migrations within and across, 11–12, 195; neighborhoods of, 194; northeast, 2–3, 17; and public health, 77; survivor solidarity across, 20, 196. *See also* citizenship; disaster citizenship; state
Boston: archdiocese of, 168, 177, 234n10; and Halifax relief aid, 122–24, 127, 157; home rule of, 64–65; labor unrest in, 80;
links to Halifax, 15, 26, 119–20, 195; and Salem fire, 72, 78
Bread and Roses Strike, 74, 86
Britain, 12, 119, 120–21, 122

Camp Hill Hospital, 22, 32–37, 39, 42
Canada: maritime provinces of, 3, 24, 110, 123, 149, 195; and Nova Scotia diaspora, 12, 41, 120–21, 123, 195; and Progressive Era, 7–8; and relations with United States, 10–11; and relief aid, 137, 139; and Salem fire, 90; as transnational, 2. *See also* border; migration; World War I
Canadian Brotherhood of Railroad Employees (CBRE), 135, 156, 158, 159–60
carpenters' union: of Halifax, 156, 158, 161, 162; of Salem, 86, 221n36
Carson, Kathleen and Johnnie, 142–43, 145–46, 157
Catholic Charities, 63, 168
Catholic Church. *See* churches
charity: balance of, 131, 194; of churches, 148, 168; hierarchy of, 106–7, 170, 193; public, 82, 86; recipients of, 141–42; and the state, 8, 42, 107. *See also* legibility; relief managers; relief workers; solidarity
Chelsea (Massachusetts), 15, 62, 65
Chicago Fire (1871), 6
churches: denominational boundaries between, 3, 36, 133, 135, 137–38, 146, 148; ecumenism among, 36, 135, 151; ethnic,

19, 63, 168, 171–73, 234n10; and Halifax explosion, 133–34, 136–37, 139, 147–54; and Salem fire, 173–82, 187; spiritual solace of, 165–67; ultramontanism of, 181–82, 239n88. *See also* charity; clergy; disaster relief; *fabrique*; French Canadians; Halifax Relief Commission; Irish; *specific churches*; social workers; United Church of Canada

citizenship: aid as a right of, 2–3, 8; and borders, 128; new forms of, 18, 20, 199; and Salem fire, 57. *See also* disaster citizenship; governance; legibility; Progressive Era; state

City Home (Halifax), 124–26, 130, 142

The City of Comrades (film), 189–90, 241n2

Civic Improvement League, 24–26

civil society, 19; mediation role of in disaster, 134; and the state, 4, 9, 193. *See also* churches; clergy; unions

clergy, 5, 10; authority of, 142–43, 146–47, 193; as extensions of the state, 137–38, 169–71; local knowledge and networks of, 142, 154–55, 157, 164, 168, 175; role in Halifax disaster relief, 36, 132–33, 135–47, 230n16; role in Salem disaster relief, 63, 168–71, 173–76, 216–17n46; as translators and brokers, 139, 140, 141, 164, 168, 174, 175. *See also* Binette, Donat; Carsons, Kathleen and Johnnie; churches; French Canadians; Mariggi, Rita; McManus, Charles; O'Connell, William; Rainville, Georges; relief managers; social workers

Cole, Charles H. (Adjutant General of Massachusetts), 61, 71–72

Colwell, Henry S. (Deputy Mayor of Halifax), 38–39, 40–41

Committee of Fourteen, 15, 63, 65, 171; and labor, 85; and tourists, 82. *See also* Committee of 100; relief managers; Salem fire

Committee of 100, 62–63, 65, 176. *See also* clergy; Committee of Fourteen; relief managers; Salem fire

Cutten, George B., 108, 125, 127, 156–57. *See also* Halifax Relief Commission

Dalhousie University: volunteers of, 22, 27, 31, 32, 33, 43

Danvers (Massachusetts), 56; French Canadian parish in, 92–93, 178–81; labor unrest in, 184, 240n106; as refuge from Salem Fire, 78, 88–91

Dartmouth (Nova Scotia): impact of Halifax explosion on, 27, 106, 147, 148, 149, 154; relief assistance in, 31, 45, 49. *See also* Dartmouth Relief Committee

Dartmouth Relief Committee, 108, 134, 139, 230n16. *See also* Halifax Relief Commission

Deacon, Byron: advice for relief workers, 12–14, 15, 107, 207n58

Dennis, Agnes, 41–44. *See also* Halifax Local Council of Women; relief managers

Department of Labor. *See under* United States

diaspora, 2; and disaster aid, 19; French Canadian, 56, 85, 91–93, 195, 206n52; Halifax, 12, 107, 119–28, 195. *See also* border; Canada; disaster citizenship; French Canadians; Massachusetts; migration; New England; Nova Scotia; survivors

disaster citizenship, 3, 10, 20, 80, 194, 196, 199; and civil society, 187; and migration, 12. *See also* Progressive Era; state; survivors; solidarity

disaster relief: contemporary lessons for, 196, 199; employers and the state as agents of, 3, 133–34; and fraud, 98–102, 175–76; institutions and ideologies of, 4–5, 12–15; politics of, 62, 66–67, 196; and religious networks, 138. *See also* churches; Deacon, Byron; labor; legibility; military; Perry, Montayne; Progressive Era; unions

disasters: citizens and, 2; and historical change, 5, 8, 18, 20; lessons of, 13, 197–99; working-class, 3, 191. *See also* disaster citizenship; governance; Halifax explosion; Salem fire

Eagan, Richard E., 58, 59, 76

Eisnor, Ralph, 156, 161. *See also* carpenters' union

emergent organizations, 37, 38, 47–48. *See also* disaster citizenship; legibility; relief managers; relief workers; solidarity; survivors

fabrique (French Canadian vestry), 176–77, 181–82. *See also* St. Joseph's parish (Salem)

Index 279

Falk, J. Howard Toynbee, 109–10. *See also* Halifax Relief Commission; social workers
family economies, 19; and boarders, 110, 111, 114–16; and disaster relief, 107, 110–15; gendered and informal, 114–19, 192. *See also*, Halifax Relief Commission; labor; legibility; survivors
Forest River Park. *See* refugee camps
Franco-Americans. *See* French Canadians
fraternal orders, 9, 91–92, 132, 133–34, 149, 176, 183. *See also* Knights of Columbus; Union St.-Jean-Baptiste d'Amérique
French Canadians: clericalism of, 235n30; contestation with "Irish" church, 168, 172–82, 187; labor movement of, 172, 183–86; networks of, 91, 102; religion of, 63–64, 165–69; in Salem, 11, 19, 57, 101, 196; self-policing of, 100–101; *survivance* of, 172–73; terminology for, 206n52. *See also* churches; diaspora; *fabrique*; Massachusetts; New England; O'Connell, William; Rainville, Georges; refugee camps; Salem fire; Salem Relief Committee; St. Joseph's parish (Salem); survivors; Union St.-Jean-Baptiste d'Amérique

Galveston (Plan), 5, 12, 20, 62
governance, 2–3, 5, 18, 194, 196; and disasters, 8, 18, 20, 23; and labor, 160; and martial law in Salem, 76–77. *See also* citizenship; disaster citizenship; military; Progressive Era; public health; relief managers; state
Grant, MacCallum (Lieutenant Governor of Nova Scotia), 39, 41
Graves, Frank A. (Colonel), 75, 76–77, 95, 194
Greeks: in Halifax, 129; in Salem, 56, 57, 101, 220n5, 224n124
Grove Presbyterian Church (Halifax), 136, 147–48, 150–53. *See also* Kaye Street Methodist Church; United Church of Canada; United Memorial Church

Haiti, 3, 198–99
Halifax, 2, 24, 206n53; as "city of comrades," 190–91; and progressive reform, 24–27, 41–42; religions in, 136; ties to United States of, 2, 119–20, 123, 143; working classes of, 109–11, 226n19; and World War I, 12. *See also* Boston; Canada; diaspora; Halifax explosion; Massachusetts; migration; Nova Scotia
Halifax explosion, 2, 17, 21–23, 29–37, 191–99; city management following, 12, 24, 37–45; destruction of, 52–53; and labor movement, 158–64; and religion, 132–33, 136–37; sources of donations after, 121–22, 125; as working class disaster, 3, 45–50, 108, 111, 136. *See also* churches; clergy; Dartmouth; diaspora; disaster relief; Halifax; Halifax Relief Commission; MacMechan, Archibald; Mariggi, Rita; Massachusetts-Halifax Relief Committee; Progressive Era; relief managers, relief workers, solidarity; survivors; unions; World War I
Halifax Labor Party, 162–63. *See also* laborism
Halifax Local Council of Women, 41, 44; and progressive reform, 24, 26
Halifax Longshoremen's Association (HLA), 29, 157–58
Halifax Relief Commission, 108, 127, 206n54, 225n9; expert knowledge of, 134, 139, 154–58, 164, 193; and housework pensions, 113, 115–19; management of Massachusetts donations, 122–28; Rehabilitation Department of, 108, 109, 123, 137–38, 155–56; and state power, 130, 134–35, 154, 160, 162, 192–93. *See also* Carson, Kathleen and Johnnie; City Home; clergy; Cutten, George B.; Halifax explosion; labor; legibility; Low, Robert Smith; Mariggi, Rita; social workers; state; survivors; unions
Halifax Relief Committee, 16, 38–40, 62, 108, 155–56, 192; and clergy, 136, 137, 148; material, 43; medical, 50; women on, 44. *See also* Halifax Relief Commission; Red Cross
Halifax Shipyards, 163, 234n96
Hurley, John F. (Mayor of Salem), 56–57, 215n36; and politics of disaster relief, 62, 65, 77, 196. *See also* French Canadians; Irish; Salem
Hurricane Katrina. *See* New Orleans
Hurricane Sandy, 3

immigration. *See* migration
Imo (ship), 21, 136
Industrial Workers of the World. *See* Wobblies
Irish: bishops, 174, 176; church, 182, 234n10; politics of, 57, 62, 64–65, 77, 192; and Salem fire relief, 11, 63, 79; workforce, 56, 185. *See also* French Canadians; Hurley, John F.
Italians: in Halifax, 128–31, 157; in Salem, 56, 57, 87, 99, 101–2, 171, 220n5

Jews: in Halifax, 150; in Salem, 56, 57, 95–96, 166, 171–72, 220n5. *See also* Ratshesky, A. C.

Kaye Street Methodist Church (Halifax), 136, 147, 150–53. *See also* Grove Presbyterian Church; United Church of Canada; United Memorial Church
Klein, Naomi, 19–20
Knights of Columbus, 62, 63, 105, 137
Knights of Labor, 67
Komagata Maru, 11

labor: and disaster relief, 3, 13, 19, 104, 109, 113, 163; domestic, 80, 84, 87–88, 110–11, 114–19, 121–22, 192; movement, 134, 156, 158–64, 183–87; regulation of following disaster, 8, 84–87, 94–95, 154–56; reproductive, 2–3, 5, 19, 79, 116, 162, 163–64, 193, 194. *See also* laborism; legibility; race; state; Trades and Labour Council; unions
laborism, 134–35, 155, 160, 162, 164
Lantz, Christian, 15, 16–17, 78, 107, 208n81. *See also* Halifax explosion; Progressive Era; Salem fire; YMCA
legibility, 8–10; of family economies, 107–8, 118; of refugee camps, 96–97; and relief management, 38, 40, 43–45, 50, 109, 196; of relief workers, 23, 29, 37; and the state, 192, 199; of the working class, 107, 118. *See also* emergent organizations; military; Progressive Era; relief managers; social workers
looting, 23, 45, 72, 73
Low, Robert Smith, 161–62, 233n91. *See also* Halifax Relief Commission

MacGillivray, Dougald, 16, 40–41, 137–38. *See also* Halifax Relief Commission
MacIntosh, Clara, 41–45. *See also* relief managers
MacKenzie, Lavinia, 124–25. *See also* diaspora; survivors
MacKinnon, Christine, 34, 37. *See also* relief workers
MacLennan, Hugh (author), 22, 24
MacMechan, Archibald, 26–27. *See also* Halifax explosion; relief managers
Mariggi, Rita, 29, 128–31, 142, 157, 192. *See also* clergy; Halifax explosion; social workers; survivors
Marine Trades and Labor Federation strike (1920), 163
Maritime provinces. *See under* Canada
Martin, Peter F. (Mayor of Halifax), 38
Massachusetts: disaster assistance of, 16; industry in, 182; links to Halifax of, 11, 12, 122–28, 195; National Guard of, 73; state government of, 66, 68. *See also* Boston; French Canadians; Halifax Relief Commission; Massachusetts-Halifax Relief Committee; Salem
Massachusetts-Halifax Relief Committee, 122, 127; Halifax committee of, 123–24, 127; labor representation on, 156, 195; relief work of, 125. *See also* MacKenzie, Lavinia; Ratshesky, A. C.
McCall, Samuel (Governor of Massachusetts), 124, 126–27
McLachlan, J. B., 116
McManus, Charles, 140, 141–42, 143–46. *See also*, Carson, Kathleen and Johnnie; clergy; St. Joseph's parish (Halifax)
McNeil, Neil (Archbishop of Toronto), 139
migration, 2, 56, 195; of Halifax refugees, 12, 119–23; and industrialism, 56–57, 67; in Progressive Era, 4, 11, 17; and race, 125–26; of Salem refugees, 56–57, 88–93, 220n5. *See also* border; citizenship; diaspora; disaster citizenship; Halifax explosion; Salem fire; state
military: and appearance of order in disaster, 45, 50–52, 59, 69, 71–77; control of Salem refugee camps, 79–83, 93–96, 175; and disaster relief, 18, 19, 38, 105, 132, 194; Na-

tional Guard, 2, 57, 73, 76–77, 79, 83, 101; survivor resistance to, 11, 87, 98, 104, 122; voluntary aid of, 22–23, 27, 29–31, 37, 53, 58–60, 191. *See also* governance; relief managers; state; solidarity; World War I
militia. *See* military
Mitchell, John Hanlon, 27, 43, 132–33. *See also* MacMechan, Archibald
Mont Blanc (ship), 21, 29, 117, 136
Moors, John Farwell, 15, 16, 17, 61, 63, 65, 122. *See also* Progressive Era; relief managers
Murray, Florence, 22–24, 35–37, 49, 190. *See also* relief workers
Murray, George Henry (Premier of Nova Scotia), 12, 155–56, 163
mutual aid. *See* solidarity

National Guard. *See* military
Naumkeag Steam Cotton Company, 54, 56, 193; strike at, 168, 182–87. *See also* French Canadians; labor; Salem fire; unions
neoliberalism, 19–20, 199
New England: Catholic Church in, 179; cross-border connections of, 3, 119–21, 123; French Canadians in, 56, 91–92, 177, 182, 195
Newfoundland, 12, 119, 121, 197
New Orleans: Hurricane Katrina and, 1–2, 3, 20, 198–99, 202n2
Nova Scotia: diaspora, 107, 121, 123, 195; and Progressive reform, 7, 25–26, 155. *See also* Canada; Halifax; Massachusetts

O'Connell, John (union organizer), 183–87
O'Connell, William (Cardinal Archbishop of Boston): struggle with Georges Rainville, 177–82
O'Keefe, Mattias J. (Mayor of Salem), 57, 214n9
Ontario, 12, 26, 106; aid offered by, 121

Pelletier, Ovid, 86–87, 97–98
Pequot Mills. *See* Naumkeag Steam Cotton Company
Perry, Montayne, 26, 57, 194; on disorder, 57, 61, 68, 71, 77, 100–101, 104; on order, 59–60, 63, 69, 81, 83, 87, 169–70

Picton (ship), 29, 128
Poles: and labor movement, 183, 185, 186; in Salem, 56, 57, 100, 101, 168, 170–71, 220n5, 224n124. *See also* Naumkeag Steam Cotton Company; O'Connell, John; St. John the Baptist parish (Salem)
Port-au-Prince, 4, 198–99
Powderly, Terence V., 67–68, 91
Prince, Samuel, 190
privacy. *See* autonomy
Progressive Era, 4–5, 7, 10–11; and disaster relief, 15, 134, 154; and expert knowledge and management, 17, 20, 26–27, 38, 41, 43, 44, 68, 164, 192–93; and ideas of domestic labor, 116; ideology of, 13–14, 20, 71, 104, 191; transnationalism of, 12, 17, 24, 195; and the welfare state, 8. *See also* Canada; Deacon, Byron; Halifax; Halifax Local Council of Women; laborism; legibility; migration; Nova Scotia; Perry, Montayne; state
Protestant Churches. *See* churches
public health, 8; concerns for during Salem fire, 76–77; reform of, 62; in refugee camps, 80, 82–83, 93–94, 94–95

Quebec: awareness of Salem fire in, 92; as destination for Salem refugees, 90–91; French institutional roots in, 12, 173, 175–76, 177, 239n88. *See also* churches; diaspora; *fabrique*; French Canadians
Queen's University (Kingston), 122

racism, 122; and labor, 159–60; and migration, 125–26; in service work, 81. *See also* migration; unions
Rainville, Georges, 168, 173–75, 178–81. *See also* O'Connell, William
Ratshesky, A. C., 15–17, 63, 65, 85, 122, 124; as relief expert, 68
Red Cross, 9; and disaster relief, 12–15; and ethnicity of disaster survivors, 56–57; and gender, 44; Salem fire response of, 64, 66–67. *See also* Bicknell, Ernest; Deacon, Byron; Dennis, Agnes; Falk, J. Howard Toynbee; Sexton, May
refugee camps, 19, 79–80, 192; abandonment of, 102–4; control of labor in, 83–88,

104; fraud in, 100–101; networks of support within, 101–2; regulation of, 80–84, 93–95; resistance to, 80, 86, 88, 95–100, 105–6. See also autonomy; disaster citizenship; disaster relief; French Canadians; military; Pelletier, Ovid; public health; Salem fire; survivors

relief managers, 23, 37–45, 61–62, 106; authority of, 64; expert knowledge of, 44–45, 109–110, 134; as middle class outsiders, 15, 26, 43, 45, 50, 44–45, 73–74, 80, 108, 194; perceptions of disorder of, 23, 50, 68–69, 71–72, 191; as transnational, 11, 17; women as, 41–44, 108. See also Archibald, Edith; Deacon, Byron; Dennis, Agnes; Halifax Relief Commission; legibility; MacIntosh, Clara; MacMechan, Archibald; military; Perry, Montayne; Progressive Era; Salem Relief Committee; Sexton, May; social workers; solidarity; survivors

relief workers, 22–23, 27, 29–37, 58–61; Christian ecumenism among, 36, 135; doctors and nurses as, 30, 31, 41, 45, 49–50, 61, 88, 105; organization among, 35–36, 40, 44–45, 58–59, 106, 191, 193. See also churches; emergent organizations; legibility; military; Progressive Era; relief managers; social workers; solidarity; state

Robie Street Methodist Church (Halifax), 147, 148

Rockhead Hospital, 32–34

Ropes, Charles F. (Lieutenant Colonel), 74, 86

Salem, 12, 56, 206n53; as "city of comrades," 191; local politics of, 56–57, 62, 64–65, 191–92; the Point neighborhood of, 54–56; social hierarchy in, 74–75, 77; transnationalism of, 11–12. See also French Canadians; Greeks; Hurley, John F.; Italians; Jews; migration; Poles; Salem fire

Salem fire, 2, 6, 58–61, 191–99, 242n12; destruction of, 70, 73–74; ethnic relief organization following, 171–73; federal role in relief of, 66–68; and French Canadian labor movement, 172, 182–87; and French Canadian political culture, 173–82; and martial law, 74, 76, 79; perceptions of order and disorder following, 61, 68–70, 71–74, 77; politics of relief during, 62–64, 65, 76–77, 79, 215n35, 215n41; refugees of, 78–81, 88–96, 93–104, 220n5, 223n94; role of expert relief managers and outsiders in, 20, 61, 65–66, 68, 170, 194; tourists and, 72–73, 82; unemployed of, 67–68; victims of, 57, 95. See also churches; clergy; Committee of Fourteen; Committee of 100; disaster citizenship; disaster relief; French Canadians; governance; Pelletier, Ovid; Perry, Montayne; Progressive Era; public health; military; refugee camps; relief managers; relief workers; solidarity; survivors; United States

Salem Rebuilding Commission, 15, 64, 192

Salem Relief Committee, 67, 76, 90

San Francisco Earthquake (1906), 6, 12–13, 15

Sexton, May, 23, 24, 26, 43–44; relief work of, 45, 194. See also Red Cross; relief managers

sinistrés. See French Canadians

social capital: importance of in disasters, 9–10, 128, 192, 197. See also civil society; solidarity

social workers: authority of, 122, 123, 129, 142, 156, 169–70, 193–94; and local knowledge, 108–10, 136, 137, 139–41, 154–55, 164, 192; professionalization of, 8, 42, 137; as relief workers, 13, 134; women as, 108. See also clergy; Halifax Relief Commission; legibility; MacGillivray, Dougald; Progressive Era; relief managers; Wisdom, Jane B.

soldiers. See military

solidarity, 4–5, 10, 189–91, 194, 197; and charity, 107; and churches, 139, 150; destruction of, 75–76, 83, 102; in disaster, 18, 20, 53, 112, 192; of relief workers, 29, 31, 34; of survivors, 23, 45, 47–48, 70–71, 100–101, 112, 131; of unions, 158–59, 197. See also autonomy; border; civil society; emergent organizations; social capital

St. Alphonsus parish (Beverly), 181

state: citizen challengers of, 20, 99, 128, 194; and disaster relief, 2–3, 6–7, 24, 45, 79, 95; domestic laborers as employees of, 116; expansion of, 5, 154, 169, 191–92; expectations of in disaster, 5, 11; and labor, 154, 162, 193; progressive, 4–5, 9, 17–18, 20, 107–8, 134–35, 194, 196; as relief organization, 24, 154, 191–92; welfare, 8, 199. *See also* borders; citizenship; civil society; disaster citizenship; disaster relief; legibility; migration; military; Progressive Era; World War I

St. John the Baptist parish (Salem), 168, 170, 171. *See also* Poles

St. Joseph's parish (Halifax), 46, 117, 133, 136–37, 147; and disaster relief, 139, 142–43; as refuge after the explosion, 49, 121. *See also*, Carson, Kathleen and Johnnie; McManus, Charles

St. Joseph's parish (Salem), 83, 166; and French Canadian community, 169, 173, 178, 181–82, 187, 193, 238n83. *See also* churches; clergy; Rainville, Georges

St. Mark's Anglican Church (Halifax), 136–37

St. Paul's Anglican Church (Halifax), 147, 190; networks of, 138, 144; as relief shelter, 126, 137, 148

survivors, 2, 19–20, 23–24, 27, 45–51, 69–71, 191, 192; homosociality among, 112; informal economies of, 107, 114–16, 191; membership in organizations of, 132–34, 136–38, 146, 153; networks and relationships of, 107, 111–13, 131, 136; transnational ties of, 12, 123, 128. *See also* autonomy; Carson, Kathleen and Johnnie; charity; churches; citizenship; civil society; disaster citizenship; emergent organizations; family economies; Mariggi, Rita; migration; Pelletier, Ovid; refugee camps; social capital; social workers; solidarity; state; unions

Trades and Labour Congress, 160, 161
Trades and Labour Council (Halifax), 149, 155–56, 162
Truro, Nova Scotia, 36–37, 138

unions: during disasters, 4, 18–19, 195–96; and Halifax explosion, 134–35, 154–58, 160–62, 193; as mutual aid organizations, 9, 29; relationship with the state, 134, 154–55, 193; and Salem fire, 86–87, 168– 69, 172, 183–87; shifting solidarities of, 158–60, 162–64. *See also* Canadian Brotherhood of Railroad Employees; carpenters union; civil society; disaster citizenship; Eisnor, Ralph; French Canadians; Halifax Labor Party; Halifax Longshoremen's Association; Halifax Relief Commission; laborism; O'Connell, John; refugee camps; state; Trades and Labour Council; United Textile Workers

Union St.-Jean-Baptiste d'Amérique (USJBA), 91, 92, 93
United Church of Canada, 150–51
United Kingdom. *See* Britain
United Memorial Church, 150–54, 193
United States: Congress of, 7, 66; Department of Labor, 67–68, 91, 103; disasters in, 197; Employment Office of, 7, 68; and federal disaster relief, 8, 66–67, 137; Post Office, 96–97; transnational region of, 2, 10–11, 17, 120
United Textile Workers, 68, 183–86

Victoria General Hospital, 35, 37, 47, 142
Victorian Order of Nurses (VON), 31, 41, 44
Voluntary Aid Detachment (St. John's Ambulance) (VAD), 32, 42, 44, 45

Wallace, William B., 108, 115–16, 129. *See also* Halifax Relief Commission
Walsh, David (Governor of Massachusetts), 15, 61, 65, 167–68
Wilson, Woodrow (President), 66
Wisdom, Jane B., 41–44, 129, 157, 194. *See also* relief managers; social workers
Wobblies, 80. *See also* Bread and Roses Strike
Women's Trade Union League, 159
World War I, 15, 176, 189; and growth of the state, 19, 154; and Halifax, 12, 24; Halifax explosion as part of, 7, 52–53, 122, 197; inflation during, 183; labor shortage during, 17. *See also* Canada; Halifax
Worrell, Clarendon Lamb (Anglican Archbishop of Halifax), 133, 137, 149

YMCA: as community gathering place, 49; Halifax, 31, 36; Salem, 15, 16, 57, 63, 78

JACOB A. C. REMES is an assistant professor of public affairs and history at the Metropolitan Center of SUNY Empire State College. He is a winner of the Herbert G. Gutman prize from the Labor and Working-Class History Association and the Eugene A. Forsey Prize from the Canadian Committee on Labour History.

THE WORKING CLASS IN AMERICAN HISTORY

Worker City, Company Town: Iron and Cotton-Worker Protest in Troy and Cohoes, New York, 1855–84   *Daniel J. Walkowitz*

Life, Work, and Rebellion in the Coal Fields: The Southern West Virginia Miners, 1880–1922   *David Alan Corbin*

Women and American Socialism, 1870–1920   *Mari Jo Buhle*

Lives of Their Own: Blacks, Italians, and Poles in Pittsburgh, 1900–1960   *John Bodnar, Roger Simon, and Michael P. Weber*

Working-Class America: Essays on Labor, Community, and American Society   *Edited by Michael H. Frisch and Daniel J. Walkowitz*

Eugene V. Debs: Citizen and Socialist   *Nick Salvatore*

American Labor and Immigration History, 1877–1920s: Recent European Research   *Edited by Dirk Hoerder*

Workingmen's Democracy: The Knights of Labor and American Politics   *Leon Fink*

The Electrical Workers: A History of Labor at General Electric and Westinghouse, 1923–60   *Ronald W. Schatz*

The Mechanics of Baltimore: Workers and Politics in the Age of Revolution, 1763–1812   *Charles G. Steffen*

The Practice of Solidarity: American Hat Finishers in the Nineteenth Century   *David Bensman*

The Labor History Reader   *Edited by Daniel J. Leab*

Solidarity and Fragmentation: Working People and Class Consciousness in Detroit, 1875–1900   *Richard Oestreicher*

Counter Cultures: Saleswomen, Managers, and Customers in American Department Stores, 1890–1940   *Susan Porter Benson*

The New England Working Class and the New Labor History   *Edited by Herbert G. Gutman and Donald H. Bell*

Labor Leaders in America   *Edited by Melvyn Dubofsky and Warren Van Tine*

Barons of Labor: The San Francisco Building Trades and Union Power in the Progressive Era   *Michael Kazin*

Gender at Work: The Dynamics of Job Segregation by Sex during World War II   *Ruth Milkman*

Once a Cigar Maker: Men, Women, and Work Culture in American Cigar Factories, 1900–1919   *Patricia A. Cooper*

A Generation of Boomers: The Pattern of Railroad Labor Conflict in Nineteenth-Century America   *Shelton Stromquist*

Work and Community in the Jungle: Chicago's Packinghouse Workers, 1894–1922   *James R. Barrett*

Workers, Managers, and Welfare Capitalism: The Shoeworkers and Tanners of Endicott Johnson, 1890–1950   *Gerald Zahavi*

Men, Women, and Work: Class, Gender, and Protest in the New England Shoe Industry, 1780–1910   *Mary Blewett*

Workers on the Waterfront: Seamen, Longshoremen, and Unionism in the 1930s   *Bruce Nelson*

German Workers in Chicago: A Documentary History of Working-Class Culture from 1850 to World War I   *Edited by Hartmut Keil and John B. Jentz*

On the Line: Essays in the History of Auto Work   *Edited by Nelson Lichtenstein and Stephen Meyer III*
Labor's Flaming Youth: Telephone Operators and Worker Militancy, 1878–1923
   Stephen H. Norwood
Another Civil War: Labor, Capital, and the State in the Anthracite Regions of Pennsylvania, 1840–68   *Grace Palladino*
Coal, Class, and Color: Blacks in Southern West Virginia, 1915–32
   Joe William Trotter Jr.
For Democracy, Workers, and God: Labor Song-Poems and Labor Protest, 1865–95
   Clark D. Halker
Dishing It Out: Waitresses and Their Unions in the Twentieth Century
   Dorothy Sue Cobble
The Spirit of 1848: German Immigrants, Labor Conflict, and the Coming of the Civil War   *Bruce Levine*
Working Women of Collar City: Gender, Class, and Community in Troy, New York, 1864–86   *Carole Turbin*
Southern Labor and Black Civil Rights: Organizing Memphis Workers   *Michael K. Honey*
Radicals of the Worst Sort: Laboring Women in Lawrence, Massachusetts, 1860–1912
   Ardis Cameron
Producers, Proletarians, and Politicians: Workers and Party Politics in Evansville and New Albany, Indiana, 1850–87   *Lawrence M. Lipin*
The New Left and Labor in the 1960s   *Peter B. Levy*
The Making of Western Labor Radicalism: Denver's Organized Workers, 1878–1905
   David Brundage
In Search of the Working Class: Essays in American Labor History and Political Culture
   Leon Fink
Lawyers against Labor: From Individual Rights to Corporate Liberalism   *Daniel R. Ernst*
"We Are All Leaders": The Alternative Unionism of the Early 1930s
   Edited by Staughton Lynd
The Female Economy: The Millinery and Dressmaking Trades, 1860–1930
   Wendy Gamber
"Negro and White, Unite and Fight!": A Social History of Industrial Unionism in Meatpacking, 1930–90   *Roger Horowitz*
Power at Odds: The 1922 National Railroad Shopmen's Strike   *Colin J. Davis*
The Common Ground of Womanhood: Class, Gender, and Working Girls' Clubs, 1884–1928   *Priscilla Murolo*
Marching Together: Women of the Brotherhood of Sleeping Car Porters
   Melinda Chateauvert
Down on the Killing Floor: Black and White Workers in Chicago's Packinghouses, 1904–54   *Rick Halpern*
Labor and Urban Politics: Class Conflict and the Origins of Modern Liberalism in Chicago, 1864–97   *Richard Schneirov*
All That Glitters: Class, Conflict, and Community in Cripple Creek   *Elizabeth Jameson*
Waterfront Workers: New Perspectives on Race and Class   *Edited by Calvin Winslow*
Labor Histories: Class, Politics, and the Working-Class Experience
   Edited by Eric Arnesen, Julie Greene, and Bruce Laurie

The Pullman Strike and the Crisis of the 1890s: Essays on Labor and Politics
   Edited by Richard Schneirov, Shelton Stromquist, and Nick Salvatore
AlabamaNorth: African-American Migrants, Community, and Working-Class Activism
   in Cleveland, 1914–45   *Kimberley L. Phillips*
Imagining Internationalism in American and British Labor, 1939–49   *Victor Silverman*
William Z. Foster and the Tragedy of American Radicalism   *James R. Barrett*
Colliers across the Sea: A Comparative Study of Class Formation in Scotland and
   the American Midwest, 1830–1924   *John H. M. Laslett*
"Rights, Not Roses": Unions and the Rise of Working-Class Feminism, 1945–80
   Dennis A. Deslippe
Testing the New Deal: The General Textile Strike of 1934 in the American South
   Janet Irons
Hard Work: The Making of Labor History   *Melvyn Dubofsky*
Southern Workers and the Search for Community: Spartanburg County, South Carolina
   G. C. Waldrep III
We Shall Be All: A History of the Industrial Workers of the World (abridged edition)
   Melvyn Dubofsky, ed. Joseph A. McCartin
Race, Class, and Power in the Alabama Coalfields, 1908–21   *Brian Kelly*
Duquesne and the Rise of Steel Unionism   *James D. Rose*
Anaconda: Labor, Community, and Culture in Montana's Smelter City   *Laurie Mercier*
Bridgeport's Socialist New Deal, 1915–36   *Cecelia Bucki*
Indispensable Outcasts: Hobo Workers and Community in the American Midwest,
   1880–1930   *Frank Tobias Higbie*
After the Strike: A Century of Labor Struggle at Pullman   *Susan Eleanor Hirsch*
Corruption and Reform in the Teamsters Union   *David Witwer*
Waterfront Revolts: New York and London Dockworkers, 1946–61   *Colin J. Davis*
Black Workers' Struggle for Equality in Birmingham   *Horace Huntley*
   and David Montgomery
The Tribe of Black Ulysses: African American Men in the Industrial South
   William P. Jones
City of Clerks: Office and Sales Workers in Philadelphia, 1870–1920   *Jerome P. Bjelopera*
Reinventing "The People": The Progressive Movement, the Class Problem,
   and the Origins of Modern Liberalism   *Shelton Stromquist*
Radical Unionism in the Midwest, 1900–1950   *Rosemary Feurer*
Gendering Labor History   *Alice Kessler-Harris*
James P. Cannon and the Origins of the American Revolutionary Left, 1890–1928
   Bryan D. Palmer
Glass Towns: Industry, Labor, and Political Economy in Appalachia, 1890–1930s
   Ken Fones-Wolf
Workers and the Wild: Conservation, Consumerism, and Labor in Oregon, 1910–30
   Lawrence M. Lipin
Wobblies on the Waterfront: Interracial Unionism in Progressive-Era Philadelphia
   Peter Cole
Red Chicago: American Communism at Its Grassroots, 1928–35   *Randi Storch*
Labor's Cold War: Local Politics in a Global Context   *Edited by Shelton Stromquist*

Bessie Abramowitz Hillman and the Making of the Amalgamated Clothing Workers
of America   *Karen Pastorello*
The Great Strikes of 1877   *Edited by David O. Stowell*
Union-Free America: Workers and Antiunion Culture   *Lawrence Richards*
Race against Liberalism: Black Workers and the UAW in Detroit   *David M. Lewis-Colman*
Teachers and Reform: Chicago Public Education, 1929–70   *John F. Lyons*
Upheaval in the Quiet Zone: 1199/SEIU and the Politics of Healthcare Unionism
Leon Fink and Brian Greenberg
Shadow of the Racketeer: Scandal in Organized Labor   *David Witwer*
Sweet Tyranny: Migrant Labor, Industrial Agriculture, and Imperial Politics
Kathleen Mapes
Staley: The Fight for a New American Labor Movement   *Steven K. Ashby
and C. J. Hawking*
On the Ground: Labor Struggles in the American Airline Industry   *Liesl Miller Orenic*
NAFTA and Labor in North America   *Norman Caulfield*
Making Capitalism Safe: Work Safety and Health Regulation in America, 1880–1940
Donald W. Rogers
Good, Reliable, White Men: Railroad Brotherhoods, 1877–1917   *Paul Michel Taillon*
Spirit of Rebellion: Labor and Religion in the New Cotton South   *Jarod Roll*
The Labor Question in America: Economic Democracy in the Gilded Age
Rosanne Currarino
Banded Together: Economic Democratization in the Brass Valley   *Jeremy Brecher*
The Gospel of the Working Class: Labor's Southern Prophets in New Deal America
Erik Gellman and Jarod Roll
Guest Workers and Resistance to U.S. Corporate Despotism   *Immanuel Ness*
Gleanings of Freedom: Free and Slave Labor along the Mason-Dixon Line, 1790–1860
Max Grivno
Chicago in the Age of Capital: Class, Politics, and Democracy during the Civil War
and Reconstruction   *John B. Jentz and Richard Schneirov*
Child Care in Black and White: Working Parents and the History of Orphanages
Jessie B. Ramey
The Haymarket Conspiracy: Transatlantic Anarchist Networks   *Timothy Messer-Kruse*
Detroit's Cold War: The Origins of Postwar Conservatism   *Colleen Doody*
A Renegade Union: Interracial Organizing and Labor Radicalism   *Lisa Phillips*
Palomino: Clinton Jencks and Mexican-American Unionism in the
American Southwest   *James J. Lorence*
Latin American Migrations to the U.S. Heartland: Changing Cultural Landscapes
in Middle America   *Edited by Linda Allegro and Andrew Grant Wood*
Man of Fire: Selected Writings   *Ernesto Galarza, ed. Armando Ibarra and Rodolfo D. Torres*
A Contest of Ideas: Capital, Politics, and Labor   *Nelson Lichtenstein*
Making the World Safe for Workers: Labor, the Left, and Wilsonian Internationalism
Elizabeth McKillen
The Rise of the Chicago Police Department: Class and Conflict, 1850–1894   *Sam Mitrani*
Workers in Hard Times: A Long View of Economic Crises
Edited by Leon Fink, Joseph A. McCartin, and Joan Sangster

Redeeming Time: Protestantism and Chicago's Eight-Hour Movement, 1866–1912
  William A. Mirola
Struggle for the Soul of the Postwar South: White Evangelical Protestants and
  Operation Dixie   *Elizabeth Fones-Wolf and Ken Fones-Wolf*
Free Labor: The Civil War and the Making of an American Working Class   *Mark A. Lause*
Death and Dying in the Working Class, 1865–1920   *Michael K. Rosenow*
Immigrants against the State: Yiddish and Italian Anarchism in America
  Kenyon Zimmer
Fighting for Total Person Unionism: Harold Gibbons, Ernest Calloway,
  and Working-Class Citizenship   *Robert Bussel*
Smokestacks in the Hills: Rural-Industrial Workers in West Virginia   *Louis Martin*
Disaster Citizenship: Survivors, Solidarity, and Power in the Progressive Era
  Jacob A. C. Remes

The University of Illinois Press
is a founding member of the
Association of American University Presses.

———————————————————

University of Illinois Press
1325 South Oak Street
Champaign, IL 61820-6903
www.press.uillinois.edu